# Regulation of Biological Control Agents

Ralf-Udo Ehlers
Editor

# Regulation of Biological Control Agents

Springer

*Editor*
Ralf-Udo Ehlers
Department for Biotechnology
and Biological Control
Institute for Phytopathology
Christian-Albrechts-University Kiel
24118 Kiel
Germany
ehlers@biotec.uni-kiel.de

**Disclaimer**

The information in this book compiled by the authors and editor has been checked for accuracy and completeness. The information is in accordance with the standard practices accepted at the time of publication. However, neither the authors, editors, and publisher, nor any party involved in the creation and publication of this work warrant that the information is in every respect accurate and complete, and they are not responsible for errors or omissions or for any consequences from the application of the information in this book.

ISBN 978-90-481-3663-6 e-ISBN 978-90-481-3664-3
DOI 10.1007/978-90-481-3664-3
Springer Dordrecht Heidelberg London New York



Printed on acid-free paper

Springer is part of Springer Science+Business Media (www.springer.com)

# Preface

Regulation is implemented by governments when human activities may cause damage to the society or the environment in order to avoid, prevent or minimise impacts. Regulation should concentrate on safety aspects and try to minimise negative consequences for trade and the economy. Biological control agents (BCAs) are generally regarded as sustainable and environmentally safe tools to manage pest insects, nematodes, weeds and diseases in agriculture, forestry and horticulture. However, no human activity is without potential risks, so regulation of BCAs is necessary to avoid potential hazards.

Plant protection products based on micro-organisms, semiochemicals and botanicals are subject to registration in all OECD countries (Organisation for Economic Co-operation and Development). Their potential for use in plant protection and substitution of hazardous chemical substances is, however, not well exploited. One reason is the stringent regulation policy that basically follows rules implemented for registration of synthetic chemical pesticides. This situation motivated the EU Commission to call for proposals for appropriate and balanced regulatory systems for BCAs. As a result, the EU-supported REBECA (Regulation of Biological Control Agents) Policy Support Action (www.rebeca-net.de) was started and gathered experts from academia, regulation authorities and industry with the objective of elaborating proposals that can accelerate the regulation process for BCAs and make it more cost-effective without compromising the level of safety for human health and the environment. Based on assessments of the potential risks of BCAs, including invertebrate agents, proposals for improvement of existing registration requirements and administration of regulation were developed.

This book summarises the results of the REBECA Action. It is also a comprehensive guide for the registration practice and requirements to apply for authorisation of biological control agents. In the first part of the book, an overview on existing regulation requirements and the general practice in OECD countries is summarised and policy aspects are reviewed and discussed. In the second part of the book, information on benefits and risks of the different biological control agents are reviewed by experienced scientists who have been working for decades in the field of biological control. This part can also be used by authorities to get an overview on the real risks related to the use of these agents. In the last part, the results of discussion among participants of the REBECA Action on how regulation of BCAs can be improved in the future is summarised by the members of the REBECA consortium.

This book will be of great help for those dealing with regulation of biological control agents in registration authorities and industry. It is also important for those who develop new products based on BCAs, as they should always have in focus the registration requirements during development of biocontrol products. Last, but not least, this book can function as the basis for future activities and discussions on how to improve existing regulation requirements. The REBECA Action was a successful platform for exchange of experience in regulation and development of possible amendments. I hope, policy-makers, scientists, member of regulatory authorities and the private sector will continue their co-operations started within the REBECA Action in order to make plant protection safer, life easier for farmers and provide healthier food produce for consumers.

The preface of this book is also a good opportunity to express my thanks to all who have contributed to the REBECA Action and to producing this book. The first acknowledgement goes to the unknown EU officials who took the initiative for the call (Sixth Framework Program of the EU. Call identifier: FP6-2004-SSP-4). Without their initiative we would today probably have to deal with more data requirements instead of fewer. Thanks are also due to the EU Commission for the financial support.

My particular thanks go to Olaf Strauch for his professional management during the Action's lifetime and to Miriam Döring and Heike Kuhlmann for their support in organisation of meetings and in administration. Thanks also to Dr. Ingmar Schmidt and Susanne Neufeldt at the Christian-Albrechts-University Kiel for keeping a scientist in line with EU administrative rules. My warmest regards go to my colleagues of the REBECA consortium, who were the backbones for success: Rüdiger Hauschild contributed his in-depth professional know-how in registration of BCAs; Anita Fjelsted managed to attract regulatory personnel and initiate fruitful networking among all stakeholders; Wyn Grant, the grey eminence, with an excellent feeling for what would be acceptable for EU and MS policy; Jeff Bale, who linked with the IOBC executives; Uli Kuhlmann, with his scientific excellence in risk assessment and links to friends of biological control all around the world; Bernard Speiser and Lucius Tamm with excellent contacts to organic agriculture and professional skills in Swiss-EU-network management; Heikki Hokkanen and Ingeborg Hokkanen-Menzler provided their expertise in socio-economics; and Hermann Strasser, who contributed the results of the previous EU projects on safety aspects (BIPESCO and RAFBCA). These were, of course, not their only qualities and I am particularly thankful to all of them for their support that made the REBECA Project a success. My gratitude also to the other authors of this book and their contributions to REBECA, in particular to Claude Alabouvette, for contributing his long-term experience in regulatory aspects and is never ending support to biological control. Thanks also to Roland Perry for proofreading and to Suzana Bernhart and Elisabete Machado (Springer) for their support.

I also want to thank all participants from biocontrol companies, universities and research organisations, regulatory authorities and consultancy companies, who came to join the workshops and discussions during the REBECA Action. We would not have been able to provide so much information within such a short time without their input. I also thank colleagues from overseas, in particular Bill Schneider and Trevor Jackson. My sincere thanks also to Ulf Heilig for provision of his expertise as a consultant and his support to our activities to inform the biocontrol industry about REBECA.

I hope this book will stimulate co-operation and activities for further improvement of regulatory policy. Finally, for those who work in biological control and have for the first time been confronted with regulation of these wonderful biocontrol techniques, please do not get frustrated; there is light at the end of the tunnel.

Kiel, Germany
August 2010

Ralf-Udo Ehlers

# Contents

# Contributors

**Claude Alabouvette** Agrene, 2100 Dijon, France, c.ala@agrene.fr

**Jeffrey S. Bale** School of Biosciences, University of Birmingham, Birmingham B15 2TT, UK, J.S.Bale@bham.ac.uk

**Gabriele Berg** Graz University of Technology, Environmental Biotechnology, Graz 8010, Austria, gabriele.berg@tugraz.at

**Wolfgang Burgstaller** Faculty of Biology, Institute of Microbiology, University of Innsbruck, Innsbruck A 6020, Austria, wolfgang.burgstaller@uibk.ac.at

**Massimiliano Cardinale** Graz University of Technology, Environmental Biotechnology, Graz 8010, Austria, m.cardinale@tugraz.at

**Christelle Cordier** Agrene, 2100 Dijon, France, christelle.cordier@agrene.fr

**Patrick De Clercq** Department of Crop Protection, Ghent University, 9000 Ghent, Belgium, patrick.declercq@ugent.be

**Ralf-Udo Ehlers** Department for Biotechnology and Biological Control, Institute for Phytopathology, Christian-Albrechts-University Kiel, 24118 Kiel, Germany, ehlers@biotec.uni-kiel.de

**Anita Fjelsted** Ministry of the Environment, Danish Environmental Protection Agency – Pesticides and Genetechnology, 1401 Copenhagen, Denmark, anfje@mst.dk

**Wyn Grant** Department of Politics and International Studies, University of Warwick, Coventry CV4 7AL, UK, w.p.grant@warwick.ac.uk

**Rüdiger Hauschild** GAB Consulting GmbH, 21769 Lamstedt, Germany, ruediger.hauschild@gab-consult.de

**Emma J. Hunt** CABI Europe-Switzerland, 2800 Delémont, Switzerland, e.hunt@cabi.org

**Stefan Hutwimmer** Faculty of Biology, Institute of Microbiology, University of Innsbruck, Innsbruck A 6020, Austria, stefan.hutwimmer@uibk.ac.at

**Ulrich Kuhlmann** CABI Europe-Switzerland, 2800 Delémont, Switzerland, u.kuhlmann@cabi.org

**Antoon J.M. Loomans** Department of Entomology, National Reference Centre, Plant Protection Service, 6700 HC Wageningen, The Netherlands, a.j.m.loomans@minlnv.nl

**Susan Mattock** Chemicals Regulation Directorate, Health and Safety Executive, York Y01 7PX, UK, sue.mattock@hse.gsi.gov.uk

**Thierry Mercier** ANSES, French Agency for Food, Environmental and Occupational Health Safety, 94700 Maisons-Alfort, France, thierry.mercier@anses.fr

**Henry Müller** Graz University of Technology, Environmental Biotechnology, Graz 8010, Austria, henry.mueller@tugraz.at

**Catherine Regnault-Roger** Université de Pau et des Pays de l'Adour-France, Pau Cedex BP 1155-64012, France, catherine.regnault-roger@univ-pau.fr

**Bernhard Speiser** Research Institute of Organic Agriculture (FiBL), 5070 Frick, Switzerland, bernhard.speiser@fibl.org

**Hermann Strasser** Faculty of Biology, Institute of Microbiology, University of Innsbruck, Innsbruck A 6020, Austria, hermann.strasser@uibk.ac.at

**Olaf Strauch** Department for Biotechnology and Biological Control, Institute for Phytopathology, Christian-Albrechts-University Kiel, 24118 Kiel, Germany, o.strauch@biotec.uni-kiel.de

**Lucius Tamm** Research Institute of Organic Agriculture (FiBL), Ackerstrasse, 5070 Frick, Switzerland, lucius.tamm@fibl.org

**Christin Zachow** Graz University of Technology, Environmental Biotechnology, Graz 8010, Austria, christin.zachow@tugraz.at

# Part I
# General Aspects of Regulation

# Chapter 1
# Regulation of Biological Control Agents and the EU Policy Support Action REBECA

**Ralf-Udo Ehlers**

**Abstract** Biological control uses living organisms like bacteria, fungi, nematodes, insects or mites (including viruses) for the control of weeds or pests and diseases of crop plants. Information on the use of these biocontrol agents and associated risks are summarized. An overview on the regulation of biological control agents and an introduction into the objectives and the organisation of the Policy Support Action REBECA is provided. The history of regulation of chemical compounds is compared with the development of regulation of biocontrol. Often the precautionary principle is consulted to justify anticipatory restrictions in regulation. A comment of the European Commission on the use of the principle is analysed and the consequences for regulation of biological control agents are discussed. The different stakeholders (academia, industry, farmers and producers, consumers and the retail sector, environmentalists organised in non-government organisations, regulatory authorities and policy makers) and their interests in regulation are described.

## Contents



R.-U. Ehlers (✉)
Department for Biotechnology and Biological Control, Institute for Phytopathology, Christian-Albrechts-University Kiel, Hermann-Rodewald-Str. 9, 24118 Kiel, Germany
e-mail: ehlers@biotec.uni-kiel.de

R.-U. Ehlers (ed.), *Regulation of Biological Control Agents*,
DOI 10.1007/978-90-481-3664-3_1, 

## 1.1 Biological Control and Regulation of Biological Control Agents

Biological control uses living organisms like bacteria, fungi, nematodes, insects or mites (including viruses) for the control of weeds or pests and diseases of crop plants. Chemical compounds of natural origin, like plant extracts and semiochemicals (molecules functioning in bio-communication), are also assigned to the group of biological control agents (BCAs).

In the European Union, the registration requirements for active ingredients of all plant protection products were laid down in the EU Directive 91/414/EEC (EU 1991). This directive was amended by Directive 2001/36/EC (EU 2001) and 2005/25/EC (EU 2005) to adapt to the special requirements for plant protection products based on micro-organisms (MBCAs). On October 21, 2009, Dir. 91/414 was replaced by EC Regulation No. 1107/2009 (EU 2009a). Registration requirements and a comparison of registration practice in different OECD (Organsiation for Economic Co-operation and Development) countries are provided in Chapter 2 (Hauschild et al., 2011).

In organic farming specific rules have been developed to define which substances are allowed for use and which are exempted. BCAs used in organic farming are not excluded from registration by the European Commissions authority DG SANCO (Directorate General for Health and Consumer Affairs) and subsequent national authorisation. Minimum requirements for organic production are laid down in EC Regulation No. 889/2008. Annex II provides a list of plant protection products referred to in Article 5(1) of the Regulation (EU 2008). In addition to the EU and national authorisation, different international and national organisations (e.g., Bioland, Demeter) review BCAs for their possible potential and applicability for organic farming within their specific system. The rules have been summarized by Speiser and Tamm (2011) in Chapter 4.

Nematodes, mites and insects belong to the group of invertebrate biological control agents (IBCAs) or macro-organisms. Nematodes used in biological control of insects belong to the genera *Steinernema* or *Heterorhabditis*. *Phasmarhabditis hermaphrodita* is used for control of slugs (Grewal et al., 2005). An overview on mites with control potential is provided by Gerson et al. (2003). The majority of parasitic insects used in biological control are in the order Hymenoptera (e.g., Wajnberg and Hassan 1994; Malais and Ravensberg 2003; Helyer et al., 2003). A comprehensive review on methods to assess the risk of introducing exotic IBCA for use in area-wide, classical biological control or commercial biocontrol was edited by Bigler et al. (2006). These marco-organisms are not subjected to registration of plant protection products of the European Commission, which were laid down in the EU Directive 91/414/EEC. However, some member states (MS), like Austria, require some fundamental data for registration of IBCAs.

Risks related to the use of IBCAs are mainly due to import and release of exotic species. These aspects are summarized in Chapter 11 (de Clercq and Bale 2011). For the use of exotics in biocontrol there is no specific legislation in any jurisdiction within Europe so far. In those European countries, where regulation of IBCA is in place, it is either in the hands of authorities or institutes dealing with plant health or nature conservation and exceptionally dealt with by pesticide registration authorities. Hunt et al. (2011) reviewed the practice of IBCA regulation in OECD countries in Chapter 3 and Bale (2011) summarized proposals of the REBECA consortium on how to organize regulation of IBCAs (Chapter 16). An overview on IBCAs widely used commercially or in classical biological control in Europe and neighbouring Mediterranean countries is provided by the European and Mediterranean Plant Protection Organisation (EPPO 2010).

Viruses, bacteria and fungi need to be registered. Table 1.1 provides a list of all strains of microbial biological control agents (including viruses) that are currently authorized by the European Commissions Directorate General for Health and Consumer Affairs (SANCO). Table 1.2 lists all strains, for which the registration is currently reviewed.

Baculo- and nucleopolyhedrosis viruses are used in biological control of insects, almost exclusively against lepidopteran pests (Shuler et al., 1994; Hunter-Fujita 1998). Because of their safety for mammals (no transmission of mammalian pathogens) insect-baculovirus expression systems have received wide acceptance in pharmacology and medical research for production of recombinant proteins (Murhammer 2007). Safety aspects of baculoviruses were summarized in the document ENV/JM/MONO(2002)1 (OECD 2002). Chapter 12 presents the proposal to the EU authority SANCO for regulation of these viruses (Hauschild 2011).

Recently, mild strains of plant pathogenic viruses, which cause mild foliar mottle but no fruit symptoms, are inoculated to healthy plants to protect the crop against more virulent virus strains; however, these viruses have not yet received a registration as plant protection organism (Desbiez and Lecoq 2003).

One of the most successful biological control agents is the entomopathogenic bacterium *Bacillus thuringiensis* (Bt) (Charles et al., 2000). Comprehensive data material is available on the safety of Bts as insecticides (Glare and O'Callaghan 2000) and the World Health Oranisation (WHO) ranks Bt as the safest existing insecticide (International Labour Organisation and United Nations Environment Programme 1999).

Bacteria are also used to control plant diseases. Of major importance are members of the Enterobacteria, *Pseudomonas* and *Bacillus* spp. (Siddiqui 2006). Much research progress was made on the understanding of the mode of action of rhizobacteria for disease control and growth and plant health promotion (Bakker et al., 2008, Boland and Kuykendall 1998). Possible risks related with the use of bacteria in biological control are summarized in Chapter 7 (Alabouvette and Cordier 2011) and Chapter 8 (Berg et al., 2011).

**Table 1.1** Microbial control agents (including viruses) listed on Annex 1 of the Directive 91/414/EEC until July 2010

| Microbial control agent | Strain | Product name | Company | Use |
|---|---|---|---|---|
| *Ampelomyces quisqualis* | AQ10 | AQ10 | Intrachem, I | Powdery Mildew |
| *Bacillus subtilis* | QST 713 | Serenade | Agraquest, USA | Fungal Control |
| *Bacillus thuringiensis* subsp. *aizawai* | ABTS-1857 | Xentari | Valent Bioscience, USA | Lepidoptera |
| | GC-91 | – | Mitsui Agriscience, J | Lepidoptera |
| *Bacillus thuringiensis* subsp. israelensis | AM65-52 | Gnatrol | Valent Bioscience, USA | Fungus Gnats |
| *Bacillus thuringiensis* subsp. kurstaki | ABTS 351 | Dipel | Valent Bioscience, USA | Lepidoptera |
| | PB 54 | Belthirul | Probelte, E | Lepidoptera |
| | SA 11 and SA12 | – | Mitsui Agriscience, J | Lepidoptera |
| | EG 2348 | Leptinoc, Rapax | Intrachem, I | Lepidoptera |
| *Bacillus thuringiensis* subsp. tenebrionis | NB 176 | Novodor | Valent Bioscience, USA | *Leptinotarsa decemlineata* |
| *Beauveria bassiana* | ATCC 74040 | Naturalis | Intrachem,I | White Flies |
| | GHA | Botanigard | Laverlam, CO | Homoptera, Thrips |
| *Coniothyrium minitans* | – | Contans | Prophyta, DE | *Sclerotinia sclerotiorum* |
| *Cydia pomonella NPV* | Mexican | Madex | Andermatt, CH | Apple Codling Moth |
| | Mexican | Granupom | Probis, DE | Apple Codling Moth |
| | Mexican | Carpovirusine | Arysta, F | Apple Codling Moth |
| *Gliocladium catenulatum* | J1446 | Prestop | Verdera, FIN | Fungal Control |
| *Lecanicillium muscarium* | Ve6 | Mycotal | Koppert, NL | Homopteran Insects |
| *Metarhizium anisopliae* | BIPESCO 5F/52 | Granmet | Agrifuture, I | Insect Pathogen |
| *Paecilomyces fumosoroseus* | Apopka strain 97 | Preferal | Biobest, BE | Homopteran Insects |
| *Paecilomyces lilacinus* | 251 | Bioact | Prophyta, DE | Plant Nematodes |
| *Phlebiopsis gigantea* | several | Rotstop | Verdera, FIN | *Heterobasidium annosum* |
| *Pseudomonas chlororaphis* | MA342 | Ceral, Cedomon | Bioagri, SE | Seed Treatment Fungi |
| *Pythium oligandrum* | M1 | Polyversum | Biopreparaty, CZ | Fungal Control |
| *Spodoptera exigua* NPV | – | Spodex | Certis, USA | *Spodoptera exigua* |
| *Streptomyces* sp. | K61 | Mycostop | Verdera | Fungal Control |
| *Trichoderma aspellerum* | ICC 012 | Remedier, Tenet | Isagro, I | Fungal Control |
| | T11 | Tusal | Newbiotechnic, E | Fungal Control |
| | TV1 | Xedarvir | Xeda, F | Fungal Control |
| *Trichoderma atroviride* | IMI 206040 | Binap T | Binap Bio-innovation, S | Fungal Control |
| | T 11 (T25) | Tusal | Newbiotechnic, E | Fungal Control |
| *Trichoderma gamsii* | ICC080 | Remedier, Tenet | Isagro, I | Fungal Control |
| *Trichoderma harzianum* | T22 | Trianum | Koppert, NL | Fungal Control |
| | ITEM908 | Micover | Agrifutur, I | Fungal Control |
| *Trichoderma polysporum* | IMI206031 | Binap T | Binap Bioinovation, SE | Fungal Control |
| *Verticillium dahliae alboatrum* | WCS850 | Dutch Trig | BTL Bomendienst, NL | Dutch Elm Disease |

**Table 1.2** Microbial control agents, including granulose- (GV) or nucleopolyhedro-viruses (NPV). Listing on Annex 1 of the Directive 91/414/EEC pending until July 2010

| Microbial control agent | Strain | Use |
|---|---|---|
| *Adoxophyes orana* GV | BV-0001 | *Adoxophyes orana* |
| *Aureobasidium pullulans* | DSM 14940 + 14941 | *Erwinia amylovora* |
| *Candida oleophila* | O | Post harvest fungal control |
| *Helicoverpa armigera* NPV | – | *Helicoverpa armigera* |
| *Paecilomyces fumosoroseus* | Fe9901 | Insect control |
| *Pseudomonas* sp. | DSMZ 13134 | Seed treatment fungi |
| *Pseudozyma flocculosa* | PF-A22 UL | Powdery Mildew |
| *Spodoptera littoralis* NPV | – | *Spodoptera littoralis* |
| *Trichoderma atroviride* | I-1237 | Fungal control |
| Zucchini Yellow Mosaik Virus | weak strain | Zuchini Yellow Mosaic |

Likewise, fungi are used to control insects and plant diseases. Fungi for insect and nematode control are in the genera *Metarhizium*, *Beauveria*, *Paecilomyces* and *Lecanicillium* (Butt et al., 2001). The major groups of fungi used in disease suppression are in the genera *Trichoderma* and *Gliocladium* (Verma et al., 2007; Kubicek and Harman 1998; Harman and Kubicek 1998), but non-virulent isolates of plant-pathogenic fungi, like *Fusarium* spp., are also used (Lemanceau and Alabouvette 1991). Of general concern during the regulation process of fungal BCAs are toxic fungal metabolites. These risks are reviewed by Strasser et al. (2011) in Chapter 9. Proposals for improvement of the regulation requirements for MBCA are summarized in Chapter 13 (Strauch et al., 2011).

Among the so called botanicals, some are highly toxic and thus are excluded from use as plant protection products (e.g., nicotine). Others, like neem or pyrethrum, are less toxic for non-target organisms and have long been used safely in integrated pest management (Regnault-Roger et al., 2005). Throughout evolution, organisms have developed semiochemicals that are involved in intra-and inter-specific communication and several molecules are currently used for monitoring insect pest populations or applied in mating disruption (sex pheromones) and others can be used as repellents (allomones) or attractants (kairomones) (Howse et al. 1998). The sex pheromones are long chain fatty acids, which are not subjected to registration when used in monitoring flight of adult insects or estimating their population density with, e.g., sticky traps; when used for area-wide control of mating (mating disruption), they need an authorisation. Safety of pheromones and other semiochemicals used for arthropod pest control has also been reviewed by the OECD (2003). Risks of botanicals and semiochemicals were reviewed by Regnault-Roger (2011) in Chapter 10 and recommendations on how to improve registration for botanicals is presented in Chapter 14 (Tamm et al., 2011) and for semiochemicals in Chapter 15 (Speiser et al., 2011).

Agricultural ecosystems benefit from the resident communities of antagonistic macro- and micro-organisms responsible for naturally occurring biological control of pest and disease species. The environmental and economic significance of biological control by far exceeds chemical control when taking into account the economic benefit of the naturally occurring antagonistic spectrum present at any agro-ecosystem. These antagonists prevent outbreaks of most of the known pest and disease populations, thus avoiding major crop damage. Only a minority of pest and disease populations need to be reduced by control measures, the majority do not exceed the economic threshold level due to the antagonistic potential of BCAs. Knowledge-based ecosystem management (Pickett and Buggs 1998) can help to preserve or even promote the positive impacts of BCAs. Under these circumstances biological control is never regulated by any authority. Whatever is endemic at a certain place is considered to be part of the natural environment.

When used by man in plant protection, BCAs are introduced or applied as an inoculative release, an augmentative or an inundative application. The application can be limited to a glasshouse or field or can be area-wide, which is typical for classical biological control. In classical biological control, natural enemies are released against introduced exotic pests, diseases or weeds. They have been imported from the place of origin of the pest. Biological control makes use of these natural resources for plant protection. BCAs are taken from natural environments. They are not synthetic. Mankind and other organisms share a long-lasting evolution with these antagonistic beneficial organisms, of which some are also used in biological control. This does not imply that biological control agents are without risks.

Regulation comes into play only when biological control agents or botanicals and semiochemicals are artificially augmented in the environment. Whether used in commercial biological control or classical biological control makes no difference.

When it comes to inundative or inoculative use of BCAs, their economic significance is small, with an overall annual turnover of 3% of the total plant protection revenues (IBMA 2008), but is growing rapidly with annual increases of between 5 and 20% (Frost and Sullivan 2001). Commercialisation of BCAs is mainly in the hands of small-and-medium-sized enterprises (SMEs).

The potential of biological control for use in agriculture, horticulture and forestry is immense. Nowadays, fewer and fewer chemical compounds make it to the market. By contrast, the exploitation of the huge biodiversity with potential for biological control has only just begun and provides an impressive reservoir for plant protection with potential to substitute many hazardous chemical control products.

## 1.2 Regulation of Biological Control Agents in Europe – the REBECA Policy Support Action

Plant protection products (PPPs) can be harmful to humans and the environment. For this reason their risks need to be evaluated and active ingredients must be authorised prior to commercial use and authorities need to develop risk management strategies to minimize possible negative effects. Authorisation for use is only given if unacceptable negative effects to humans and the environment can be excluded. Registration of PPPs based on BCAs follows rules originally developed for the risk assessment of synthetic chemical compounds. Although the data requirements for micro-organisms have been adapted twice to facilitate the registration process, the requirements still are one of the major hurdles for BCAs to reach the market. The stringent regulation policy for BCAs, based mainly on registration rules for synthetic chemical pesticides, has hampered the development and use of biological control in Europe.

The current situation for registration of BCAs is as follows:

- Considering the market potential, costs are too high (between 0.5 and 2.5 million €)
- The market size often cannot support costs, consequently few products are available
- BCA registration takes too long, sometimes exceeding 9 years for Annex 1 inclusion
- A major obstacle is the subsequent member state authorisation (additional 2 years)
- Countries vary in interpretation of guidelines
- Mutual recognition is not well implemented
- Guidelines/requirements are not set up for BCAs
- With a lack of knowledge and experience, regulation adopts the precautionary principle
- Efficacy trials are more difficult and costly for BCAs
- Regulation authorities and SMEs often have limited knowledge on BCA registration
- Registration is a blackbox that cannot attract venture capital and investment
- Registration is a major barrier of entry for SMEs

Much investment went into research and development of BCAs in the public and private sector. Despite these activities, progress in exploitation of BCAs in agriculture has been limited. This motivated the EU Commissions´ General Directorate for Research to publish the following call for proposals: "Despite considerable research efforts on BCAs the number of such products on the market in Europe is currently still extremely low. BCA cannot be treated like synthetic chemicals and need different approaches for registration purposes". After 15 years of disappointing results with registration of biocontrol agents following Dir. 91/414, the need for a review of regulation procedures for BCAs was realized.

The result of an application to this call was the EU Policy Support Action REBECA (Regulation of biological control agents in Europe), which gathered all stakeholders in biocontrol in Europe to build a network for exchange of information and a platform for discussions on how to improve regulation of BCAs in Europe. The Action was supported by valuable contributions from experts from overseas.

The Action first wanted to review possible risks of biocontrol agents. In parallel, experts compared regulation in the EU with rules in other OEDC countries. The results were then presented in a first joint conference. The next activity was to work on the development of proposals for alternative or improved regulation rules. The proposals were then presented during the final conference held in Brussels. The progress was reviewed by an Action Steering Group, which gathered members of science, policy, regulation and non-governmental organisations. The flow chart of the REBECA Action is presented in Fig. 1.1.

The work was divided into the following work packages (WP), which were managed by different REBECA participants.

- WP 1: *Management and co-ordination* was in the hands of Olaf Strauch, Miriam Döring and Ralf-Udo Ehlers (Christian-Albrechts-University, Kiel, Germany)
- WP 2: *Comparison of current legislation practice* was divided into two tasks, the review on IBCAs managed by Ulrich Kuhlmann (Commonwealth Agriculture Bureau International, Delemont, Switzerland) and all other agents organised by Rüdiger Hauschildt (GAB Consulting GmbH, Lamstedt, Germany) and by Bernard Speiser and Lucius Tamm (FIBL, Research Institute for Organic Agriculture, Frick, Switzerland)
- WP 3: *Risk assessment of microbial biocontrol agents* organised by Hermann Strasser (University Innsbruck, Austria)
- WP 4: *Risk assessment of botanicals and semiochemicals* organised by Lucius Tamm (FIBL, Switzerland)
- WP 5 RA: *Risk assessment of macrobials* organised by Jeffrey Bale (University of Birmingham, UK)
- WP 6: *Risk trade-off and cost-benefit analysis of regulation* organised by Heikki Hokkanen (University of Helsinki, Finland)
- WP 7: *Measures to accelerate regulation* organised by Anita Fjelsted (Danish Environmental Protection Agency, Copenhagen, Denmark)

The objectives of the Actions were to elaborate proposals that could help to

- develop less bureaucratic and more efficient regulation procedures
- develop more balanced regulation according to potential hazards
- maintain the same level of safety for human health and the environment
- accelerate market access
- lower registration costs
- define "low risk products", which might be exempted from registration
- propose alternative regulation systems

The results of the Action and much additional information on regulation requirements and biocontrol safety information were disseminated on the webpage http://www.rebeca-net.de, which also made available the reports and deliverables.

## 1.3 History of Biocontrol Registration

In Europe, PPP regulation was introduced in the 1960s. On the initiative of the chemical industry, governments gave authorisation exclusively for those pesticides, for which evidence for their efficacy was provided. Environmental aspects were

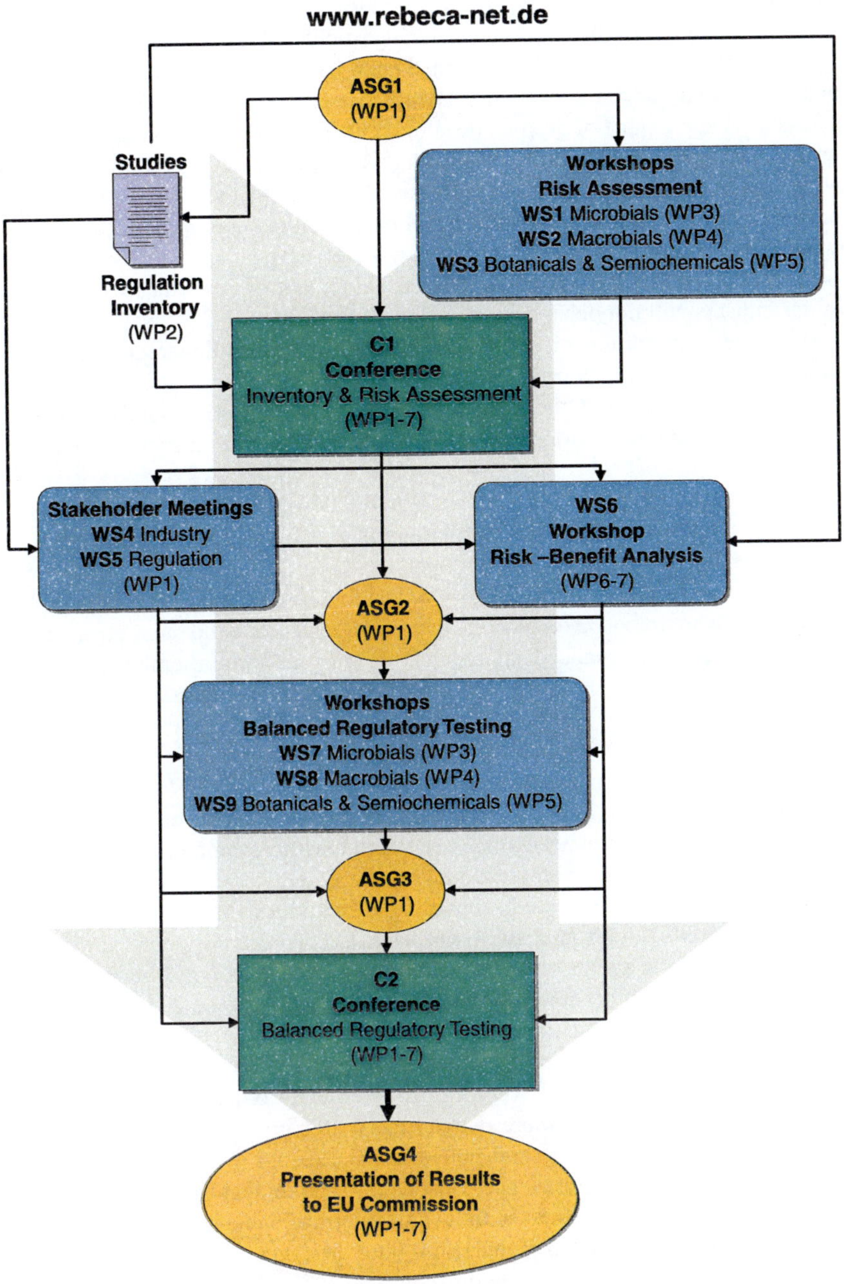


**Fig. 1.1** Flow chart presenting the organisation of the REBECA EU Policy Support Action. AGS: Action Steering Group; WP: Work Package; WS: Workshop

only considered and included in the registration process in response to concerns about accumulation of the organochlorine insecticide DDT in the food chain. Since then PPPs posing unacceptable risks have been banned and/or substituted, and the chemical industry adapted to the increasingly strict standards by monitoring safety aspects at an early stage of product development. The history of regulation has been a process of replacement of one chemical group by another, which often exhibited another set of problems. This process was accompanied by the development of more and more stringent rules taking into account scientific reports of damage caused by synthetic compounds and anticipated risks of new compounds. Governments responded to reports of damage with the development of new rules to ensure that similar impacts will not occur with new compounds.

Since the introduction of regulation in Europe, registration requirements and guidance documents had always been developed in consultation with multinational agrochemical companies. Other than regulation of synthetic compounds, regulations for biological plant protection products have not evolved within such a process:

- Regulation of biological PPPs was not a gradual evolution involving industry
- Regulation was not based on scientific reports of damages, as there are hardly any reports on damage of BCAs
- There is no evolution of regulatory rules for BCAs; the rules for synthetic compound were imposed on biocontrol without consulting the biocontrol industry
- Adapted and more balanced approaches existing in some member states were even rolled back with the introduction of Dir. 91/414 as a consequence of better harmonisation.

For example, in Germany, before implementation of Dir. 91/414, the requirements for PPP based on insect viruses were much reduced after the first file (*Cydia pomonella* GV) had been processed. With the implementation of Dir. 91/414, applicants had to provide a complete data set again.

Although not a good example for handling even minor risks, for many years Italy had no regulation for microbial BCAs in place. Companies only needed to use the scientific name of the agents on their products. *Bacillus thuringiensis, Trichoderma harzianum* and many other micro-organisms had been marketed without evaluation of safety data until 2006. No damage was recorded.

With the introduction of the EU regulation old active ingredients had to undergo the process of re-registration. According to EU policy objectives, this process is targeted at the substitution of more risky PPPs. With increasing knowledge and scientific evidence about damage and potential risks of old synthetic compounds, a re-registration is a logic consequence. However, for biological control agents, which have been safely used for decades without any reports of damage and for which more and more knowledge has been gathered proving their safety, such a re-registration seems unnecessary. The re-registration requirement was the consequence of handling biologicals like synthetic PPPs and was not based on scientific information on damage and risks. Many biological control agents, for which re-registration has not been applied, are now out of the market. It does not mean that they are risky. The market is too small to justify the registration costs. Farmers have lost safe natural products due to policy decisions, which aim to limit negative effects of synthetic compounds; as the same rules are implemented also for the safe alternatives, the effect is counter-productive.

Compared with the chemical industry, the participation of the biocontrol industry in defining regulatory rules was minor. One reason certainly was the rudimentary representation and ineffective group organisation of the comparatively young biocontrol enterprises. Another was the limited knowledge and experience available in these companies and also on the side of regulation authorities. Only a few years ago, the OECD asked for industry participation when discussing guidance documents for micro-organisms and invertebrates, but it was only with the start of the REBECA Action that an intensive dialogue between all stakeholders in regulation of biological control agents was introduced. The Action was very well attended and resulted in a better dissemination of knowledge and experience among all stakeholders. The policy aspects of regulation are reviewed by Grant (2011) in Chapter 5.

With the limited economic importance of biocontrol during the time of implementation of Dir. 91/414, one can understand why little emphasis was given to specify regulation for BCAs. However, this situation has now changed. Problems with chemical control compounds increase and growers in Europe are starting to realise the potential of BCAs. The biocontrol industry is flourishing with up to 20% increase in annual sales. Growers start to realize that BCAs have the potential to close control gaps and substitute some of the environmentally risky synthetic PPPs. In order to protect consumers more effectively from residues of synthetic PPPs, avoid hazards for users of synthetic PPPs and preserve agro-ecosystems, a rapid market access for biological products would be desirable. A better adapted regulation procedure would help to reduce restrictions and ease the market access for environmentally sound biocontrol PPPs.

In view of the history of regulation of BCAs, the REBECA consortium proposes to

- continue the dialogue between all stakeholders
- critically review the existing regulatory practice
- develop new and innovative strategies for BCA regulation
- consider more adapted regulatory measures according to the real risks of BCAs

## 1.4 The Precautionary Principle in Risk Assessment

The precautionary principle is the basis of European risk management and is thus also applied to biological control agents. It is often mentioned that BCAs might possibly pose risks similar to synthetic PPPs or pose unknown risks that have not yet been identified. These "unknown unknowns" are often a justification for the execution of the precautionary principle on BCAs and why rules similar to those developed for chemical compounds are applied.

The decision making in regulation is based on data from investigations and applying experimental models for assessment of potential risks. Data are used to predict hazards and quantify the probability of occurrence and the development of risk management strategies. However, the system could not always prevent hazards to the environment. Atrazines, for instance, were detected in the ground water and their use had to be banned. Only recently, tolyfluanid-containing fungicides were banned because the compound is metabolised in the soil to dimethylsulfamid (DMS), which is displaced into the ground water.

These failures of the regulatory system to prevent hazards to the environment, have resulted in it becoming customary to demand the application of the precautionary principle for regulation of PPPs, including those of biological origin. This new approach is forming the basis of the European regulatory systems and is reflected also in the Rio Declaration (1992): "in order to protect the environment, the precautionary approach shall be widely applied by states. Where there are threats of serious or irreversible damage, lack of full scientific certainty shall not be used as a reason for postponing cost-effective measures to prevent environmental degradation". The REBECA consortium could not identify major threats with severe consequences for humans and the environment related with the use of currently registered BCAs or invertebrate BCAs.

Within the EU Commission, the interpretation of the precautionary principle treats the principle less like a dogma but more as the beginning of a serious analysis of how to approach risks within the authorities dealing with risk assessment and management. The Commission published a communication on the precautionary principle (European Commission 2000) outlining the EU Commission's approach to use the principle and establishing guidelines for application. The Commission clearly states "that recourse to the precautionary principle presupposes that potentially dangerous effects... have been identified and that scientific evaluation does not allow the risk to be determined with sufficient certainty." Is this an argument to demand the application of the precautionary principle for the regulation of BCAs? Risks related with the use of BCAs have been described and in many cases their dimension has been scientifically assessed. The RAFBCA project (QLK1-CT-2001-01391) worked on fungal antagonists and the ERBIC project (FAIR5-CT97-3489) on invertebrate BCAs. Both projects identified potential risks and also concluded on their dimension and probability of occurrence. Together with the results gathered and summarized by the REBECA Action (www.rebeca-net.de: Safety information) or the biopesticide fact sheets provided by the Environmental Protection Agency in the USA (http://www.epa.gov/pesticides/biopesticides/#factsheet) much information is available to conclude that regulation of BCAs can be based on scientific evidence and that we do not need to apply the precautionary principle. Thus, we do not have so many "unknown unknowns" but rather a set of known risks with limited dimension.

The Commissionś communication further outlines the general principles of risk management measures (COM (2000)1):

- proportionality
- non-discrimination
- consistency,
- examination of the benefits and costs of action or lack of action
- examination of scientific developments

Measures should be proportional to the desired level of protection and should not be discriminatory in their application. A comparable situation should not be treated differently and different situations should not be treated in the same way. Taking this principle literally, we must analyse whether the reduced risks related to biological PPPs now paves the way for the separation of the risk assessment practice of biological and synthetic products.

The Commission demands that "measures should be consistent with the measures already adopted in similar circumstances or using similar approaches." Biological PPPs often only share their use in plant protection with synthetic compounds. Many other comparable agricultural practices are not regulated like BCAs. The use of organic fertilizers (containing a much higher amount of micro-organisms than used in biological control) is not regulated. Nitrogen-fixing *Rhizobium* bacteria are applied to seeds and are not regulated. In many countries the plant-growth promoting products are subject to lower level regulation. Even in the food industry alternative approaches are successfully used. The "qualified presumption of safety" (QPS) concept provides a generic assessment system for micro-organisms deliberately introduced into the food chain (see also Chapter 17). This system allows for experience to be introduced into the assessment and should be further elaborated for the assessment of plant protection products.

In addition, the Commission states that "measures.... shall be re-examined and if necessary modified depending on the results of the scientific research and the follow up of their impact." As much more scientific information is now available this seems to be a good opportunity to review the legislation of BCAs and develop more balanced, better adapted and more cost-effective regulation procedures for BCAs.

The REBECA Action was a starting point to produce a network of all stakeholders involved in regulation of BCAs. Within the time frame of the Action, the activities concentrated on providing proposals for a short term improvement of conditions. Further activities in the analysis of the risks and the development of innovative regulation strategies must now follow to provide the appropriate conditions for a faster development of biological control measures in European agriculture. The rules defined by the Commission need to be applied also to BCAs.

Reviewing the Commissions' communication of the precautionary principles the REBECA consortium proposes to

- treat BCAs in a non- discriminative way
- consider their lower risk compared with synthetic compounds
- take into consideration experience and available data from comparative use
- re-examine measures based on new scientific results on the safety of BCAs

## 1.5 Stakeholders

The REBECA Action tried to get as many competent stakeholders as possible to participate in the Action. In the area of regulation of BCAs, stakeholders are in academia and industry, and farmers and producers are affected; stakeholders also include consumers and the retail sector, environmentalists organised in non-government organisations (NGOs), regulatory authorities and policy makers.

### *1.5.1 Academia*

Scientists, who are working in development of BCA in public entities, are interested in successful implementation of their R&D results. Most of the BCAs currently in the market originate from the activity of public research organisations, institutes of higher education or governmental research organisations. Much research into the safety of BCAs is also undertaken by these research organisations, a motivation for the academic sector for more research activities into scientific assessment of risks and risk analysis. Often these activities result in more rather than less registration requirements.

### *1.5.2 Industry*

An important stakeholder is the biological control industry. The structure of bio-control industry is diverse. The large (transnational) chemical companies have no major interest in BCAs for several reasons. Most products have a short shelf life and thus do not fit well into the distribution logistics of the chemical companies. BCAs are often more expensive than the synthetic compounds in their portfolio. Marketing strategies for BCAs are more difficult to develop and biocontrol products would be competing with their own synthetic products. On the other hand, this industry has huge R&D and registration departments, which involve tremendous costs and the usually smaller markets of BCAs cannot show a financial return on the investment. The chemical companies prefer to go for "blockbusters" rather than for niche products, like BCAs. However, since the concerns regarding pesticide residues increase and are constantly highlighted in the media, chemical companies are currently developing interest in the biocontrol sector. For example, Bayer Cropscience (Monheim, Germany) is testing *Bacillus firmus* for nematode control and BASF (Limburgerhof, Germany) and AgraQuest Inc (Davis, CA, USA) have entered into a license, supply and distribution agreement for Serenade®, a bio-fungicide based on *Bacillus subtilis*. Syngenta Bioline Ltd (Little Clacton, Essex, UK) are producers of natural beneficial insects, mites and bumblebees for integrated pest management in horticulture.

In the past the economic significance of biological control was negligible but since the biocontrol industry has become the major supplier for PPP in the glasshouse sector and has now expanded applications into out-door crops, the biocontrol industry has become a small, but important, competitor. As a consequence this might also motivate competitive interests rather than support of activities to ease registration requirements for BCAs.

Most biocontrol companies are small and medium-sized enterprises (SMEs). Several biocontrol companies are spin-offs of public research organisations. These start-up companies usually lack capital for larger investment into registration. Investment capital is difficult to obtain from the financial markets as business plans appear unattractive, due to the unpredictable duration of the registration process. Authorisation also of biocontrol products can last for more than 10 years and can involve costs exceeding 2 million €. Consequently, companies were either successful when they were marketing IBCA (insects, mites and nematodes), which are usually exempted from registration (e.g., Koppert, in The Netherlands or Biobest, Belgium) or when they were able to attract venture capital to support the registration (e.g. Agraquest). Some of these companies now have smaller product portfolios. Others were able to start joint ventures with, or were acquired by, larger companies in the food and agriculture sector who supplied the necessary financial resources for product registration (e.g. Bioagri AB in Sweden).

The biocontrol industry is organized within the International Biocontrol Manufacturers´Association (IBMA) (http://www.ibma.ch) and the BioPesticide Industry Alliance (BPIA) (http://www.biopesticideindustryalliance.org). Within the REBECA Action the IBMA was often represented by Ulf Heilig, a private consultant in registration support, who contributed a lot to the discussions and elaboration of proposals.

The interests of industry in the Action were quite diverse. On the one hand, larger companies, who run experienced registration departments and had registered products in the market, were more reluctant about reduction of the registration requirements. They had gone through the mill, why should other have an easier run? Other companies, who were new in the business and had not yet registered their results of R&D or had products in registration, were more open to support the Action. In the area of IBCA regulation the larger companies were the driving forces to define Europe-wide regulation rules and smaller companies did not participate in the work, due to lack of expertise and personnel.

Working with biocontrol industry one must always have in mind that registration is a possibility to protect markets and exploit competitive advantages. In the biological control sector innovation is not easily protected. Living organisms cannot be patented and the same is also the case for protection of results of genetic improvement by selective breeding. The biocontrol industry is trying to keep intellectual property in-house. Under these circumstances an authorisation for a biological control agent is of larger value than for a well protected chemical compound.

### *1.5.3 Farmers and Producers*

Users of BCAs are found in the conventional and organic agricultural and horticultural sector. Forestry is increasingly moving away from plant protection, but in some countries produces considerable demands for BCAs, particularly for *B. thuringiensis* based products. As an increasing number of synthetic chemical compounds have not been defended (re-registered) by the chemical industry or have been withdrawn due to environmental concerns, the agricultural sector is lacking alternative PPPs. Of the previously existing active ingredients of PPPs listed in Annex 1 of Dir.

91/414, 67% were not defended, 7% were rejected and 26% approved within the re-registration process (Richardson 2009). Biological control would be able to fill part of this gap; however, the sparse financial input into registration resulted in limited product availability. As a consequence, the majority of the farmers and producers have not considered these products as realistic alternatives. The image of the early biocontrol industry was bad. In the past, the products were considered to be of low quality, too expensive and lower in control efficacy than chemical compounds. With this image of biocontrol products, producers did not lobby for biological alternatives to be supported by governments. The chemical paradigm (knock-down effect, cheap, easy-to-use, preventive treatment) is difficult to change and biological control products had major problems in persuading the conventional sector to use their products.

This has changed, not radically, but in small steps, since the conventional sector has experienced successful replacement of synthetic compounds by BCAs (e.g., in the greenhouse sector in Mediterranean countries, the use of CpGV against codling moth in apple orchards, *B. thuringiensis* products against lepidopterans with resistance to synthetic insecticides). The lobby of farmers still is more in favour for chemical compounds, however, the door has been opened and in the future producers might advocate more for political support of biological alternatives.

### *1.5.4 Consumers and Retail Sector*

The debate about pesticide residues in food produce was one of the driving forces for the development of biological control. For a long time non-government organisations (NGOs), like Greenpeace, made public residues in vegetables and fruit and offered residue-free shopping lists on their webpage, without any major impact on the use of synthetic compounds. It was only when the NGOs began to search for residues in produce sampled from the shelves of different retailers that the campaign began to have an impact on the purchasing policy of the retailers. Suppliers are today put on contracts, in which they have to guarantee that pesticide residues in their produce would not exceed retailers specifications, which are below what governments allowed and which is limited to only two or three substances. Although the retail sector, in the beginning, had just implemented these rules without discussing alternative control strategies with the suppliers, this policy made many producers switch to alternative and residue-free control strategies in the horticulture sector. The further development of new products to supplement the PPP portfolio suddenly is of increasing interest and cooperation between the biocontrol and retail sector should in the future be intensified to enhance the confidence in the quality and potential of biological control strategies. Thus, the policy of the retail sector has become the major driving force for implementation of biological control. Whether the retail sector will support the activities to reduce registration requirements is doubtful as they are advocates for the safety of the consumers and have little expertise in judging risks of BCAs or comparing these risks with risks resulting from the use of chemical PPPs.

### 1.5.5 Environmentalists Organised in NGOs

PAN (Pesticide Action Network) and Greenpeace are two NGOs active in the assessment of risks related with the use of chemical PPPs. Their activities related to pesticide residues in food have resulted in an increasing implementation of biological control in the past decade. However, so far they have not participated in the discussion on risks and regulation of BCAs. Asked to participate in the activities of the REBECA Action, they confessed that they lack expertise as their focus is on chemical control substances. Criticism is more powerful when better alternatives can be offered. Consequently, the biological control sector should increase their efforts to integrate and cooperate with NGOs.

### 1.5.6 Regulatory Authorities

Regulation in Europe is a two-phased process. The active ingredient is authorised by the EU Commission DG SANCO and the formulated product is still a matter of national authorisation. However, one Rapporteur Member State (RMS), which is usually selected by the applicant, is in charge of putting together the data requirements and producing the Draft Assessment Report (DAR) for submission to SANCO. Northern Europe countries share a well developed infrastructure for pesticide registration; several Southern European member states have caught up, but new and smaller member states still lack the resources and expertise. The EU-wide harmonisation of registration rules was a necessary political step to exclude competitive advantages in the agricultural sector and improve on the safety for the consumers. As a result of the "mad-cow-disease", the EU created another European organisation, the European Food Safety Authority (EFSA), which is now advising the Commission in questions of pesticide safety. This organisation is building expertise and hopefully will, in the future, also provide excellence in reviewing risks of BCAs.

The re-registration of the PPPs was a tremendous workload for the authorities. Now the work is done and at the same time fewer chemical products are being developed and only a few make it to the market. As companies can select the RMS, authorities in Europe will face competition for submissions and might run into shortage to keep their departments busy and maintain the expertise. Some countries have already implemented guidance programmes to support authorisation of BCAs (GOENOG in NL and Biopesticide Scheme in the UK). The aim is to bring more biological products to the market, facilitate the initial contact between companies and authorities and help industry through the approval process (see Chapters 5 and 17). As it will reduce costs for the evaluation, it is a useful strategy to attract companies to those member state authorities that provide this support.

Regulators administrate the rules set by policy makers. They are dealing with the dossiers and transfer regulation into practice. The progress of the REBECA Action depended greatly on the contribution of regulatory personnel. Their expertise was very valuable as they were open to provide information and actively participated.

In the beginning of the Action, we anticipated much more input and innovative proposals for change from the biocontrol industry and less from regulation authorities. During one workshop a participant regulator mentioned that the job of a regulator was to regulate and not have visions about future solutions to ease the registration of BCAs. For several reasons the contribution from industry was less compared with the regulators, who contributed to the discussions and provided input for improvements.

### *1.5.7 Policy-Makers*

Policy-makers did not participate in REBECA, possibly because the REBECA consortium was not able to attract their attention or because their awareness of biological control was/is remote, which is probably due to the rudimental level of representation of the biocontrol industry at the EU and MS administrations and its limited resources to support lobbying. In the past, the biocontrol industry had no lobby and thus was not noticed. This becomes apparent when analysing the decisions of European policy-makers on the reduction of the use of pesticides (EU 2009b). The European Parliament decided on a "Thematic Strategy on the Sustainable Use of Pesticides", stating that low pesticide-input farming needs to be promoted, priority should be given to non-chemical methods and meaningful support to organic farming. MS should be required to set up National Action Plans for reducing pesticide use and the development of plant protection products with a low risk profile should be encouraged. It is obvious that neither Parliament Members nor politicians on the MS level considered that the use of biological control agents would result in a significant reduction of chemical pesticides, otherwise they would have recommended the use of BCAs in their documents. In this aspect, policy is not meeting its own objective, which is to reduce pesticide use. Their support for biological control, with few exceptions, has always been of minor impact and was limited to support of research projects.

Another problem is that policy-makers are usually not aware that decisions taken to restrict the use of chemical pesticides have, at the same time, negative effects on biological control. As BCAs are covered within the same legislation as synthetic compounds, restrictions on the use of synthetic compounds automatically apply also for BCAs. These trade-off effects are often neglected by policy-makers.

Within the REBECA Action representatives of almost all stakeholders contributed to the success of the activities. Besides the provision of reviews on existing regulatory practice and the proposals made to improve the regulation of biological control agents, which are all summarized in this book, a significant success of the REBECA Action was the organisation of a platform for exchange of information and opinions for all stakeholders This initiated a Europe-wide discussion on regulation of BCAs, which will also lead, hopefully one day in the near future, to a further improvement of regulation for biological control agents and thereby accelerate the provision of environmentally friendly plant protection products for the agricultural sector.

## References


Alabouvette C, Cordier C (2011) Risks of microbial biocontrol agents and regulation: are they in balance?. In: Ehlers R-U (ed) Regulation of biological control agents. Springer, Dordrecht, pp 157–173

Bakker PAHM, Raaijmakers JM, Bloemberg G, Höfte M (2008) New perspectives and approaches in plant growth-promoting rhizobacteria research. Eur J Plant Pathol 119:1–126

Bale JS (2011) Regulation of invertebrate biological control agents in Europe: recommendations for a harmonised approach. In: Ehlers R-U (ed) Regulation of biological control agents. Springer, Dordrecht, pp 232–373

Berg G, Zachow C, Cardinale M, Müller H (2011) Bacterial ecology and human pathogenicity of plant-associated bacteria. In: Ehlers R-U (ed) Regulation of biological control agents. Springer, Dordrecht, pp 175–189

Bigler F, Babendreier D, Kuhlmann U (2006) Environmental impact of invertebrates for biological control of arthropods: methods and risk assessment. CAB International, Wallingford, CT

Boland GT, Kuykendall LD (1998) Plant-microbe interactions and biological control. Marcel Dekker, New York, NY

Butt TM, Jackson CW, Magan N (2001) Fungi as biocontrol agents: progress, problems and potential. CABI Publishing, Wallingford, CT

Charles JF, Delécluse A, Nielsen-le Roux C (2000) Entomopathogenic bacteria: from laboratory to field application. Kluwer, Dortrecht

De Clercq P, Bale JS (2011) Risks of invertebrate biological control agents – *Harmonia axyridis* as a case study. In: Ehlers R-U (ed) Regulation of biological control agents. Springer, Dordrecht, pp 243–255

Desbiez C, Lecoq H (2003) Zucchini yellow mosaic virus. Plant Pathol 46:809–829

EPPO (2010) List of biological control agents widely used in the EPPO region. EPPO Standards on Safe use of Biological Control – PM 6/3 http://archives.eppo.org/EPPOStandards/biocontrol_web/bio_list.htm. http://archives.eppo.org/EPPOStandards/biocontrol_web/bio_list.htm

EU (1991) 1991/414/EEC: Council Directive of 15 July 1991 concerning the placing of plant protection products on the market Official Journal of the European Communities L 230

EU (2001) 2001/36/EC: Commission Directive of 16 May 2001 amending Council Directive 91/414/EEC concerning the placing of plant protection products on the market. Off J Eur Commun L 164/1

EU (2005) 2005/25/EC: Commission Directive of 14 March 2005 amending Annex VI to Directive 91/414/EEC as regards plant protection products containing micro-organisms. Off J Eur Union L 90/1

EU (2008) Regulation (EC) No 889/2008 of 5 September 2008 laying down detailed rules for the implementation of Council Regulation (EC) No 834/2007 on organic production and labelling of organic products with regard to organic production, labelling and control. Off J Eur Union L 250/1

EU (2009a) Regulation (EC) No 1107/2009 of the European Parliament and of the Council of 21 October 2009 concerning the placing of plant protection products on the market and repealing Council Directives 79/117/EEC and 91/414/EEC. Off J Eur Union L 309/1

EU (2009b) Directive 2009/128/EC of the European Parliament and of the Council of 21 October 2009 establishing a framework for Community action to achieve the sustainable use of pesticides. Off J Eur Union L 309/71

European Commission (2000) Communication from the Commission on the precautionary principle. Commission of the European Communities COM 2000–1

Frost and Sullivan (2001) European biopesticides market http://www.frost.com/prod/servlet/report-toc.pag?ctxixpLink=FcmCtx3&searchQuery=European+biopesticide+market+2001&repid=3905-01-00-00-00&bdata=aHR0cDovL3d3dy5mcm9zdC5jb20vc3JjaC9jYXRhbG9nLXNlYXJjaC5kbz9xdWVyeVRleHQ9RXVyb3BlYW4rYmlvcGVzdGljaWRlK21hcmtldCsyMDAxJnBhZ2VTaXplPThAfkBTZWFyY2ggUmVzdWx0c0B%2BQDEyOTEwNDAxOTcxOTE%3D&ctxixpLabel=FcmCtx4. www.frost.com

Gerson U, Smiley RL, Ochoa R (2003) Mites (Acari) for pest control. Wiley-Blackwell, Hoboken, NJ

Glare TR, O'Callaghan M (2000) Bacillus thuringiensis – biology, ecology and safety. Wiley-Blackwell, Hoboken, NJ

Grant W (2011) Policy aspects of regulation. In: Ehlers R-U (ed) Regulation of biological control agents. Springer, Dordrecht

Grewal PS, Ehlers R-U, Shapiro-Ilan DI (2005) Nematodes as biological control agents. CAB International, Wallingford, CT

Harman GE, Kubicek CP (1998) Trichoderma and Gliocladium, Vol 2. Enzymes, biological control and commercial application. Taylor and Francis, London

Hauschild R (2011) Facilitations in the regulation of plant protection products containing baculoviruses. In: Ehlers R-U (ed) Regulation of biological control agents. Springer, Dordrecht, pp 259–266

Hauschild R, Speiser B, Tamm L (2011) Regulation according to EU Directive 91/414: data requirements and procedure compared to regulation practice in other OECD countries. In: Ehlers R-U (ed) Regulation of biological control agents. Springer, Dordrecht, pp 25–77

Helyer N, Brown K, Cattlin ND (2003) A color handbook of biological control in plant protection. Manson Publishing, Portland, OR

Howse P, Jones O, Stevens JM (1998) Insect pheromones and their use in pest management. Chapman and Hall, London

Hunt EJ, Loomans AJM, Kuhlmann U (2011) An international comparison of invertebrate biological control agent regulation: what can Europe learn?. In: Ehlers R-U (ed) Regulation of biological control agents. Springer, Dordrecht, pp 79–112

Hunter-Fujita FR, Entwistle PF, Evans HF, Crook NE (1998) Insect viruses and pest management. Wiley, Hoboken, NJ

International Labour Organisation and United Nations Environment Programme (1999) Bacillus thuringiensis: environmental health criteria. Series No. 217, World Health Organisation, Geneva

Kubicek CP, Harman GE (1998) Trichoderma and Gliocladium, vol 1. Basic biology, taxonomy and genetics. Taylor and Francis, London

Lemanceau P, Alabouvette C (1991) Biological control of *Fusarium* diseases by fluorescent *Pseudomonas* and non-pathogenic *Fusarium*. Crop Protec 10:279–286

Malais MH, Ravensberg WJ (2003) Knowing and recognizing: the biology of glasshouse pests and their natural enemies. Koppert, Berkel en Rodenrijs

Murhammer DW (2007) Baculovirus and insect cell expression protocols, 2nd edn. Humana Press, Totowa, NJ

OECD (2002) Consensus document on information used in the assessment of environmental applications involving baculoviruses. Series on Harmonization of Regulatory Oversight in Biotechnology, No.20. ENV/JM/MONO(2002)1

Pickett CH, Bugg RL (1998) Enhancing biological control: habitat management to promote natural enemies of agricultural pests. University of California Press, Berkley

Regnault-Roger C (2011) Risks of biocontrol agents containing compounds of botanical origin or semiochemicals. In: Ehlers R-U (ed) Regulation of biological control agents. Springer, Dordrecht, pp 215–242

Regnault-Roger C, Philogène BJR, Vincent C (2005) Biopesticides of plant origin. Lavoisier Publishing, Paris

Richardson DM (2009) Assessment of the impact of crop protection by the "cut-off criteria" in a new regulation for authorisation of plant protection products. Comm Agric Appl Biol Sci 74:3–8

Shuler ML, Wood HA, Grandos RR, Hammer DA (1994) Baculovirus expression systems and biopesticides. Wiley-Liss, New York, NY

Siddiqui Z (2006) PGPR: biocontrol and biofertilization. Springer, Dortrecht

Speiser B, Tamm L (2011) Regulation of plant protection in organic farming. In: Ehlers R-U (ed) Regulation of biological control agents. Springer, Dordrecht, pp 113–125

Speiser B, Tamm L, Mattock S (2011) Proposals for regulation of semiochemicals. In: Ehlers R-U (ed) Regulation of biological control agents. Springer, Dordrecht, pp 305–321

Strasser H, Hutwimmer S, Burgstaller W (2011) Metabolite toxicology of fungal biocontrol agents. In: Ehlers R-U (ed) Regulation of biological control agents. Springer, Dordrecht, pp 191–213

Strauch O, Strasser H, Hauschild R, Ehlers R-U (2011) Proposals for bacterial and fungal biocontrol agents. In: Ehlers R-U (ed) Regulation of biological control agents. Springer, Dordrecht, pp 267–288

Tamm L, Speiser B, Mercier T (2011) Proposals for regulation of botanicals. In: Ehlers R-U (ed) Regulation of biological control agents. Springer, Dordrecht, pp 289–304

Verma M, Brar SK, Tyagi RD, Surampalli RY, Valéro JR (2007) Antagonistic fungi, *Trichoderma* spp.: panoply of biological control. Biochem Engin J 37:1–20

Wajnberg E, Hassan SA (1994) Biological control with egg parasitoids. CABI, Wallingford, CT

# Chapter 2
# Regulation According to EU Directive 91/414: Data Requirements and Procedure Compared with Regulation Practice in Other OECD Countries

**Rüdiger Hauschild, Bernhard Speiser, and Lucius Tamm**

**Abstract** The data requirements and the administrative procedure needed for the registration of biological plant protection products and their active ingredients (micro-organisms including viruses, plant extracts, and semiochemicals) are described for the European Union, the USA, Canada and Australia and compared between these systems. Experiences from registration procedures are compared. While data requirements and formalities are rather similar in all systems considered, the time span needed for evaluation and the possibility to predict this time span are quite different. Alternative regulatory measures existing in different regulatory systems are described and initiatives for the facilitation are presented.

## Contents



R. Hauschild (✉)
GAB Consulting GmbH, Hinter den Höfen 24, 21769 Lamstedt, Germany
e-mail: ruediger.hauschild@gab-consult.de

R.-U. Ehlers (ed.), *Regulation of Biological Control Agents*,
DOI 10.1007/978-90-481-3664-3_2, 

## 2.1 Introduction

Invertebrate and microbial biocontrol agents, plant extracts, and semiochemicals (here collectively called BCAs) are interesting alternatives to conventional pesticides for the control of plant pests and diseases. Their use is in most cases safer to humans and the environment when compared with conventional plant protection products.

Nevertheless, the industry complains that the current registration system for BCAs in the EU is costly and time-consuming. For the industry, long registration periods are a severe problem because they delay the onset of the returns for the investments made during research and development. In addition, longer registration periods result in shorter periods of sale under patent protection.

Apparently, registration times are far shorter in the USA than in the EU. However, a comparison between the time needed for evaluation for microbial BCAs in the EU and the USA is very difficult because of differences in the procedures. The fact that the time needed for listing in Annex I of Directive 91/414 in the EU was longer than product registration in the USA is only one aspect (Table 2.1). Times summarized in the table include both times needed for evaluation of the dossier by authorities and for the generation and provision of additionally required studies or information by applicants. Provisional national registrations can be applied for and granted before Annex I listing, so some of these products were already on the market in some EU member states before Annex I inclusion. Times required to obtain provisional national registrations differ considerably between member states and products, and no complete, detailed information could be obtained. On the other hand, not all member states granted provisional registrations before Annex

**Table 2.1** Time periods for selected microbial BCAs between submission of the dossier and Annex I inclusion in the EU or national registration in the USA

| Organism – product | EU period (month, year) | EU Annex I inclusion time frame (months) | USA registration time frame (months) |
|---|---|---|---|
| *Paecilomyces fumosoroseus* – Preferal® | 5.94–6.01 | 85 | 60 |
| *Coniothyrium minitans* – Contans® | 11.98–8.03 | 57 | 15 |
| *Pseudomonas chlororaphis* – Cedomon® | 1.96–4.04 | 99 | - |
| *Ampelomyces quisqualis* – AQ10® | 2.96–10.04 | 104 | ? |
| *Gliocladium catenulatum* – Prestop® | 3.99–10.04 | 67 | 13 |
| *Bacillus subtilis* – Serenade® | 5.00–2.07 | 81 | 14 |
| *Spodoptera exigua* NPV – Spodex® | 7.97–8.07 | 121 | 12 |
| *Paecilomyces lilacinus* – Bioact® | 4.06–4.08 | 72 | 21 |
| Average time frame | | 86 | 23 |

The period is indicated from the month of dossier submission to the month of inclusion on Annex I of Directive 91/414/EEC, or granting of national registration. Some of the products have obtained provisional registrations and were already on national markets before the active ingredient was listed in Annex I. In the EU, times needed to generate further studies and provide them by the applicants are included

I inclusion so far. Originally, it was assumed that longer registration periods reflect greater data requirements and thus higher costs for dossier preparation.

To date (March 2010), eight micro-organisms are listed in Annex I of Council Directive 91/414/EEC as new active substances. Inclusion of 40 strains belonging to 17 species or subspecies, which were evaluated as "existing active substances", was published in December 2008 and came into force in May 2009. A peer review (cp. below) will be organised by EFSA until 2012. Plant protection products containing these micro-organisms were continuously and are still on the market. Another 11 micro-organisms (strains) are being evaluated as new substances. An overview on the number of microbial strains used in plant protection products in the EU, USA, Canada, and Australia can be found in Table 2.2

Comparisons between the EU and the USA, Canada or Australia for the products containing micro-organisms that are available on the market are extremely difficult for several reasons. Names for the same product may differ between countries. Product availability changes for economical reasons (supply, distribution). Not all products that are registered are currently available. On the other hand, products used to control plant pests or pathogens may still be on the market under different labelling. Furthermore, a national registration in a single EU member state would not be comparable to a national registration in e.g. the USA with regard to market size.

Until today, only one plant extract ("botanical") is included in Annex I of Directive 91/414 as new active substance: Laminarin is a purified substance from brown algae (*Laminaria* sp.). Annex I inclusion for eight plant extracts and plant oils (pyrethrum, extracts from tea tree, garlic extract (oil not included), citronella

**Table 2.2** Number of strains available in plant protection products in different countries

| EU Annex I (strains) | | EU member states (strains available in products) | USA | Canada | Australia | New Zealand |
|---|---|---|---|---|---|---|
| Included as new substances | 8 | | | | | |
| Included as existing substances | 40 | > 60 | 70 | 17 | 24 | 22 |
| Under evaluation as new substances | 11 | | | | | |

Differences between strain numbers for Annex I inclusion and for strain numbers available in products is due presence or absence of provisional national registrations during the Annex I evaluation process. Some strains registered in the US for use in plant protection products are commercialised in Germany as Plant Resistance Improvers (PRI, cp. Section 2.6.5) and included in the number for EU member states. In the USA, no distinction is made like that in the EU between plant protection products and biocides. Strains only used in products considered to be biocides are not included in the list

oil, clove oil, rape seed oil, spearmint oil, pepper) came into force in September 2009 in the course of the 4th stage of re-evaluation (so-called "List 4 substances") with products containing these active ingredients still on the market.

32 semiochemical compounds and blends all belonging to the "Straight-Chained Lepidopteran Pheromones" (SCLP) were evaluated as "existing active substances" and inclusion into Annex I of Directive 91/414 was published in December 2008, entering into force in September 2009.

The list of biochemical pesticides in the USA currently includes 160 registered products, with most of them being pheromones (50), followed by repellents (29, some of them are not considered as plant protection products in the EU, but as biocides), plant growth regulators (phytohormones, 21), and attractants for traps (14). 18 products are intended for insect, nematode, or plant pathogen control based on physical modes of action. None of these products are, or contain, classical plant extracts.

This report contains a description of the registration system in the EU, with emphasis on aspects that could be improved. Where adequate and available, the EU system is compared with non-EU systems. The main focus is on registration of active ingredients under Directive 91/414/EEC, resulting in inclusion in Annex I. However, some attention is also given to registration of plant protection products, which is under national authority. Invertebrate biocontrol agents are not discussed in this chapter.

## 2.2 Methodology and Terminology

This report is based on a review of data requirements inside and outside the EU. In addition, the practical experience of the industry and regulators is also considered. The report covers microbial biocontrol agents (fungi, bacteria, viruses), plant extracts (crude, purified), microbial extracts, and pheromones. These products do not have much in common except that all of them are of natural origin and that all of them are very different from the typical, synthetic pesticides.

The following legislation was considered for the comparison of data requirements:

**EU**: Council Directive 91/414/EEC, with the Commission Directive 2001/36/EC, and Commission Directive 2005/25/EC for "Uniform Principles for Evaluation" included in the Council Directive.

**USA**: Federal Insecticide, Fungicide, and Rodenticide Act (FIFRA); Code of Federal Regulations (40 CFR 158) which was in force until the end of 2007. Differing requirements according to the new rule (Federal Register, Part III, Environmental Protection Agency, 40 CFR Parts 158 and 172, Pesticides; Data Requirements for Biochemical and Microbial Pesticides, October 26, 2007) are included in the comparison. The new rule was proposed in March 2006 and commented afterwards. These regulations also cover uses that are not considered as plant protection in the EU, e.g. insect repellents for humans or biocides.

**Canada**: Regulatory Directive DIR 2001-02 (Guidelines for the Regulation of Microbial Pest Control Agents and Products) and Regulatory Directive DIR2002-02 (The PMRA Initiative for Reduced-Risk Pesticides).

**Australia**: "Manual of Requirements and Guidelines" and "Guidelines for the Registration of Biological Agricultural Products" from the Australian Pesticides and Veterinary Medicines Authority.

**Switzerland**: Data requirements in Switzerland are regulated by Pflanzenschutzmittelverordnung (PSMV, SR 916.161, published 18 May 2005). These data requirements are identical to those in the EU and are therefore not mentioned explicitly in this comparison.

**New Zealand**: In New Zealand, data requirements for microbial plant protection products are not separated from those for chemical plant protection products, and are defined case-by-case according to the particular micro-organism, the product, and the uses.

## 2.3 Legal Framework and Regulatory Procedures

### *2.3.1 Regulatory Procedures in the EU*

In Europe, registration of active substances and plant protection products is performed in a two-step system. The active substance first has to be evaluated at EU level. The dossier containing all information on the active substance and on at least one representative product and one representative use is submitted to a member state, the designated Rapporteur Member State (RMS). Authorities of the RMS first check the completeness of the dossier, then evaluate the dossier and distribute the Draft Assessment Report (DAR) to the other member states, the applicant, and EFSA. EFSA, the European Food Safety Authority, is a decentralized agency of the European Union, financed by the Community budget. EFSA provides risk assessments and risk communication on all matters related to food safety, including plant protection products and their residues. A peer review process is then initiated which involves all member states and EFSA. This process includes a written procedure as well as meetings. Following this evaluation, an EFSA scientific report with conclusions of the peer review is released. Finally the EU Commission (DG SANCO)[1] prepares a draft directive which aims at inclusion (or non-inclusion) of the active ingredient into Annex I of Directive 91/414/EEC, for which the Member States vote in a committee. When all the existing active substances have gone through the EU re-evaluation process, this annex will contain the active ingredients for all plant protection products registered in the EU. No guarantee is given regarding the length of the evaluation process. Annex I inclusion is valid for 10 years.

Plant protection products are regulated at the national level of member states. Formally, applications for provisional national registrations can be submitted when the dossier for Annex I inclusion is declared complete by all RMS and when completeness is published. In some member states, especially if application for product evaluation is made to the RMS, evaluation of the dossiers for Annex I inclusion and for national registration can be treated in parallel, which accelerates the process for national registration. However, some member states refuse provisional registrations after publication of the completeness of the dossier and await release of the DAR or even Annex I inclusion before they evaluate the dossier for national registration.

In addition to a complete dossier on the active ingredient and the product, data on efficacy have to be provided in most member states of the EU for national registration of a plant protection product. In case the application concerns a product for field application, the efficacy data have to be specific for the country or at least the climatic zone. This restriction is normally not relevant for products that are only intended for use in greenhouses. Efficacy data for biological plant protection products are reduced in some countries when compared with the requirements for chemical plant protection products. This reduction refers both to the number of successful studies that have to be submitted and to the extent of efficacy obtained in the trials.

In addition, specific forms, most frequently in the national language, have to be filled in. In some member states, summaries in the dossier (Documents M and N) have to be submitted in the national language, in others only parts of the dossier (e.g. Document N) have to be in the national language. In practice, the national distributor for the product often prepares the required documents in the national language.

The organisation of the regulatory bodies within the member states varies greatly. In some member states, the regulation is mainly carried out within the ministry of agriculture, in others within the ministry of environment, the ministry of health, or

[1]DG Sanco is the Health and Consumer Protection Directorate General

related agencies. Often, several regulators from several ministries or agencies are involved in the evaluation of the active substances and the plant protection products. In several member states both the risk assessment and risk management part is carried out within the regulatory agencies. However, in many other member states external experts are involved or even responsible for the risk assessment. Some, but not all, member states have pre-submission meetings with applicants.

#### 2.3.1.1 Fourth Stage of Re-evaluation

Substances that were already registered in an EU member state when Dir. 91/414 came into force can stay on the market, but are subject to re-evaluation. The re-evaluation was divided into four stages. The 4th stage comprised among other groups microbials, plant extracts and semiochemicals. Commission Regulation 1112/2002 required that substances had to be notified by autumn 2002. Commission Regulation 2229/2004 assigned the notified substances to RMS. Dossiers had to be submitted to RMS by 30 June/30 November 2005. Currently, evaluations within the 4th stage are ongoing. Inclusion of all micro-organisms into Annex I was published on December 8, 2008 and came into force on 1 May 2009. Similarly, Annex I inclusion of all semiochemicals and eight plant extracts was published on December 18, 2008 and came into force on 1 September 2009. A peer review to clarify open issues will be organised in 2011/2012. Cooperation between RMS until Annex I inclusion was working well for microbials and less pronounced for botanicals. All semiochemicals were reviewed by one RMS, Austria.

### *2.3.2 Regulatory Procedures Outside the EU*

#### 2.3.2.1 Regulatory Procedures in the USA

Contrary to the practice in the EU, both the active substance and the end use product are being evaluated by a centralised authority in the USA. US pesticide regulation is under the authority of the US Environmental Protection Agency (EPA; registration) and the US Food and Drug Administration (FDA; MRL enforcement). After a pesticide is registered by EPA, states can register pesticides under specific state pesticide registration laws. A state may have more stringent requirements for registering pesticides for use in that state. States may also register an additional use of a federally registered pesticide product or a new end-use product to meet special local needs. EPA reviews these registrations and may disapprove them under certain circumstances. In practise, additional data are mainly required by authorities in California.

The Biopesticides and Pollution Prevention Division was founded in 1995 within the EPA Office of Pesticide Programs. It comprises the Microbial Pesticide Branch (20 staff, responsible for microbial pesticides and plant incorporated protectants, foreign genes introduced in transgenic plants) and the Biochemical Pesticide Branch (23 staff).

The “Biochemical Classification Committee” was formed in 1995. It decides if an active substance can be a classified as a Biochemical Pesticide, based on literature, SAR analysis, whether or not it is a significant food component or if it has other uses, etc. The Committee has the options to decide that an active substance (1) is a biochemical pesticide, (2) is not a biochemical pesticide, but eligible for review using the reduced data set, (3) is a conventional chemical pesticide, (4) is not a pesticide. The Committee has evaluated 212 chemicals, classifying 42 as conventional pesticides.

Prior to submission of a dossier, the US Environmental Protection Agency (EPA) invites applicants to one or more pre-submission meetings, during which the applicant is advised what studies are necessary for the particular product in question based on the applicant's preliminary identification of the product and whatever data are available from the literature or other sources. The applicant then submits a summary of the meeting to the agency for comment and approval. The dossier has to contain all the required studies, but no summary dossier is requested in the US.

Since the implementation of the Pesticide Registration Improvement Act (PRIA) of 2003, which established fees for registrations, time limits are set to process a registration application. US EPA is supposed to register a microbial pesticide in 16 months from receipt of a complete application. Therefore, inadequate applications are often denied unless negotiations with the company will allow them to be delayed to wait for additional data. Registrations can be granted as “conditional” registrations (1 year), if data are missing or classified as “supplementary”, and the risk is low enough to market the product (e.g. field testing to verify lack of effects on non-target organisms). Data submitted later will be classified as confirmatory data. Unconditional registrations are valid forever, but re-examined every 15 years to ensure that the original assessment is still valid (re-registration).

### Biopesticides (USA)

Biopesticides are divided into two groups: Microbial pesticides include microbial entities such as bacteria, fungi, viruses, and protozoa (protista). Biochemical pesticides include, but are not limited to, products such as semiochemicals (e.g. insect pheromones), hormones (e.g. insect juvenile growth hormones), natural plant and insect regulators (attractants, repellents), and enzymes. In the regulation valid until end 2007, biochemical and microbial pesticides are generally distinguished from conventional chemical pesticides by “their unique modes of action, low use volume, target species specificity or natural occurrence”. This definition was written about 30 years ago and has now been revised to better describe how it has always been interpreted for biochemical pesticides (see “new rule”, below).

Financial aspects: Grants for research projects needed for the registration of biopesticides, especially for minor uses, can often be obtained from the US Department of Agriculture’s IR4 program. Small businesses and government are exempt from fees paid to EPA for review (PRIA). The new pesticide data requirement rule attempts to reduce the number of studies and data waiver paperwork needed for biopesticides.

New rule: Since the 1970s, the US EPA encourages the registration of BCAs through exemptions, explicit footnotes, tiered data tables and by facilitating data waivers. After two decades' experience, data requirements have been revised. The new data requirements are included in Federal Register, Part III, Environmental Protection Agency, 40 CFR Part 158, Pesticides; Data Requirements for Biochemical and Microbial Pesticides from October 26, 2007. The new rule formalizes pre-submission meetings to reach agreement on what data are needed and to provide EPA assistance on data waivers. The NAFTA procedures for joint review of biopesticides describe how the US-EPA and the Pest Management Regulatory Agency in Canada will coordinate the joint pre-submission consultation to agree on data needed. Exemptions are based on the following legislation: 40CFR 152.20: All Biocontrol organisms except microbial pesticides and plant incorporated protectants (nematode symbiont policy: exempt, unless bacteria grown separately and/or genetically engineered). 40CFR 152.10: Products to attract pests for survey or detection and physical barrier products. 40CFR 152.500: Devices. 40CFR 152.25: Specific exemptions for: (1) pheromones used in pheromone traps, (2) "foods" used to attract pests, (3) "natural cedar" chips, panels, etc, (4) "Minimal Risk Pesticides" (see below).

The new rule contains revised definitions for microbial and biochemical pesticides that better describe the established interpretations of the old definitions. A biochemical pesticide is characterized as follows: (1) It is a naturally-occurring substance or structurally-similar and functionally identical to a naturally-occurring substance; (2) it has a history of exposure to humans and the environment demonstrating minimal toxicity, or in the case of a synthetically derived biochemical pesticide, is equivalent to a naturally-occurring substance that has such a history; and (3) it has a non-toxic mode of action to the target pest(s). Non-toxic modes of action include: (i) lures, attractants, repellents, irritants; (ii) systemic acquired resistance induction; (iii) growth/developmental changes (IGRs, PGRs); (iv) physical modes of action (suffocation, desiccation, coatings). The naturally-occuring part of the definition allows for pre-registration analysis of the product to determine if any adverse effects have been identified as a result of its human and environmental exposure. The non-toxic mode of action makes it less likely that the more toxic naturally-occurring chemicals, might not have sufficient data submitted to support an adequate risk assessment. The Biochemical Classification Committee is authorized to accept well known substances with toxic modes of action for review using the biochemical pesticide reduced set of data requirements if it can be justified as appropriate for that chemical, but those are not allowed to be called "biochemical pesticides".

### Minimal Risk Pesticides

In the USA, there is a list of substances that can be used as pesticides without any registration, although they still need a residue limit, or exemption, for food or feed uses. These substances are called Minimal Risk Pesticides, as described in the US Code of Federal Regulation, 40CFR 152.25(f). The list contains many essential oils[2]. All inerts must be on EPA's 4A inert list, all ingredients must be identified on the label, and the label may not contain false or misleading claims. This regulation was developed by an EPA workgroup in 1994 and revised in accordance with public comments for a final Federal Register publication in 1996 (61 FR 8878, March 6, 1996). A public petition that expresses concern about the potential lack of efficacy for some of these ingredients when used as public health pesticides (which would be classified as biocides and not plant protection products in the EU) has recently been submitted to EPA and is under review. Another enforcement problem has been with identifying exactly what chemical substances are included under the names listed. Currently, CAS numbers are used to describe the substances on the EPA inert substance classification lists.

#### 2.3.2.2 Regulatory Procedures in Canada

In Canada, applicants are asked to meet with the authorities before the submission of a dossier, to define data requirements applicable for the active ingredient and the product. Canada accepts dossiers in their own PMRA format as well as in US EPA or OECD formats. The dossier is evaluated by the Pest Management Regulatory Agency (PMRA) and is scheduled to be finished within 12 month. Additionally, a 52 days front-end and a 30–75 days tail-end administrative phase has to be considered, resulting in a maximum time between receipt of the application and the decision of 16 months. If data are incomplete and additional information is requested, the "review clock" is reset to 0, starting with the submission of the additional information. Full registration is granted for 5 years, with the possibility for 2 renewal periods of 5 years. After 15 years, the product is subject to re-evaluation. Temporary or conditional registrations are possible to allow the applicant to collect further data required for full registration.

##### Reduced-Risk Pesticides Initiative

The Canadian Regulatory Directive DIR2002-02 is the basis for the "PMRA Initiative for Reduced-Risk Pesticides". This initiative is not restricted to a particular group of substances, but may apply to all groups of products including chemicals, botanicals, micro-organisms and semiochemicals. Data are required for the technical active ingredient and at least one product. Data requirements are very similar to those for "biochemicals" in the USA. In this initiative, there is emphasis on the replacement of more risky plant protection products by reduced-risk plant protection products.

---

[2]Currently, the list includes the following substances: castor oil, cedar oil, cinnamon and cinnamon oil, citric acid, citronella and citronella oil, cloves and clove oil, corn gluten meal, corn oil, cottonseed oil, dried blood, eugenol, garlic and garlic oil, geraniol, gernanium oil, lauryl sulfate, lemongrass oil, linseed oil, malic acid, mint and mint oil, peppermint and peppermint oil, 2-phenethyl propionate (2-phenylethyl propionate), potassium sorbate, putrescent whole egg solids, rosemary and rosemary oil, sesame (includes ground sesame, plant) and sesame oil, sodium chloride (common salt), sodium lauryl sulfate, soybean oil, thyme and thyme oil, white pepper and zinc metal strips.

#### 2.3.2.3 NAFTA Joint Review Process

If applicants apply in both the USA and Canada, these countries may use a joint review process starting with a joint pre-submission meeting to determine the actual data requirements. For microbials, the initial evaluation is based on the taxonomy and a literature research. Evaluation of the active ingredient and the product are done in parallel. The evaluation period normally takes 12 months, with additional national administrative periods. Using this procedure, applicants simultaneously get access to both markets.

#### 2.3.2.4 Regulatory Procedures in Australia

Pre-submission meetings are recommended to determine the applicable data requirements. For microbials, the taxonomy and a literature search on possible risks are used for initial evaluation. Time frames are set to process the evaluation within 12 months, but missing data can delay the process.

## 2.4 Analysis of Formal Data Requirements

### *2.4.1 Formal Dossier Requirements*

The authorities in the EU now require dossiers according to the OECD format. Distinction is made between data requirements for the active substance, in this case the micro-organism or MPCA (microbial pest control agent), and the formulated product, or MPCP (microbial pest control product). Dossiers are organized in 7 sections (see Table 2.3) and include Tier I, Tier II, and Tier III summaries for studies, literature information and risk assessments for toxicity and the environmental impact. These summaries and risk assessments are not required in the USA and Australia. In Canada, formal requirement for summaries is a quite recent introduction, but dossiers without summaries are equally accepted. Most dossiers are submitted in "national" formats, but the OECD data format is generally accepted in USA and Canada.

Summary documents are highly appreciated by regulators because information concerning the data point from literature or studies is presented in a short and

**Table 2.3** Dossier organization according to the OECD Guidance Document

| Section number | Active agent – Annex II | Formulation – Annex III |
|---|---|---|
| Section 1 | Identity, biological properties/physical–chemical properties | Identity, formulation |
| Section 2 | Analytical methods | Analytical methods |
| Section 3 | Human health | Human health |
| Section 4 | Residues | Residues |
| Section 5 | Fate and behaviour in the environment | Fate and behaviour in the environment |
| Section 6 | Effects on non-target organisms | Effects on non-target organisms |
| Section 7 | – | Efficacy |

compact form. All essential literature can be revised and summarized. Detail information in publications can be pointed out by the applicant and is thus more easily accessible to the evaluators. These presentations can lead to considerable time savings for evaluating authorities, but it has to be taken into account that the applicant needs sufficient expertise to provide these summaries.

Dossiers for national registration of plant protection products in some EU member states must be submitted in the national language. At least specific forms have to be provided. In some member states, summaries in the dossier (Documents M and N) have to be submitted in the national language, in others only parts of the dossier (e.g. Document N).

### 2.4.2 "Waivers"

In the USA, Canada and Australia, certain data requirements may be met with a "waiver". The applicant has to apply for a waiver, by providing a scientific argument (mostly derived from published literature and own data of the applicant). If the waiver is granted, no study has to be provided.

This "waiver system" does not formally exist in the EU. However, the summaries in the M-Documents give space for the same scientific argumentation to fulfil the data requirement using published literature and data from the applicant instead of a study. No formal waiver is necessary in the EU to replace a study with a "reasoned case" based on existing, public or non-public, information. However, it seems that waivers are accepted more easily in the USA, Canada and Australia than reasoned cases are accepted in the EU. Furthermore, the scientific justification might be evaluated differently in different EU member states, which may lead to additional data requirements during the evaluation process.

At the scientific level, the arguments why a study is not necessary in a particular case are very similar in the EU and the USA, but the formal procedure is different. For reasons of simplicity, the term "waiver" is used in this document for all systems.

### 2.4.3 Data Requirements for Microbials

"Microbials" are a heterogenous group including fungi, bacteria, viruses, and protozoans (protista) used in biological plant protection products. Data requirements are formulated to account for all groups, even if some points are not applicable for particular groups (e.g. metabolite production for baculoviruses).

#### 2.4.3.1 Definitions of Microorganisms

**EU**: A "micro-organism" according to the Council Directive 91/414/EEC is defined as "A microbiological entity, cellular or noncellular, capable of replication or of transferring genetic material. The definition applies to, but is not limited to, bacteria, fungi, protozoa (protista), viruses and viroids."

**USA** old version 40 CFR 158.65 (p. 86): Biochemical and microbial pesticides are treated together: "Biopesticides include naturally occurring substances (biochemical pesticides), micro-organisms (microbial pesticides), and pesticidal substances produced by plants containing added genetic material (Plant-Incorporated Protectants, PIPs)." Only living entities are considered as microbial pesticides, while dead micro-organisms are included in biochemical pesticides. Contrary to the EU, genetically modified plants are as well covered by this regulation.

- Microbial pesticides consist of a micro-organism (e.g., a bacterium, fungus, virus or protozoan) as the active ingredient.
- Plant-Incorporated-Protectants (PIPs) are pesticidal substances that plants produce from genetic material that has been added to the plant. The protein and its genetic material, but not the plant itself, are regulated by EPA.
- Biochemical pesticides are naturally occurring substances that control pests by non-toxic mechanisms. Biochemical pesticides include substances such as insect sex pheromones as well as various scented plant extracts that attract insect pests to traps.

**USA new regulation**: Biochemicals are separated from microbials. Extracts from plants or from micro-organisms with toxic properties continue to be treated as conventional chemicals. "Microbial pesticides" includes all living or dead microbial pesticides: "*Microbial pesticide*" is a "microbial agent intended for preventing, destroying, repelling, or mitigating any pest, or intended for use as a plant regulator, defoliant, or desiccant, that

- is a eukaryotic micro-organism including, but not limited to, protozoa, algae, and fungi;
- is a prokaryotic micro-organism, including, but not limited to, Eubacteria and Archaebacteria; or
- is a parasitically replicating microscopic element, including but not limited to, viruses."

**Canada**: In Canada, a "microbial pest control agent" is defined as "a micro-organism (bacterium, alga, fungus, protozoan, virus, mycoplasma or rickettsia and related organisms) and any associated metabolites, to which the effects of pest control are attributed."

**Australia**: In Australia, microbial plant protection products are classified as "Biological agricultural chemical products". "A biological agricultural chemical product is an agricultural chemical product where the active constituent comprises or is derived from a living organism (plant, animal, micro-organism, etc), with or without modification." Micro-organisms are included in Group 3: "microbial agents (e.g. bacteria, fungi, viruses, protozoa)". Biological agricultural chemical products include, among others, also natural chemicals (pheromones, hormones etc. (Group 1), plant extracts and oils (Group 2), and microscopic insects (Group 4).

Decisions on actual data requirements are made on a case-by-case basis; special emphasis is on environmental expression and suppression of indigenous species. Five micro-organisms are excluded from the requirements of APVMA approval as constituent in plant protection products: *Helicoverpa zea* NPV, *Metarhizium anisopliae*, *Paecilomyces lilacinus* strain 251, Rabbit calicivirus, and *Trichoderma harzianum* Rifai strain T-39.

Regulation of micro-organisms is done on strain level in all regulatory systems. Each "new" variety, subspecies, or strain of an already registered microbial pest control agent must be evaluated. In the EU, strains can be treated together, if they are "similar". In the USA and Australia, non-indigenous micro-organisms are subject to additional data requirements. All regulatory systems follow a case-by-case approach in the evaluation of micro-organisms for the use in microbial plant protection products.

In all regulatory systems, additional data have to be provided for genetically modified organisms (GMOs). In the EU, GMOs are evaluated also under the Council Directive 2001/18/EC. On the other hand, no GMOs are expected for the use in plant protection products in the EU in the near future as they will not be accepted by the market. Therefore, GMOs are not further considered in this comparison.

#### 2.4.3.2 Data Requirements for the Active Ingredient, the Microorganism (MPCA)

Section 1

According to the OECD numbering system, the identity of the micro-organism including criteria for the identification have to be described under point 1. This includes literature information and identification studies on the species and the strain and represents the central information for evaluation of all aspects considered in the application. The question whether the organism is indigenous or not to the area of application is required in the EU, USA, Canada, and particularly emphasised in Australia. The composition of the technical grade of the active substance (the micro-organism after fermentation) and of the material used for manufacture of the end use product is required in all systems. Production methods and quality control data are demanded as well. Differences occur for Point IIM 1.4.5 "The formation, presence, and/or impact of unintentional ingredients": a "theoretical discussion" on impurities and other unintentional ingredients on human health or product quality, and appropriate quality criteria have to be provided in the USA and Canada, but not in the EU and Australia. Likewise, physical and chemical properties of the technical product are required if the manufacturing product is stored before formulation of end-use products in the USA and Canada. The international regulatory status is to be provided in Australia and Canada. Samples of the micro-organism and analytical standards have to be provided on request in the EU, USA, and Australia. Additionally, reference substances for relevant impurities may be required in the EU. The patent status of the MPCA is only required in Canada.

Under Point IIM 2, biological properties of the micro-organism shall be described in a very similar way in the EU, USA, Canada, and Australia. For all regulatory systems, information is to be provided on strain level where possible. A point that is particularly important for bacteria and fungi is the potential of the micro-organism to produce metabolites that are of concern for human health and/or the environment. This information is required in the EU, Canada (with a special emphasis on toxins and genotoxins), and Australia. Information on physiological properties and on genetic stability of the micro-organism have to be provided in the EU and Canada, but not in the USA and Australia. The USA and Canada ask for a description of extrachromosomal elements involved in pesticidal activity, pathogenicity, or toxicity. This information is in practice also required in the EU, if extrachromosomal elements are involved in the mode of action. Information on resistance or sensitivity of the micro-organism to antimicrobial agents is required in the EU, Canada, and Australia, but not the USA. It is required only in Canada, that the relationship of the micro-organism to known human dermatophytes has to be discussed.

Under Point IIM 3, information on function, mode of action and handling of the micro-organism is summarized. Data requirements are again similar, with differences for the information on resistance or cross-resistance (only required in the EU and Australia). Material Safety Data Sheets and procedures for the decontamination of water are only required in the EU and Canada, while measures to render the micro-organism harmless are only required in the EU.

Section 2

Data demanded in Section 2 "Analytical methods" are very similar for maintenance of the master seed stock, the production processes, and methods of detection, differentiation between similar strains, and the determination of contaminants, pathogens, or metabolites. However, post registration monitoring methods to determine and quantify residues from different matrices are required in the EU and the US according to the new rule, if residues (the micro-organism itself or its metabolites) are considered to be relevant.

Section 3

Section 3 comprises "Toxicological and Exposure Data and Information on the Microbial Pest Control Agent". In general, a tiered test system is applied in all systems. Testing is started with basic studies (acute testing) and, if negative effects occur, continued with further tests. These depend on the outcome of the tests where negative effects were observed.

A summary on the potential of the microbial pest control agent to be hazardous to humans with consideration of its pathogenicity, infectivity and pattern of clearance, and its toxicological effects is demanded in the EU, in Canada, and in Australia, but not in the USA. Occupational health is a central point in all regulatory systems and reports on sensitisation, allergies, or hypersensitisation, especially of workers, have to be provided. Similarly, information on persons with increased susceptibility and literature information on clinical cases has to be supplied. However, the sensitisation properties of micro-organisms are still a matter of debate. As tests systems used for determination of dermal and inhalative sensitization of chemicals are not appropriate for microbials, no proper test system is available. In the EU and in Canada, tests are therefore not required. In the EU, "all micro-organisms should be regarded as potential sensitisers" until appropriate test methods are available (Commission Directive 2001/36/EC, p.13). However, applicants may submit data to demonstrate that the micro-organism does not have a sensitizing potential. In practice, all microbials are currently classified as potential sensitizers in the EU member states. In the USA, studies on the sensitization properties are "required if commonly recognized use practices will result in repeated human contact by inhalation or dermal routes". According to the new USA rule a skin sensitisation study is not required, but hypersensitivity incidents have to be reported.

Studies on acute infectivity, toxicity, and pathogenicity have to be performed for oral, intratracheal or inhalative, and intravenous or intraperitoneal administration in the EU, USA, Canada, and Australia. In these studies, effects on the test animals as well as the clearance have to be determined. The US guidelines define the requirements in more detail: oral tests have to be performed with the manufacturing product and technical active ingredient. In the new proposal only studies for the technical active ingredient, but not for the end-use product or manufacturing product are required.

"Acute intratracheal/inhalation infectivity, toxicity and pathogenicity" in the USA has to be covered by studies only if a possibility of inhalation exists. Methods for the determination of intravenous/intraperitoneal infectivity are detailed according to the nature of the micro-organism: the administration should be intravenous for bacteria and viruses (no longer required for viruses in the new rule), intracerebral for viruses and protozoa (only according to the old rule, no longer required in the new rule), and intraperitoneal for fungi or protozoa. In Canada and Australia, intravenous administration is recommended for bacteria and viruses, and intraperitoneal for fungi or protozoa. Additionally, in Australia intracerebral studies are required for neurotropic agents. All studies shall be performed as single studies with a single high dose. Further studies are required if negative effects are observed.

The genotoxic potential has to be determined in the EU for purified metabolites if the micro-organism produces exotoxins, or for the entire micro-organism. If production of exotoxins is not known, expert judgement is advised to decide on the necessity of a test with broken cells of the micro-organism. Information requirements include bacterial gene mutation assays (Ames test), clastogenicity tests, and gene mutation tests in mammalian cells. The risk of insertional mutagenesis has to be discussed for a virus. In Canada, an appropriate and sensitive analytical test (e.g. HPLC) must be performed to detect the presence of possible genotoxins in the technical product, if a related fungus or actinomycete produces a genotoxin. In the USA, genotoxicity tests are only currently required in Tier II and have been deleted in the new rule. Genotoxicity testing on appropriate extracts of the micro-organism is also required in Australia.

Cell culture studies have to be provided in the EU in Tier I for micro-organisms which are able to replicate intracellularly. In the USA, tissue culture studies with the TGAI are only necessary for virus products. In Canada, cell culture assays are required in Tier I for virus only. Studies have to be performed to assess infectivity and toxicity in a human cell line, a primary cell type, or a primate continuous line. If infection occurs and the virus is replicated in the mammalian system, a cell transformation assay is required.

Information on short-term toxicity is required in the EU unless the information already provided is sufficient to assess human health effects. A study is not necessary if it can be explained why there is no risk, e.g. exposure is not likely to occur, and if there is enough information for a proper risk assessment. The route of administration depends on the exposure for humans. In a short-term toxicity study (28 days minimum) pathogenicity and infectivity have to be assessed and clearance from different organs must be determined. Similarly, short-term studies are also required in Australia. In the USA and Canada, short term toxicity studies are only required in Tier II, if significant negative effects were detected in Tier I acute studies. Toxicity assays on toxins can be demanded in all systems in Tier II, if concerns arise in acute toxicity tests. If needed, also further toxicity studies may be required if concerns arise from Tier I studies.

Section 4

Information on "Metabolism and Residue Studies on the Microbial Pest Control Agent" is summarized in Section 4. In the EU, no studies are required if literature data are available and if no negative effect on human health is known. Information on persistence of the micro-organism or relevant metabolites after application has to be provided. If significant persistence above naturally occurring levels occurs, the full data set as for chemicals may be required. In the USA and Canada, residue data are only necessary if Tier II or Tier III toxicity data are required, i.e. if Tier I toxicity test results gave reasons for concern by the micro-organism or its metabolites. Studies are not necessary for indigenous species. If toxins are produced by the micro-organism, the same residue testing scheme as for chemicals is applicable. In Australia, residue data are generally not necessary for microbials, but information has to be provided on the natural occurrence.

Section 5

In Section 5 "Fate and Behaviour Studies on the Microbial Pest Control Agent in the Environment", the information requirements can normally be covered using literature information. Studies are only needed if mammalian toxins are produced or if studies on toxicity, pathogenicity, infectivity, or on ecotoxicology showed adverse effects. In the EU, mobility studies for the micro-organism are required if negative effects were observed in toxicology studies.

## Section 6

Section 6 includes "Ecotoxicological Studies on the Microbial Pest Control Agent (Effects on non-target organisms)". In the EU and Canada, studies listed according to the OECD numbering scheme are required unless certain non-target organisms are not exposed. In the USA, the extent of ecotoxicity testing depends largely on the use of the product and thus on exposure of potential non-target organisms, and conditions under which studies are required are explicitly defined. For Australia, studies have to be provided and the host range for the micro-organism must be described. Studies are not required if the micro-organism does not survive in the Australian environment, if certain non-target organisms are not exposed, or if high host specificity is demonstrated. Higher Tier tests are required if negative effects on the test organisms are observed.

In particular, data requirements differ for avian toxicity. In the USA, oral and injection tests are required, except if the product is only applied in greenhouses. In the new rule for USA, the avian injection test is replaced by an inhalation test, which is only required if there is any indication that the microbial pesticide or its toxins may be pathogenic to birds. Canada requires toxicity tests on birds by oral and pulmonary (inhalation or injection) administration, whereas in the EU, only a study using oral administration is required. Toxicity tests with wild mammal species are only required in the USA, Canada, and Australia, and only if tests conducted on laboratory animals to assess human toxicity, pathogenicity, and infectivity, are inadequate or inappropriate for assessment of hazard to wild mammals.

Information on effects on fish and aquatic invertebrates is required in all systems. Studies assessing the effect of the micro-organism on algal growth are required in the EU and Australia, but not in the USA and Canada. Studies using aquatic plants or terrestrial plants have to be provided alternatively in the EU. Aquatic plants have to be tested in Canada and in Australia if aquatic exposure is anticipated, but not in the USA.

Effects on terrestrial plants have to be assessed in USA and Australia. In Canada, this test has only to be provided if the MPCA is related to a plant pathogen (this is also the condition in the new rule in USA).

Effects of the micro-organism on bees and other terrestrial arthropods have to be tested in all systems, but not for greenhouse use in USA. In the new USA rule this requirement is reduced, and arthropod tests are only required if the micro-organism acts through infecting insects. Effects on earthworms need to be determined in the EU and Australia, but not in Canada and the USA. Studies on effects on other terrestrial invertebrates have to be provided in Canada only if the micro-organism is intended to control non-arthropods, and in Australia. Effects on soil micro-organisms have to be assessed in the EU and Australia, and can be required in Canada, depending on the micro-organism. Studies on soil micro-organisms are not required in the USA. The new rule explicitly says as a response to comments received to the proposal: "Current knowledge indicates that the inherent variability in physical and biological environments, the adaptability of microbes, and redundant degradation pathways in microbial and mesofaunal communities, leads to no significant or lasting impact on ecosystems from introduction of pesticidal microbes even where changes to these populations can be meaningfully tracked (...). Moreover, microbial ecosystems are highly variable. Any transitory, limited, effects from the introduction of a typical microbial pesticide into the environment would be very difficult to detect and analyze."

Generally, the strategy in USA and Canada is directed towards a maximum hazard testing as the first step. If no hazard is detected, no further evaluation is accomplished. Exposure is only evaluated if a hazard is detected. In the EU effects and exposure are compared from the beginning. The exposure is calculated and a risk assessment is provided by the applicant (cp. Annex III data requirements below).

### 2.4.3.3 Data Requirements for the Product (MPCP)

Section 1

Information on the product as required in Section 1 include the identity, physical, chemical, and technical properties of the product, and data on application. Data requirements are widely the same for the EU, USA, Canada, and Australia.

Section 2

Data demanded for methods of analysis, manufacturing, and quality control are again the same in all systems and equivalent to the data required in Annex II. Methods for the determination of residues, if relevant, are only required in the EU, and are in most cases identical to the information for the determination of the micro-organism provided in Annex II.

Section 3

"Toxicological Studies and exposure data" required in Section 3 are similar between the EU, the USA, Canada, and Australia. Acute oral and dermal (percutaneous) toxicity tests using the formulated product are required in the EU unless justification is provided, in the USA and Australia. In the new USA proposal, acute oral toxicity tests are no longer required for the end-use product or manufacturing product, but may be combined with the limit dose infectivity/pathogenicity testing for the technical active ingredient. Acute inhalation toxicity studies are demanded generally (Australia), or conditionally if inhalative exposure can occur (EU, USA). The new USA rule specifically recognizes that all the acute toxicity tests on the end-use and manufacturing products can be waived if the inerts are not likely to pose significant risk. Skin and eye irritation studies are required as well, unless the micro-organism or formulants are already classified as "sensitizing". For details on classification as "sensitizer", refer to Annex II information. Monitoring data for operator and bystander exposure are always demanded.

Section 4 and Section 5

Information on residues and on fate and behaviour in the environment can be based on published literature. Studies are only required if negative effects were observed in toxicology or ecotoxicology studies. In most cases residues only derive from the micro-organism and not from other ingredients. Therefore, the argumentation is in general similar to that in Annex II.

Section 6

Ecotoxicity tests are frequently performed with the unformulated micro-organism or the technical product. Therefore, testing of the formulated product is in most cases only necessary if the formulation is suspected to modify the effect on non-target organisms. Point IIIM 11 "Summary and evaluation of environmental impact" includes a summary of all data relevant to environmental impact and an environmental risk assessment. For this risk assessment, effects of the micro-organism on different non-target organisms as determined in the studies required in Annex II are compared with application rates for the product under the appropriate conditions of use. This summary with the risk assessment is to be provided by the applicant in the EU, but not in the USA, Canada, and Australia.

Section 7

"Efficacy Data and Information (including Value Data) for the Microbial Pest Control Product" is not required in the EU for listing of active substances in Annex I of Commission directive 91/414. However, these data are necessary for national registration of the product in most EU member states. Efficacy data are also required in Canada and Australia, but are not required to be submitted for review in the USA. Efficacy data are only required for review in the USA if the product is to control a public health pest, but these products are considered to be biocides and not plant protection products in the EU.

#### 2.4.3.4 Differences Between Data Requirements in the EU and in Non-EU Countries

The major differences in formal data requirements for micro-organisms as active ingredients in plant protection products concern methods for residue analysis if relevant metabolites are expected, the demand for information on short-term toxicity, infectivity, and pathogenicity (justifications based on lack of relevant exposure can be accepted for not providing a study), and some ecotoxicty tests. For a summary of the main differences, please refer to Table 2.4.

It should be realised that in the EU the data requirements ask for information, which can be provided from published literature, internal information from the applicant, or from studies. A justification for not submitting a study can always be submitted. The acceptance of such justifications differed in some cases between different member states, but is expected to be better harmonised by the different

**Table 2.4** Summary of the major differences in the formal data requirements between the EU, the USA, Canada, and Australia

| Section | Subject | Major differences |
|---|---|---|
| 1 | Biological, physical-chemical and technical properties | A theoretical discussion on effects of impurities and other unintentional ingredients on human health or product quality is required in the USA and Canada, but not in the EU and Australia<br>Physical and chemical properties of the technical product are required if the manufacturing product is stored before formulation of end-use products in the USA and Canada, but not in the EU and Australia |
| 2 | Analytical methods | Methods for the determination of residues are only required in the EU if relevant residues are expected |
| 3 | Human health | Information on short term-toxicicty in Tier I is required in the EU and Australia, not in the USA and Canada |
| 4 | Residues | EU: general information is required on persistence, multiplication, population dynamics, mobility (soil/water/air) of the microorganism and eventually occurring metabolites. Information can be derived from published literature<br>USA/CAN: data are only required if toxicology or ecotoxicology studies showed negative effects. Australia: no data required |
| 6 | Effects on non-target organisms | USA: Tests depend on application of the product<br>EU: Effects on non-target organisms have to be determined, if these are exposed. Evaluation of exposition and effects, calculation of exposure, risk assessment<br>Studies to assess effects on birds are required in the EU, USA and Canada, but not in Australia |
| 7 | Efficacy | Information is required for national registration in member states of the EU, in Canada, and Australia, but not required to be submitted and reviewed in the USA |

Member States within the EU based on the experiences with the 4th list substances (see also 5.2)

#### 2.4.3.5 Revision of Data Requirements in the USA

In the USA, data requirements were recently revised. New data requirements are included in "Federal Register, Part III, Environmental Protection Agency, 40 CFR Part 158, Pesticides; Data Requirements for Biochemical and Microbial Pesticides, October 26, 2007". Major differences to the existing rules are more flexible data requirements depending on the use of the product, the group of micro-organisms, and the particular species or strain. In particular, toxicity tests are further adapted to the micro-organism. The hypersensitivity studies are no longer required, but hypersensitivity incidents have to be reported. Avian inhalation tests are only conditionally required if the presence of toxins is suspected. Studies on the toxicity to plants are only required if the micro-organism is related to a plant pathogen.

## 2.4.4 Data Requirements for Botanicals

"Botanicals" include a vast diversity of substances or mixtures. Plants used for the extraction of "botanicals", might be known for food, feed, or medicinal uses, or are restricted in their use to plant protection purposes. Extraction methods and the extent of purification can vary as well, from crude extracts to purified single substances. Active constituents may be well-characterized or unknown. In addition, as for any biological products, variation occurs from one batch to another. "Botanicals" also differ in their modes of action. In general, substances with direct toxicity to the target pest are treated as conventional chemicals. Modes of action include induction of systemic resistance, repellence and growth regulation. Definitions differ between the regulatory systems considered in this comparison.

### 2.4.4.1 Definitions

**EU**: Plant extracts or "botanicals" are not defined in the EU legislation, and no separate data requirements exist. Therefore, the situation is different when compared to micro-organisms. Formally, they have to be registered like chemicals according to the data requirements outlined in Directive 91/414. Reduced data requirements are described in a SANCO Draft guidance document (SANCO Draft 10472/2003/rev.5) but this document covers only a small proportion of plant extracts. The SANCO draft document includes a reference list of plants for which these reduced data requirements shall apply. This list is based on experience with these plants in food or feed use, or as herbal drugs in European pharmacopeia. The document is furthermore restricted to water and ethanol extractions. Extracts derived from plants which are not mentioned in the document, or which are extracted with other solvents are treated like conventional chemicals. All data points have to be addressed, but information does not necessarily have to be provided by studies, but might be presented from published literature. In the comparison below, data requirements in the SANCO draft document and in the Directive 91/414 are considered.

**USA**: In the USA, naturally occurring substances are treated as "biochemicals" if they have a non-toxic mode of action. If Tier I testing shows toxic effects, the substance may be treated like a conventional chemical. "Biochemical and microbial pesticides are generally distinguished from conventional pesticides by their unique modes of action, low use volume, target species specificity or natural occurrence. Biochemical pesticides include, but are not limited to, products such as semiochemicals (e.g., insect pheromones), hormones (e.g., insect juvenile growth hormones), natural plant and insect regulators, and enzymes."

The new rule in the USA (in force since December 26, 2007) separates biochemicals from microbials and presents a new definition: A biochemical pesticide is a pesticide that...

(1) is a naturally-occurring substance or structurally similar and functionally identical to a naturally-occurring substance;

(2) has a history of exposure to humans and the environment demonstrating minimal toxicity, or in the case of a synthetically derived biochemical pesticides, is equivalent to a naturally-occurring substance that has such a history; and
(3) has a non-toxic mode of action to the target pest(s).

Extracts from plants or from micro-organisms with toxic properties are treated as conventional chemicals.

Canada does not have a particular set of data requirements for plant protection products based on plant extracts. However, for many plant extracts the "PMRA Initiative for Reduced-Risk Pesticides" (Regulatory Directive DIR2002-02) with reduced data requirements when compared to "conventional" chemicals might be applicable. "Reduced Risk" criteria are applicable for all other groups of products, including chemicals, micro-organisms, and semiochemicals. Data are required for the technical active ingredient and at least one product. Data requirements are very similar to those for "biochemicals" in the USA.

In **Australia**, plant protection products containing plant extracts and oils as active ingredients are classified as Group 2 "Biological agricultural chemical products". Decisions on actual data requirements are made on a case-by-case basis. Biologically derived chemicals that have direct toxicity to the target species are exempted from reduced data requirements and treated like chemicals. Unpurified or purified plant extracts can be included, as long as purification is incomplete and composition not fully characterised. Examples are "pyrethrum", consisting of a mixture of related pyrethrins, and "neem oil" and "neem extract", consisting of mixtures of characterised and uncharacterised components. If purification and full identification of a plant-derived substance is possible, these substances are as well treated as conventional chemicals and not as biological agricultural products (examples: nicotine, strychnine and ivermectin). Several plant extracts are excluded from the requirements of APVMA approval as constituents in plant protection products: cabbage extract, canola oil, *Capsicum oleoresin*, chilli extract, citronella oil, cypress wood oil, derris dust, eucalyptus oil, garlic extract, garlic oil, lanolin oil, lavender fragrance, lime oil, orange oil, pine oil, pyrethrins, pyrethrin I, pyrethrin II, quassia, rotenone, salicylic acid, sesame, tea tree oil and thymol.

#### 2.4.4.2 Data Requirements for the Active Ingredient and the Plant Protection Product

Requirements for plant extracts according to the SANCO Draft 10472/2003/rev.5 concerning information on toxicology and effects on non-target species are similar to the data requirements for micro-organisms. Data requirements refer to the active substance (if it can be purified at all) and to the plant protection product.

If components other than the active substance(s) are not considered to affect human or animal health, environmental behaviour, or have effects on non-target species, data requirements for Sections 2 and 3 to 6 can be covered using information on the active substance.

Section 1

According to Directive 91/414/EEC, the SANCO draft document, and to US/Canada and Australian data requirements the active ingredient has to be described. As a starting point, the plant species and cultivar as well as growth region and conditions, the plant organ and growth stage, have to be specified. If any substance has been identified, its chemical name according to IUPAC, CAS Number, structural formula, and ISO name have to be provided. For the description of physical-chemical properties, vapour pressure, partition coefficient, hydrolysis and photolysis have to be determined. The composition of the extract has to be described as far as possible. Concentration ranges have to be provided for all known substances included in the extract. Maximum limits have to be given for any substances that are relevant for human or animal health and the environment. If the active substance(s) is (are) not identified, a representative marker, i.e. a chemical naturally present in a known proportion in the plant extract has to be defined in order to identify the plant protection product. Five production batches collected over several periods have to be analysed for their contents. Mode of action and specificity as well as likely biological effects arising from use have to be declared. In Australia, special emphasis is placed on natural occurrence and distribution of the source organism in Australia, on the natural occurrence of the chemical or relationship to the form occurring in an organism.

A full list of ingredients of the plant protection product has to be submitted, including a precise quantity or an upper and lower limit of the extract and other ingredients. The plant protection product's trade name, physical state and function must be specified. Physical and chemical properties of the plant protection product, data on application, and further information on the plant protection product have to be given as for all other plant protection products. Information on shelf-life is as well required.

Section 2

According to Directive 91/414/EEC, validated analytical methods have to be provided for the active ingredient and for all "impurities" present in quantities $\geq 1$ g kg$^{-1}$. According to the SANCO draft document, validated analytical methods for the determination of contents of the active substances have to be provided if the substances are identified. If the active substances are not identified, a validated method of analysis of the marker in the plant protection product should be available. A validated method for analysing the active substance in water, soil and air is necessary. If exposure of the concerned compartment is not likely or the contribution compared to natural background levels is not substantial, such methods do not need to be provided. If any substances that are relevant for human or animal health and the environment are detected in the plant protection product, validated methods of analysis must be provided.

Section 3

Council Directive 91/414/EEC requires thorough testing of the active substance and/or the product, including acute, short-term, and long-term toxicity, carcinogenicity, reproductive toxicity, and delayed neurotoxicity tests. According to the SANCO draft document, toxicology testing refers to the product rather than to the active substance, to take all possible impurities into account. Information is required for acute oral, dermal, and inhalation toxicity, skin and eye irritation, and skin sensitisation in the form of studies or literature data. Oral toxicity studies may be waived if the plant is also used for food or feed. Risk assessments for the operator and worker must be addressed. Further toxicological testing is needed, if effects are observed in first tier studies.

In the USA, data on acute oral, dermal, inhalation toxicity, cutaneous and eye irritation and genotoxicity are required. Further toxicological testing is needed if effects are observed in Tier I studies. In Australia, data on acute oral, dermal and inhalation toxicity, genotoxicity, and short-term toxicity are required. Furthermore, information is required on Occupational Health and Exposure.

Section 4

The data requirements in Section 4 are similar between Directive 91/414/EEC and the SANCO draft working document.

Data on residues in or on treated products, food and feed are required according to the exposure to the plant extract's components due to the use as plant protection product, and have to be compared with the exposure due to consumption of the plant itself or to natural exposure to the plant itself. Residue data are required if human exposure to residues from products based on plant extracts is higher than from consumption of the plant itself. Supervised field trials only have to be carried out if human health or ecotoxicology are concerned. This requirement is different from microbials, where residue trials are in general not necessary. US EPA demands residue data depending on the use rate of the product. If the application rate is below 0.7 ounces active substance per acre (corresponding to 52.5 ml $ha^{-1}$), no residue studies are required. However, the proposed new rule requires residue data for biochemical pesticides only if Tier II or Tier III toxicology data were required. In Australia, residue studies are normally not required, and information demands can be covered using published literature.

Section 5

According to Council Directive 91/414/EEC, pathways and kinetics of degradation, adsorption/desorption, and mobility in soil have to be determined. Data on abiotic and biotic degradation in water, on volatility, and on photolytic degradation are required to assess the persistence of the substance in water and air.

The SANCO draft document and USA/Canada, Australia demand provision of available information from literature on natural background levels if, depending on the use of the product, exposure of water, soil or air is likely to occur. More information may be required based on expert judgement, if there is a substantial increase.

Section 6

According to Directive 91/414/EEC, the effects of the active substance and/or the product have to be determined on terrestrial vertebrates (acute, short-term dietary, and long-term reproductive toxicity to birds and mammals), aquatic organisms (acute, short-term, chronic toxicity to fish, invertebrates, sediment dwelling organisms, and algae, bioaccumulation), arthropods (acute oral and contact toxicity to bees and other arthropods depending on the use of the product), earthworms (acute and reproductive toxicity), and soil microorganisms according to Council Directive 91/414/EEC. SANCO/10472 only requires submission of "all available ecotoxicological information" which might primarily be based on published literature. If classification as dangerous substance according to Directive 67/548/EEC or 1999/45/EC is applicable, the following studies must be provided: acute effects on fish, daphnia and algae. Depending on the use, non-target arthropod testing is needed for those groups that are exposed to the product. In the USA, information is required on effects on birds (acute oral and dietary exposure), fish, and freshwater invertebrates. Depending on the use, non-target arthropod testing is needed. Data required in Australia depend on the use of the product.

Section 7

Efficacy data have to be submitted for national registrations of plant protection products in EU Member States, Canada and Australia, but are not required to be submitted for review in the USA. If the application concerns a product for field application, the efficacy data have to be specific for the country or at least the climatic zone. This restriction is normally not relevant for products that are intended for use in greenhouses.

#### 2.4.4.3 Revision of Data Requirements in the USA

Compared with the preceding version, the new rule in the USA contains changes to define when a data point is applicable. Toxicology testing is dependent on the expected human exposure through the use of the product. Exposure data have to be provided and hypersensitivity incidents have to be reported (which are new requirements). Data requirements for genotoxicity testing are more clearly described. Residue data are required depending the use of the product (and no longer the use rate).

### *2.4.5 Data Requirements for Semiochemicals*

Semiochemicals present a particular case among active ingredients used in plant protection products, as they are the only pesticides not intended to kill the pest organism. Semiochemicals can be used for (i) mating disruption, (ii) mass trapping,

(iii) monitoring and (iv) attract and kill. Semiochemicals are considered as pesticides in the first two uses. In the last two uses, they are not considered as pesticides, and are therefore exempt from registration. This section deals only with the use for mating disruption.

Semiochemicals have a high specificity for the target species. Their efficacy is not related to population effects, which makes assessment of efficacy difficult. One problem in fulfilling formal data requirements is the low production rates of most substances. Some substances are not even produced once every year; thus, request for data from a 5-batch analysis is very difficult to fulfil. On the other hand, exposure for applicators, bystanders and the environment is very low, because the released quantities are very low when compared with other pesticides. One group among semiochemicals is particularly well characterised: the Straight-Chained Lepidopteran Pheromones (SCLPs). SCLPs represent a homogenous group with low toxicity to non-target (and target-) organisms including mammals.

#### 2.4.5.1 Definitions

**EU**: Formally, there are no separate data requirements for semiochemicals in the EU; thus, they have to be registered according to the data requirements outlined in Directive 91/414. The OECD developed the "Consensus Document No 12" on "Guidance for registration requirements for pheromones and other semiochemicals used for arthropod pest control". Many EU member states choose to implement OECD 12. This document defines semiochemicals as follows: "Semiochemicals are chemicals emitted by plants, animals, and other organisms – and synthetic analogues of such substances – that evoke a behavioural or physiological response in individuals of the same or other species. They include pheromones and allelochemicals. This report pertains only to semiochemicals that affect the behaviour of arthropods".

**USA**: Semiochemicals are included in the group of "Biochemicals" according to 40 CFR 158. Data requirements are therefore formally the same, but vary according to the uses. For the definition, refer to Section 2.4.3.1.

**Canada**: Registration of products containing semiochemicals as active substances is possible as reduced-risk pesticide (Regulatory Directive DIR2002-02: The PMRA Initiative for Reduced-Risk Pesticides). Data are required for the technical active and at least one product. Synthesized pesticides can be considered as pheromone or other semiochemical pesticides, if it is demonstrated that they are structurally similar and functionally identical to a naturally occurring pheromone or semiochemical. Data are required according to OECD 12.

In **Australia**, semiochemicals in plant protection products are classified as Group 1 "Biological agricultural chemical products". Decisions on actual data requirements are made on a case-by-case basis; special emphasis is on environmental expression and suppression of indigenous species. Most pheromones are exempted from data requirements for the active ingredient for chemistry, manufacture, and biological properties. Metabolism and kinetics data are only required if the concentration of the active ingredient results in levels that can be differentiated from background levels. Residue studies are usually not required.

Environmental studies have to be provided as for chemicals, but the actual requirements are determined case-by-case. The following semiochemicals are excluded from the requirements of APVMA approval as active constituents (Pheromone use only): 4-(p-acetoxyphenyl)-2-butanone (Cue-lure), 4-(p-hydroxyphenyl)-2-butanone (Frambinone), 8,10-Dodecadiene-1-ol, 8-Dodecen-1-ol, 8-Dodecen-1-ol acetate (cis-isomer), 8-Dodecen-1-ol acetate (trans-isomer), German Cockroach Pheromone, Isomate LBAM, Isomate OFM Rosso, (E)-2-Octadecenal, (E,Z)-2,13-Octadecadienal, and (Z)-9-Tricosene.

#### 2.4.5.2 Data Requirements for Semiochemical Active Ingredients and Products

Date Requirements as in OECD 12: Sections 1 and 2

According to OECD 12, the mode of action of a semiochemical product should be explained in terms of its function in modifying the behaviour of the target pest, and information should be provided to support the claim that the active ingredient is a naturally occurring arthropod semiochemical. Qualitative information is required on the pest species life cycle, and the nature and extent of damage it causes. Other useful information includes the compatibility of semiochemicals with IPM programs and their contribution to risk reduction.

Identification of the active ingredient has to be provided as well as specific physical and chemical characteristics. Identity data are used to determine whether an active ingredient is identical or structurally similar to another active ingredient or a naturally occurring substance. The manufacturing process has to be described with the starting materials. The possible formation of impurities has to be discussed and upper and lower certified limits for each active ingredient component, and upper limits for impurities have to be given. Supporting analytical data including component identity confirmation are required.

For the characterisation of the product, identification of the active ingredient, formulants, and impurities of toxicological concern in the plant protection product has to be provided. Starting materials and the formulation are to be described including upper and lower certified limits of the technical grade active substance and formulants. An enforcement analytical method for each active ingredient component has to be provided. If the formulation process introduces or enhances the presence of impurities of toxicological concern, this must be identified along with upper limits and a corresponding enforcement analytical method.

Section 3

Sufficient information to identify potentially hazardous products is always required, including information on irritation, dermal sensitisation, acute toxicity, mutagenicity, and medical data. Studies of teratogenicity and subchronic exposure can generally be waived if long-term exposure above background levels can be excluded, or if a substance is a member of a well-characterized group, such as

SCLPs, for which toxicological concerns have already been addressed. Less information is available on the toxicity of other forms of semiochemicals containing ketone, epoxide, lactone, terpenoid, pyrazine, pyran and other aromatic structures. If they have the toxicological characteristics of other chemicals with these substructures or functional groups, they may be more toxic than the SCLPs and might potentially require long-term tests.

Section 4

For semiochemicals, residue data may not be required, if detectable residues on the consumable commodity are unlikely to occur, if residue levels are unlikely to exceed natural background levels during outbreaks of the pest, or if residues are not toxic. In Canada and the EU, applicants are encouraged to provide a scientific rationale for waiving residue data based on the low potential risk of any residues on a treated crop.

The US EPA has established an exemption from the requirement of a food tolerance (i.e., MRL) for most uses of arthropod semiochemicals, namely

- in retrievably sized polymeric dispensers used at a rate no more than 375 g active ingredient (a.i.)./ha/year;
- at a rate of no more than 50 g a.i./ha per application regardless of formulation, provided no potentially adverse effects are observed during the tier I toxicity testing;
- SCLPs at rates up to 375 g a.i./ha/year, regardless of the mode of application.

Sufficient information is required to characterize occupational and bystander exposure potential. This would include consideration of application method and rate and appropriate physical–chemical properties. For those substances with significant exposure potential and for those with toxicological concerns, additional exposure data would be required. In the EU, the USA and Canada, residue studies are conditionally required if toxicity data indicate concern. In contrast to this, residue data are generally not necessary for semiochemicals in Australia.

At EU level the exemption from MRLs setting is foreseen (not implemented so far) and exempted substances will be listed in Annex IV of Regulation 396/2005 on residues. Semiochemicals are among the candidates to be listed there.

Section 5

According to the OECD document, assessment of the environmental fate of a semiochemical (e.g., stability in air and water) is required, based on available information. Test data on a compound will only be required if its use will result in environmental contamination exceeding natural background levels. Application rates of up to 375 g SCLP/ha/year are generally understood to result in exposure levels that are comparable to natural emissions and safe for non-target species. This threshold may or may not be applicable for other kinds of semiochemicals; applicants are invited to request waivers of environmental testing, based on information that indicates that application rates are comparable to natural emissions. If ecotoxicity data or public literature indicate a hazard to biota, data on the persistence of a semiochemical and its transport from the site of application to another site or medium may be required.

If the data indicate that significant persistence and transport of these agents occurs in any part of the environment such that significant exposure to non-target organisms could be expected, then additional environmental testing will be necessary.

Determination of the estimated environmental concentration (EEC) is performed with a simple mass-balance analysis of the pesticide, taking into consideration

- the pesticide application parameters (i.e., rate, frequency, and site of application)
- initial tests that measure transport properties (volatility, dispenser water leaching, vapour pressure, and water solubility).
- persistence testing (hydrolysis, aerobic soil metabolism, aerobic aquatic metabolism, soil photolysis, aquatic photolysis, adsorption–desorption, and octanol–water partition coefficient), each of the transformation processes should be expressed as a half-life for the particular environment, or as a rate constant for the environmental process, depending on the test.

Estimated environmental concentrations can then be calculated for different times using these data and the field application rate of the pesticide. Aquatic use patterns and non-dispenser pesticides will require mass-balance analysis following persistence tests.

## Section 6

Fewer tests are required for semiochemicals when compared with chemical pesticides, and the number of organisms per test is reduced because of the non-toxic mode of action of semiochemicals and limited exposure of non-target organisms. Avian dietary toxicity is only of concern for formulations that might be ingested, e.g., granules. No wild mammal testing is required. Non-target terrestrial plant studies would only be required if effects are suspected. Aquatic invertebrate and fish toxicity data are required for direct application to aquatic sites for all semiochemicals. One species of fish (rainbow trout), an aquatic invertebrate (*Daphnia magna*) and (in Europe) an algal species should be tested. Aquatic testing is not required for fixed-point dispensers applied over land. Non-target arthropod testing is not required if no adverse effects were observed during efficacy testing (in particular on predators or parasites of the target organism, closely related species and pollinators). Following OECD 12, and depending on the use also in the USA, testing of effects on bees and terrestrial arthropods other than bees is required, while effects on earthworms and soil micro-organisms only have to be determined according to OECD 12, if the product is applied to the soil and can accumulate in soil. It is only required if the exposure exceeds natural background levels (e.g. at >375 g a.i./ha/yr for SCLP). Available information has to be provided and discussed. Ecotoxicity data are generally not necessary for semiochemicals in Australia.

Section 7

In the EU (only for national registrations, not for Annex I listing) and Canada, data from scientifically conducted efficacy trials are required to support pest control claims on the product label and to demonstrate how a product may be used most effectively. Sufficient efficacy data are required to confirm the performance. At least one study should evaluate a range of rates to demonstrate the lowest effective rate of application. In conjunction with the efficacy trials, information on any adverse effects on the crop or site should be reported, including phytotoxicity and effects on non-target arthropods.

Because the use of semiochemicals can involve specialist techniques, they require adapted trial protocols. The UK PSD has recently developed draft efficacy guideline 220 on efficacy testing of mating disruption products. This has been submitted to the EPPO fungicides and insecticides panel as a draft guideline for discussion.

#### 2.4.5.3 Revision of Data Requirements in the USA

The new data requirements according to the "Federal Register, Part III, Environmental Protection Agency, 40 CFR Parts 158 and 172, Pesticides; Data Requirements for Biochemical and Microbial Pesticides" introduce a change for semiochemicals: No toxicity data are required for SCLPs, and no non-target organism and environmental fate data are required for all arthropod pheromones, if they are applied at less than 360.66 g $ha^{-1}$.

## 2.5 Practical Experience with the Regulatory Process

### *2.5.1 Case Study: Registration of* Paecilomyces lilacinus *in the EU and the USA*

As an illustration of current regulatory practice, studies submitted for the registration of the plant protection product based on the micro-organism *Paecilomyces lilacinus* strain 251 in the USA and for the inclusion of the same strain in Annex I of Council Directive 91/414/EEC were compared. Both applications regarded the same use of the product. As the same studies were available for both dossiers when initial applications were submitted, most of the submitted studies were identical. Some studies already existed, and were therefore submitted even if they might not have been necessarily required. It must be emphasized that this section is based on the experiences with a single micro-organism, and not on a representative sample of substances.

**Data requirements:** Data submitted in the EU but not in the USA include studies on the genotoxic potential and a cell culture study. In the EU, a study to assess acute intratracheal/inhalation infectivity, toxicity and pathogenicity of the formulated product including analysis of the clearance of the micro-organism was additionally requested during the evaluation process. Furthermore, studies on effects on terrestrial arthropods other than bees and on effects on earthworms were additionally demanded.

**Waivers:** No studies to assess short-term toxicity, pathogenicity, or infectivity were submitted in either of the two systems, because no signs of acute toxicity, pathogenicity or infectivity were observed in acute tests. Furthermore, no data were submitted on effects of the micro-organism or the formulated plant protection product on birds, aquatic and terrestrial plants, and bees, because these non-target organisms are not exposed to the micro-organisms with the intended uses.

**Time for registration:** The comparison of the time needed for registration in the USA and for Annex I listing in the EU is extremely difficult. The time needed by the applicant to generate data that were additionally requested in the EU is counted as well as the time needed by authorities (RMS, other member states, and EFSA) for the evaluation. Evaluation in the USA took 21 months starting from the submission of the dossier to the registration of the product. A provisional national registration for the product BioAct was obtained in Italy before Annex I listing of *P. lilacinus* in December 2005. The application for the inclusion of *P. lilacinus* strain 251 into Annex I was submitted in April 2002 and Annex I inclusion was published in April 2008.

### *2.5.2 Data Requirements and "Waivers"*

Whether specific data or studies are required or whether a data requirement can be waived is better defined in the USA and in the Canadian system than in the EU. Also, the industry reports that the registration authorities in the EU are in general less willing to accept "waivers" than the authorities in the USA. The background for this is that (i) in the EU, the registration authorities take more responsibility when they register a substance, while in the USA, the responsibility stays largely with the applicant; (ii) within the US EPA, a separate unit of staff is concerned exclusively with the registration of "biopesticides", and has built up expertise with this kind of substances. Also, many biopesticides have been applied for in the USA which has given the US EPA considerable experience in assessing these kinds of products. In the EU few applications for active substances and products were submitted and evaluated under the current legislation before the EU review programme. Furthermore, these applications have been assessed by different RMS (producing the DAR). Therefore, the registration authorities within each RMS have built up less experience with such products. With this background of knowledge, the US EPA can accept waivers more easily than their European counterparts. (iii) Even if the RMS accepts a waiver, a study may later be requested, when the other member states and EFSA are evaluating and giving comments to the DAR. It is possible that some RMS rather require the maximum number of studies which might be required, in order to avoid such requests in the commenting phase. The 4th stage of re-evaluation will give the European regulators' community a chance to collectively improve their expertise in evaluating BCAs, and to harmonize their interpretation of data requirements in this field. This experience is expected to result in a number of EU guidance documents/"lessons learned documents", which will facilitate the application process for future applicants as well as the future assessments carried out within the regulatory authorities.

#### 2.5.2.1 USA

In the USA, more BCAs are registered than in the EU (see Table 2.2), and registration is faster (see Table 2.1). Given the aims of the REBECA project, the US system is more successful. This is only partly due to lower data requirements than in the EU (for details see Section 4). However, other factors contribute equally to the success of this system. (i) The process is organized much more simply than in the EU, with only one authority in charge of registration (except for some states that require separate registration). (ii) There are strict timelines for registration. (iii) Within the EPA, a separate unit is concerned with biopesticides, thus building up expertise with this kind of products.

US EPA recently conducted a survey on the data that were effectively required for the evaluation of 9 different bacteria and 11 fungi for the use in microbial plant protection products between 1997 and 2004. Data on chemical identity and technical properties were required in all cases. No data on residues or methods for residue analysis were required. For assessment of toxicity, acute studies were required, but cell culture studies or studies on subchronic toxicity/infectivity/pathogenicity were not demanded. Studies on ecotoxicity (non-target-organisms) were limited to organisms that were apparently exposed or at risk. No data on environmental fate and behaviour were required. Data requirements for the only baculovirus that was evaluated during the selected timeframe were limited to information on chemical identity and technical properties, a study on acute oral toxicity/pathogenicity, acute inhalation toxicity, and acute eye irritation. Further information on toxicity, ecotoxicity or residues was not required.

During recent evaluation of five different Straight-Chained Lepidopteran Pheromones (SCLP) by the US EPA, data on chemical identity and technical properties were required. However, no data on residues or methods to determine residues, no toxicity studies, and no studies on ecotoxicity and environmental fate and behaviour were required. No comparable survey is available for the EU, Canada, or Australia.

**Microbial pesticides:** 78 microbial pesticides are currently registered: bacteria: 17 *Bacillus thuringiensis* subspecies, 10 other *Bacillus* species, 10 *Pseudomonas* species, 2 *Agrobacterium* isolates; 29 fungi; 7 baculoviruses; 2 yeasts; 1 protozoan.

**Biochemical pesticides:** 160 biochemical pesticides are currently registered: 50 semiochemicals (pheromones), 4 insect growth regulators (e.g. azadirachtine), 21 plant growth regulators (e.g. indole-3-acetic acid), 3 herbicides (e.g. corn gluten meal), 29 repellents (e.g. capsaicin from red pepper), 14 floral attractants and plant volatiles, 18 products for insect and nematode control (e.g. soybean oil) and 21 products for plant pathogen and microbial control (e.g. sodium bicarbonate).

The registration procedures in the USA for the registration of BCAs are more predictable than those in the EU. These results in a much larger number of products registered in the USA (see introduction). The proposed new rule, which is based on long-term experience with a large number of products, will formalize the current practice of the US EPA. This procedure applies to biochemicals, microbials, semiochemicals, but also to chemicals if certain conditions are fulfilled. Likewise, applications restricted to minor uses can be treated as reduced risk pesticides.

#### 2.5.2.2 EU

No comparable survey is available for the EU, Canada or Australia. However, in the EU a similar survey was carried out for all the 4th list micro-organisms (approximately 30 strains belonging to 16 species) and is expected to be finalized. The survey will also include information regarding studies being accepted even though they were performed on closely related strains and not the specific strain in question. However, from such surveys it is not possible to see if some studies were submitted even though a waiver may have been accepted. In the EU, practice demonstrates that assessment of clearance in studies on acute toxicity, pathogenicity, and infectivity in many cases is only required for one out of the three studies. In many cases, these studies and studies on non-target organisms only have to be performed with the active ingredient or the formulated product and not with both. Furthermore, short-term toxicity studies were only required for a smaller part of micro-organisms already listed in Annex I. Tier II toxicity tests are normally not required, as results with micro-organisms in Tier I tests do not raise concern for human or environmental aspects.

So far, relatively little experience exists with the current European data requirements. Since the first micro-organisms were evaluated as "new active substances" for inclusion in Annex I, data requirements have been amended (Commission Directive 2001/36/EC) and the uniform principles for evaluation were defined (Commission Directive 2005/25/EC). The so called "old active substances", including micro-organisms, which were used in plant protection products before 1993, are subject to the 4th stage of re-evaluation. As this is still ongoing and is expected to be finalised after the peer review in 2010, there is no experience with the 4th stage available.

A common interpretation of data requirements has not yet evolved in the EU (e.g.: what is a "relevant metabolite"?). In this situation, many applicants have chosen to submit "minimum data packages", in order to save costs for studies that might not necessarily be required. This has led to demands for further studies not initially considered, which has considerably lengthened the process. Such studies are especially expensive and time consuming, if new methods have to be developed.

Little experience exists with the registration of botanicals in the current system. To date (January 2009), only one botanical has been evaluated as a new active substance in this system and is included in Annex I of Directive 91/414.

The active substance, laminarin, is a polysaccharide purified from the brown algae Laminaria sp.. Due to the chemical properties and the indirect mode of action of laminarin towards fungal plant pathogens, only a reduced set of data when compared to chemical substances was required during evaluation of the active substance and a corresponding product for inclusion into Annex I. Data requirements were reduced for human health, environmental fate and behaviour, and ecotoxicology. Several plant extracts, some of them with direct effects on the target organisms, have been evaluated as List 4 substances and will be included in Annex I in September 2009 whereas evaluation for some other plant extracts is still ongoing. No detailed information on the data used for the risk assessment is available to date.

Also for semiochemicals, relatively little practical experience exists regarding the applicability of the data requirements in the EU. EU evaluation of the so called "old substances" including the majority of semiochemicals that are used in plant protection products was just finalized in December 2008.

### *2.5.3 Experience with the Structure of the Registration Process in the EU*

The registration process under Directive 91/414 is structured in a complex way and involves a large number of authorities, such as The European Commission (mainly DG SANCO, which is responsible for the coordination of the process and the legal aspects), the registration authorities of the Rapporteur Member State (RMS) and the European Food Safety Authority (EFSA) which is responsible for coordinating a peer review process of the active substances (see Fig. 2.1). This process takes time. In addition, this structure is intrinsically conservative and may in many cases hinder the acceptability of waivers, since even though accepted by some member states, others may not accept the waiver during the peer review process. In some cases, RMS have to request data even if they do not consider them essential for the evaluation themselves. Compared with the EU system, the registration processes in the USA, Canada, Australia, or Switzerland are simpler, mainly because they involve only one country.

Historically, the current registration procedure has grown. Originally the plant protection products were only regulated by the national authorities in each of the member states. In 1991, Directive 91/414 established an EU-wide system for registration of active substances. The EFSA was created in 2002.

The Commission's proposal for revision of Dir. 91/414 of 12.07.2006 contains other important changes such as the transition from a Directive to a Regulation, and mutual recognition of plant protection product registration within each of three climatic zones.

Conclusions: The current EU registration system is complex. It reflects the structure of the political organization of the EU, which is a compromise between central administration and independence of member states (subsidiary). The system is dynamic and likely to change in the future. However, changes cannot be made solely from a technical and practical viewpoint, but must be embedded in a wider political context.

### 2.5.4 *Time Span for Registration*

As the examples in Table 2.1 show, it takes much longer for an active substance to be included in Annex I of Directive 91/414, than it takes for a substance and the corresponding product to be registered in the USA. Furthermore, in most EU member states Annex I inclusion must be followed by registration of plant protection products at national level, before commercialization can begin. However, some member states have given provisional authorisation of products prior to Annex I inclusion. In the USA, the registration process is at present guaranteed to be completed within 18 months. As can be seen from Table 2.1 this was not always possible in the past.

For the industry, long registration periods are a severe problem, because they delay the onset of the returns for the investments made during research and development. In addition, longer registration periods result in shorter periods of sale under patent protection.

Conclusions: The length of the registration period in the EU is one of the main obstacles for the industry to obtain registration of their products. The long process is mainly caused by the complicated structure of the decision-making process.

### 2.5.5 *Fees*

Fees demanded by the authorities vary widely. EU fees for evaluation of dossiers for substances of the 4th list varied between 11,600 € and 215,000 € (information was not available for all member states). In most member states, fees for BCAs were considerably reduced compared with the fees demanded for conventional chemicals as active substances. In some member states, fees are intended to recover the costs for evaluation but, in others, fees are lower than actual costs.

The fees for new active substances also differ between member states (0–215,000 €). Some countries do not require any fees for new substances, e.g. if a national registration of a product is applied for simultaneously (which also applies for new chemical substances).

In Canada, no cost recovery-fees are demanded for the evaluation of microbial plant protection products, in order to encourage the registration of biological control products. Only a label review fee of 252 CND$ is requested. In the USA, fees for products containing microbials are 25,000 US$, but fees can be reduced or waived for small enterprises. In Australia, fees depend on the data that are actually required, since the fee has to cover the costs for evaluation.

### 2.5.6 Pre-submission Meetings

Pre-Submission Meetings between the applicant and the evaluating authorities are formally recommended in the USA, in Canada, and Australia and proved to be very effective. They are also informally recommended in the EU. Practice in the EU demonstrated that these meetings were in some cases not offered by the authorities or not used by the applicants. The structure of the registration process sets a limit to pre-submission meetings in the EU: Besides the rapporteur member state, other member states and EFSA are later on also involved in the evaluation of an active ingredient, but they do not participate in pre-submission meetings.

In pre-submission meetings, the substance is briefly discussed and it is outlined what kind of data are likely to be needed, and where waivers are likely to be acceptable. Applicants can thereby better target their application dossiers to the requirements of the authorities, and save expenses on the production of unnecessary data and consequently speed up the preparation of the dossier. Authorities report that the improvements of dossier quality save labour during evaluation.

### 2.5.7 EU Harmonization with Biocide Regulation

Some BCAs, mainly pheromones and botanicals, can be used both as plant protection products and as biocides. The industry reports that the two registration processes are not harmonised to a large extent. For example, dossiers normally have to be presented in different formats. Regulatory fees are higher for biocides than for plant protection products (authorization and evaluation of dossiers) in all EU member states. However, the data requirements for microbial biocides were recently revised mainly in order to harmonise these with the data requirements for microbial plant protection products (the biocide data requirements can be found in Commission Directive 2006/50/EC and further information can be found in the Technical Notes for Guidance). Also, the Uniform Principles of the microbial plant protection products have been used as a basis when a similar and nearly identical document was recently prepared for the biocides.

Conclusions: BCAs are highly specialized products, for which the evaluation methodology differs greatly from conventional pesticides. Thus, the plant protection and biocide use of a specific substance (e.g. a certain lepidopteran pheromone) share many common aspects. The evaluation of plant protection and biocide uses should be harmonized as much as possible, to avoid further fragmentation of the regulatory community, and to save costs on unproductive tasks such as reformatting of dossiers. Increased harmonization has the potential to benefit the use of BCAs in both areas.

### 2.5.8 Level of Annex I inclusion

In workshops on semiochemicals which were held during 2006 in the framework of the REBECA project, the question was raised how the active ingredients will be listed in Annex I. Listing could be in one of the following ways:

- "Straight-Chained Lepidopteran Pheromones" (SCLPs) could be listed collectively, or
- each substance could be listed separately, or
- each blend of substances, with their proportions, could be listed separately.

The mode of listing will influence the registration requirements for future applications. In particular, it determines whether or not the active ingredient must be registered. Under the first option, new plant protection products based on old and on new SCLPs would not require Annex I inclusion, but only national registration of the plant protection product. Of course, the data of the SCLP notifiers in the 4th stage re-evaluation would have to be protected. Under the second option, new plant protection products based on old SCLPs would not require Annex I inclusion, while new SCLPs would require Annex I inclusion. Under the third option, even new blends of old SCLPs would require Annex I inclusion. Industry would prefer one of the first two options, because this facilitates the development of new end-use products.

Plant extracts are complex mixtures of numerous substances, which can be characterized to a variable degree, but rarely to 100 %. Therefore, similar problems may arise with the listing of plant extracts in Annex I.

Micro-organisms are listed in Annex I at strain level. A working group within the framework of the REBECA project has proposed facilitations in the registration of plant protection products containing baculoviruses. Based on conclusions of the "OECD Consensus document No 20 on information used in the assessment of environmental applications involving baculoviruses" (see 6.1.1), on the high host-specifity of baculoviruses, and on the high similarity of baculoviruses with regard to effects on humans, the environment, or non-target organisms, the authors propose to include members of the family Baculoviridae on the level of species. This proposal was implemented in the EU as "Guidance Document on the assessment of new isolates of baculovirus species already included in Annex I of Council Directive 91/414/EEC" (SANCO/0253/2008 rev. 2, from 22 January 2008) and reduces the data requirements for new isolates of a given baculovirus species.

## *2.5.9 Efficacy Data*

For Annex I inclusion of the active substance, efficacy data are not required. For registration of the plant protection product, efficacy data must be submitted for all uses applied for. By contrast, efficacy data are not required for registration in the USA under the preceding legislation, but will be required under the new rule.

Not all EU member states require efficacy data, and EU member states differ in their practice of how easily they accept efficacy data from other countries. Also, the practice for issuing experimental permits and the requirements for crop destruction differ across the EU.

Experimental permits: Practice for issuing experimental permits differs widely in member states, in terms of crop destruction and area that can be treated. No mechanism is foreseen to harmonize experimental approval systems, and there is no legal basis for this. However, industry and Member States could take the initiative for harmonization.

Conclusions: These aspects are under national authority and therefore cannot be harmonized with EU legislation at the moment. Nevertheless, the costs related to submission of efficacy data are often high, and add to the total costs for market introduction of a new BCA and should therefore be kept as low as possible. Currently, there are plans for mutual recognition of registration within climatic zones. If this concept will come into force, such problems would be greatly reduced.

## 2.6 Initiatives Taken to Facilitate the Registration of BCAs in the EU

Section 6 of this report contains an inventory of documents, initiatives and concepts aiming to improve the registration of BCAs. These include guidance documents at EU and OECD level (Section 6.1), legislation and regulation at national level and national initiatives for supporting SMEs in the registration of BCAs (Sections 6.2, 6.3, 6.4, 6.5, 6.6). This section also briefly discusses the new categories of "low risk substances" and "basic substance" mentioned in the Commission's proposal for a new Directive on plant protection products, as well as new concepts for risk assessment (Sections 6.7, 6.8, 6.9). These initiatives act at different levels. Some of these documents, initiatives and concepts may serve as models or as inspirations for improvements of the registration process in the EU.

### *2.6.1 Guidance Documents*

#### 2.6.1.1 Baculoviruses: OECD 20

The "OECD Consensus document No 20 on information used in the assessment of environmental applications involving baculoviruses" dates from January 2002. With respect to human health, it is stated that "Baculoviruses are naturally occurring pathogens of arthropods. Their host range is exclusively restricted to arthropods. No member of this virus family is infective to plants or vertebrates" and it concludes that "No adverse effect on human health has been observed in any of these investigations indicating that the use of baculovirus is safe and does not cause any health hazards." This guidance document summarises the current knowledge on baculoviruses with respect to risk assessment and is a valuable argument to avoid the production of additional studies.

#### 2.6.1.2 Botanicals: SANCO/10472

The draft guidance document SANCO/10472/2004 proposes data requirements for some plant protection products based on plant extracts. Data requirements are described in detail under Section 2.4.3.2. The data requirements apply only to plant protection products made from the edible parts of plants used for animal or human feed, from parts of plants mentioned explicitly, or from parts of plants currently authorised as herbal drugs in European pharmacopoeia and known traditionally for plant protection properties (also listed explicitly). Further, the data requirements apply only to plant extracts made with water and/or ethanol. For plant protection products made from other plants or plant parts or with other solvents, data requirements will be established case-by-case in a pre-submission meeting, based on the available information.

As pointed out in Section 4.3, the document is still a draft and is not legally binding. During a REBECA meeting a working group on botanicals recommended that SANCO/10472 should be reviewed and amended, based on the experience of applicants and regulators involved in the 4th stage re-evaluation.

- The document should also include extracts prepared with other extraction methods, such as $CO_2$, pressure or food-grade oils.
- The list of plants in the annex should remain indicative («plants such as»).
- The definitions for "products" and "extracts" need specification and should be used consistently throughout the document.
- The working group recommended to encourage purification of plant extracts, but to allow declaration of a range of composition for actives and "impurities". The degree of detail required should be justified by toxicological relevance.
- At the moment, there is still very little experience with this document. With more experience, the document should become an official guidance document.

#### 2.6.1.3 Semiochemicals: OECD 12

The OECD Consensus document No 12 on "Guidance for registration requirements for pheromones and other semiochemicals used for arthropod pest control" dates from February 2002. The document suggests reduced data requirements, particularly for straight-chained lepidopteran pheromones (SCLPs). Data requirements are presented in detail in Section 2.4.4.2. The rationale is that

- The application rate is typically low and probably comparable to natural emissions.
- Volatility and rapid environmental transformation minimise residues in crops and exposure of non-target organisms.
- SCLPs are of low toxicity to mammals.

Although it has no official status within the EU, several registration authorities are ready to use it as a basis for evaluation. The working group "semiochemicals" suggested to formalize its use within the EU, and also to adopt its use for registration under the biocide directive 98/8.

#### 2.6.1.4 Plant Strengtheners (EU Draft Working Document)

There have been attempts to elaborate data requirements for plant strengtheners at EU level. Draft working document SANCO/1003/2000 rev.3, 21/06/2001 was intended to be a guidance for a reduction in data requirements for "low risk products", when compared with data required for chemicals according to Council Directive 91/414/EEC. The document refers to products which are expected to be "low risk" for humans, animals and the environment from their composition and their uses.

During the drafting of "Commission Regulation (EC) No 2229/2004 laying down further detailed rules for the implementation of the fourth stage of the programme of work referred to in Article 8(2) of Council Directive 91/414/EEC" Guidance Documents were developed on Plant extracts (Sanco 10472/2003-rev. 5; 6.7.2004) and Chemical substances (Sanco 10473/2003-rev. 4; 6.7.2004). In a certain sense these documents replace the working document on plant strengtheners. The document on Plant Strengtheners was not followed up.

As active substances are not characterised or cannot be purified for many plant strengtheners, data requirements refer basically to Annex III data. Evaluation follows a tiered approach. If during evaluation of Tier 1 data it turns out that the product is not "low risk", the full data requirements are applicable. Requirements are similar to those for microbials, with some exceptions.

Sections 1 and 2

Information on the identity and composition of the product, methods of manufacture, its physical-chemical and technical properties, and application is required. For plant extracts, details on the origin of the plant, harvest season, plant parts, extraction methods and main components are demanded. In addition to data required for micro-organisms, a method to qualitatively identify the characteristics of the extract is required for plant extracts. On the other hand, analytical methods for determination of residues are not required.

Section 3

Acute toxicity studies are required for the product as for products containing micro-organisms. Short term toxicity studies and genotoxicity testing are also required. A literature summary on operator exposure based on available information has to be provided. Assessment of bystander and worker exposure is not considered to be relevant.

Sections 4 and 5

No data are required.

Section 6

Data requirements are identical to those for micro-organisms.

Section 7

Efficacy data are limited to qualitative effects and possible long-term effects. Information can be derived from literature.

### 2.6.2 Genoeg (NL)

The "GENOEG" (Gewasbeschermingsmiddelen van Naturlijke Oorsprong Effectief Gebruiken) project was initiated in The Netherlands in 2002. Its aim was to get more natural pesticides registered, to learn about their low risk profiles and to apply this knowledge for registration purposes. The project is funded by the Dutch Ministry of Agriculture, Nature and Food Quality and the Product Board for Horticulture. It is carried out in collaboration between the Board for Authorization of Pesticides (CTB), the Plant Protection Service, the Agricultural University, the Dutch Organisation for Agriculture and Horticulture (LTO), the National Institute for Public Health and the Environment (RIVM) and CLM.

Applicants were supported in registration procedures and financially by partial funding of the costs required for generation of data and for registration (co-financing of up to a maximum of 50%, with a limit of EUR 100,000). Furthermore, scientific and administrative expertise was provided to facilitate registration.

Initially, the potential "low risk profile" of the products was evaluated by experts based on information provided by the applicants. Products were also selected according to their use, efficacy, and relevance for particular growing conditions in integrated agriculture in The Netherlands. Evaluation was done in contact with the applicants. Data requirements were partially answered with statements based on literature, expert judgement, or information from registration procedures in other countries for matters of identity of the active ingredient, residues, ecological effects, and efficacy. Basic information that has to be provided is the composition of the product, the application rates and frequencies, and toxicology studies. Administrative procedures were facilitated by guidance through the registration process. Despite this support, the main responsibility remained with the applicant.

Economic risks for registration remained high due to unpredictable time scales for registration. At the same time, the market for biological pesticides is still very small, as the use of these products requires adapted culture conditions for many crops. Experience with biological pesticides is increasing among regulatory bodies. During the project, risk assessment and evaluation were adapted to biological pesticides and statements were accepted more frequently. Case-by-case evaluation again turned out to be essential for biological pesticides. By using statements instead of performing studies, costs could be saved. From the project, the following steps were considered necessary: a reduction in formal data requirements, taking into account the specific risk of biologicals, predictable registration costs, and improved communication between applicants and authorities.

Since 2002, GENOEG has supported the registration of two plant protection products with active ingredients in the 4th stage of re-evaluation (Trianum

[*Trichoderma harzianum*]; Botanigard [*Beauveria bassiana*]), one product with the active ingredient listed in Annex I (Preferal [*Paecilomyces fumosoroseus*]) and of two new substances, the registration of which is not yet completed. All five initial applicants claimed that they would have withdrawn their applications if they had not been supported by GENOEG. In 2004, this activity was scaled up with 10 new products. For more information, see www.genoeg.net. Source of information: "Five case studies on the registration of natural pesticides in The Netherlands", GENOEG, March 2006.

### *2.6.3 Rub (NL)*

The Dutch pesticide law recognizes the so-called "Regulation Exemption Pesticides" (Regeling Uitzondering Bestrijdingsmiddelen, abbreviated "RUB"). RUB are products with such a low risk for man and environment that the usual procedures for authorization are considered unnecessary. Contrary to the other procedures, where the decisions on authorization are taken by the CTB (on behalf of the government), the government itself decides about RUB authorizations, advised by the CTB. Examples of RUB: milk as a viricide and against mildew in courgette, sugar as a fungicide against a specific disease in arboriculture and several plant oils against pests and diseases, and potassium phosphonate as a fungicide in glasshouse culture. Sometimes, authorization is only given for a specific application method. For example, heavy oils like coconut oil and sunflower oil are allowed for spraying, while lighter, volatile oils may only be used for dipping and pouring. A number of 4th list substances have been registered under the RUB procedures in the past years. However the RUB procedure will no longer apply when the substances of the fourth list have been evaluated.

Source of information: R. Boeringa and M. Trapman (2004) Plant protection products in organic farming in The Netherlands. In: *Current Evaluation Procedures for Plant Protection Products Used in Organic Agriculture.* Proceedings of a workshop held September 25–26, 2003 in Frick, Switzerland.

**Conclusions:** Given that the low-risk status of a plant protection product is established, the RUB procedures allow fast registration. A number of substances have been registered under the RUB procedures in the past years, demonstrating that this pragmatic approach is effective in practice. The RUB procedure might be taken as a blueprint for similar procedures in other countries, and possibly also at EU level.

### *2.6.4 PSD Pilot Scheme and Biopesticide Scheme (UK)*

In June 2003, the UK Pesticides Safety Directorate (PSD) introduced the "pilot scheme" for BCAs. The overall aim was to increase the availability of BCAs in the UK. Within this programme, recommendations and guidance were given to applicants, starting with support early in the product development. Data requirements were not modified, but the procedure was facilitated. Pre-submission meetings were held to create confidence between the authorities and the applicants, and to increase the awareness of the regulation process among the applicants. Fee structure and success rate depend on dossier quality. An identification of "low risk" profile was done beforehand in initial pre-submission meetings. An authorisation within 42 weeks after completeness is guaranteed. Within the pilot scheme, "Exosex CM" (Codlemone, under evaluation for Annex I inclusion as existing substance), "Curbit" (Zucchini Yellow Mosaic Virus; ZYMV, under evaluation for Annex I inclusion as new active substance) and "Contans WG" (*Coniothyrium minitans*, included in Annex I) were registered, and several other biologicals are at various stages of evaluation.

On the 1st of April 2006, the pilot scheme was completed and the "Bio-pesticides Scheme" was launched as its successor. It comprises the following categories: semiochemicals, micro-organisms (bacteria, fungi, protozoa, viruses), natural plant extracts and "other novel products" on a case-by-case basis. Fees are £22,500 for biologicals/plant extracts, £13,500 for semiochemicals plus £7,500 for EU consideration (Annex I listing). A biopesticide "champion" is assigned to act as contact point for general enquiries. Pre-submission support and advice on the registration process and data requirements are free. Close contact and cooperation between the applicant and the authorities shall be maintained throughout the evaluation. Special guidance can be given by specialists. An internal biopesticides e-mail group was established, to keep awareness of current issues and to disseminate information, and an area for biopesticides was dedicated on the PSD website.

Source of information: L. Moakes (2006) Experience with the UK pilot scheme and consequences for the biopesticide scheme. In: *Review of potential risks of botanicals and semiochemicals*. Minutes of a workshop held on 13–14 June 2006 in Brussels.

### *2.6.5 Plant Resistance Improvers (DE)*

Plant resistance improvers (PRI) – sometimes also called plant strengtheners (literal translation of the German term Pflanzenstärkungsmittel) – form their own category of products in Germany. They are regulated by the German Plant Protection Act. Plant resistance improvers are defined in Article 2 no. 10 as substances which are

- solely intended to enhance the resistance of plants to harmful organisms, or
- intended to protect plants against non-parasitic impairments, or
- intended for use on cut flowers.

This means that plant resistance improvers may not have any biocidal effects or effects that are covered by the definition of a plant protection product (e.g. as growth regulators or repellents). The mode of action should be induction of systemic resistance, increase in nutrient uptake resulting in reduced susceptibility towards pests or parasites or, for micro-organisms, competition between the "active ingredient" and the parasite. A product that primarily increases growth and not resistance of the plant is classified as a fertilizer. Active agents in PRIs may be micro-organisms, plant extracts, organic or inorganic compounds.

The indirect mode of action of a micro-organism used in PRIs implies that this micro-organism is not infective towards other organisms, including man and other non-target organisms and that it does not produce harmful metabolites. Likewise, lack of a direct effect on pests and parasites from a plant extract or a chemical is assumed to imply a lack of detrimental effects on humans or the environment. The risk assessment in the course of the listing procedure is based on material safety data sheets for all the ingredients and additional data given by the applicant, and follows the pathway of answering the questions: is there any risk inherent in one of the substances or in the product? (1) The environmental risk assessment furthermore gives special interest to the predicted concentration of the substances in the environment. If certain applications are too risky for the environment, they might be excluded but, unlike for plant protection products, no sophisticated risk management is accepted for plant resistance improvers, i.e. the use has to be safe without risk mitigation. Otherwise a listing of such a product is not feasible. (2) For the risk assessment in toxicology, the criteria are similar to those for home and amateur gardening, i.e. toxic substances are not allowed in plant resistance improvers. The use of irritant products is restricted to professional users (only very few, exceptional cases). For products containing micro-organisms, the criteria of the directive 91/414/EEC are adopted. (3) The assessment in efficacy covers mainly the question of whether any of the ingredients acts as an active substance in the sense of a plant protection product with the exception of products for use on cut ornamental flowers: they may contain plant hormones. Only the plausibility of the mode of action is checked, but studies on efficacy are not required.

Recently, data requirements were changed for PRI containing living micro-organisms. Data required for the evaluation of health effects for plant protection products and biocides containing micro-organisms are now also required for PRI containing micro-organisms. The applicant has to prove that the micro-organism does not produce and secrete toxic metabolites, and has to provide a literature search proving that the micro-organism has no deleterious effects on human health. Information on the identity of the strain and production of mycotoxins is required as well as data on exposure of users, workers and bystanders, medical data, toxicity, pathogenicity, infectivity, and the sensitization potential of the micro-organism. Furthermore, information on viable or non-viable residues on treated plants is required. A description of the production process has to be provided, including information on quality control during production and determination of contaminants and content of the active ingredient. For the assessment of toxicity, pathogenicity, infectivity, and the sensitization potential of the micro-organism, information from animal experiments is currently not required and data may be derived from published literature.

The majority of microbial strains used in plant protection in Germany are listed as PRI, and a minority are registered as plant protection products (Table 2.5). Listing as a PRI represents an interesting alternative to registration as a plant protection product, provided that

**Table 2.5** Strains of micro-organisms in plant protection in Germany, under plant protection products and plant resistance improvers

| Group | Species | Strain | Target pest/use |
|---|---|---|---|
| Strains registered as plant protection product | | | |
| V | *Adoxophyes orana* GV | Swiss | *A. orana* |
| B | *Bacillus subtilis* | QST713 | *Venturia spp.* |
| B | *Bacillus thuringiensis* ssp. *aizawai* | ABTS-1857 | Lepidoptera |
| B | *Bacillus thuringiensis* ssp. *kurstaki* | HD-1 | Lepidoptera |
| B | *Bacillus thuringiensis* ssp. *tenebrionis* | NB 176 | *Leptinotarsa decemlineata* |
| F | *Coniothyrium minitans* | CON/M/91-08 | *Sclerotinia* spp./*Sclerotium* spp. |
| V | *Cydia pomonella* GV | Mexican | *Cydia pomonella* |
| B | *Pseudomonas chlororaphis* | MA 342 | *Fusarium spp., Pyrenophora graminea, P. teres, Septoria nodorum, Tilletia caries, T. foetida*) |
| Strains listed as plant resistance improver | | | |
| F | *Aureobasidium pullulans* | | *Erwinia amylovora* |
| F | *Aureobasidium pullulans* | | Foliar fungal pathogens |
| B | *Bacillus subtilis* | | *Erwinia amylovora* |
| B | *Bacillus subtilis* | FZB-24 | Soilborne fungi |
| B | *Bacillus subtilis* | B2g | Soilborne fungi |
| B | *Pseudomonas* sp. | | Soilborne fungi |
| F | *Pythium oligandrum* | | Soilborne fungi |
| F | *Trichoderma harzianum* | T-22 | Soilborne fungi |
| F | *Trichoderma harzianum* | | Increase of resistance, abiotic diseases |
| F | *Trichoderma harzianum* | | Soilborne fungi |
| F | *Trichoderma harzianum* | T 39 | *Botrytis* spp. |
| F | *Trichoderma harzianum* | | Soilborne fungi |
| F | *Trichoderma harzianum + T. polysporum* | | Soilborne and floral fungi |
| F | *Ulocladium oudemansii* | | *Botrytis* |

V = virus, B = bacterium, F = fungus (Status December 2006)

- it has an indirect mode of action against the pest or parasite, and
- production of toxic metabolites can be excluded.

However, determination of the mode of action is often difficult, particularly for living organisms. For further information on listed products please see: www.bvl.bund.de.

Apart from Spain, where a similar procedure was installed recently, Germany is the only EU member state with separate legislation for PRI, but PRI listed in Germany can be marketed in Austria as well. As mentioned in 6.1.4, there have been attempts at EU level to define data requirements for plant resistance improvers with low risk profile (Draft working document SANCO/1003/2000 rev.3, 21/06/2001). However, these activities have been discontinued. Plant resistance improvers are products of low risk, but they are not congruent with the planned low risk category of the Directive 91/414/EEC.

### 2.6.6 Italian Presidential Decree 290 (IT)

In Italy, Presidential Decree 290 dated April 23rd 2001, established a general, simplified procedure for authorization of plant protection product and adjuvants. Specifically, Article 38 dealt with substances for organic and biodynamic agriculture. It allowed the commercialization and use of several products traditionally used in organic farming but not registered/authorized in Italy (e.g. oils, lecithine, herbs, quassia). These substances could only be marketed under the technical name of the active ingredient, but not under a brand name. Source of information: C. Micheloni (2004) Plant protection products in Italy. In: B. Speiser & O. Schmid, *Current evaluation procedures for plant protection products used in organic agriculture.* Proceedings of a workshop held September 25–26, 2003 in Frick, Switzerland. The Presidential Decree 290 is no longer in force.

### 2.6.7 QPS

The development of a "QPS" (Qualified Presumption of Safety) concept was initiated in 2003 by a working group consisting of members of the former (EC) scientific committees on animal nutrition, on food and on plants. It is now continued within an EFSA working group involving the panels on micro-organisms used in animal feed, on preservation of animal feed, on plant protection products, on novel foods regulation and on GMO. The aim is to develop a scheme that would make the approval procedure for micro-organisms more consistent. For this, a more generic approach instead of a full case-by-case assessment is envisaged. This could allow the generic listing of micro-organisms, provided that certain criteria are met, e.g. absence of acquired antibiotic resistance factors. QPS should be similar in concept and purpose to the GRAS (Generally Recognised As Safe) concept used in the USA, but not identical to GRAS. The main difference is that QPS refers to the species and GRAS to the particular application. A further aim is the harmonisation of the safety assessment of micro-organisms throughout the food chain, making better use of assessment resources by focussing on those organisms that present greatest risk or uncertainties, and which would need a case-by-case risk assessment. A major advantage of the QPS status for the notifier may be the ability to change production conditions (media etc.), with only a requirement for notification rather than generating a need for an additional full safety assessment.

The EFSA Scientific Committee recommended that EFSA should develop a strategy for the introduction of an assessment system based on the QPS concept. As a first step, this should be limited to micro-organisms deliberately introduced into the food chain or used as production strains for food/feed additives. If the robustness and value of such a system has been shown in practice, its application to micro-organisms used in plant protection products may be considered. The Committee recommended considering non-spore forming gram positive bacteria, Bacillus spp., yeasts and commonly encountered filamentous fungi. The conclusions of the working group should be made available for public consultation.

A consultation on documents for the four groups of organisms took place during February 2007. The documents made available for consultation suggested that all strains belonging to the *Bacillus cereus* sensu lato group (e.g. *Bacillus thuringiensis*) should not be given a QPS status, since it is known that the vast majority of strains within this group are toxin producers and thus cannot meet the required qualifications. In the document regarding filamentous fungi (which includes e.g. *Trichoderma*) it is concluded that no filamentous fungi can be proposed for a QPS status. The main reason is the difficult taxonomy of these fungi and the lack of knowledge concerning production of toxic compounds

Bacteria directly consumed by humans only qualify for QPS status if they are free of acquired resistance to antibiotics of importance in clinical and veterinary medicine. Furthermore, all bacteria capable of toxin production should be demonstrated to be free of any toxigenic potential. It is important to stress that QPS would carry no legal status.

QPS status of an organism should be established by risk assessors, as a result of an assessment. Establishment of QPS relies on four aspects: (1) taxonomy, (2) familiarity and body of existing knowledge, (3) pathogenicity to humans and animals, (4) end use. Some safety issues like carrier, particle size and dust need to be addressed at product level on a case-by-case basis, while hazards like sensitisation are more amenable to a generic approach. The assumption could be that all products are potential sensitisers, unless otherwise demonstrated. There should be no implication that micro-organisms considered unsuitable for QPS status are less fitted for introduction into the food chain.

**Taxonomy:** QPS should be sought at the highest taxonomic level that is practicably possible by using the mechanism of qualifications to exclude undesirable strains. If the identity of a micro-organism cannot be established, a full assessment is required.

**Body of knowledge/familiarity:** The body of knowledge of the group of organisms seeking QPS must be sufficient to provide adequate assurance that any potential adverse effects in humans, livestock or the wider environment is understood and predictable.

**Pathogenicity:** Any grouping of micro-organisms in which the majority of members produce any form of adverse effect would automatically be excluded from consideration for QPS status. If only a few strains produce adverse effects, there is the option to use qualifications to exclude the undesirable strains, provided the means exist to do so (e.g. B. subtilis).

**End use:** The end use determines, to which degree end users come into contact with the micro-organisms (intended to enter food chain/not intended to enter food chain, but possibility of unintentional introduction [plant protection product]/production strains [end product free of micro-organisms]). This will influence the nature and extensiveness of the body of knowledge needed to determine whether the taxonomic unit is suitable for QPS status. The body of knowledge made available for QPS status for one use may not be sufficient to be able to extrapolate and give QPS status for another use.

The following data would be required for organisms within a taxonomic unit with QPS status: (1) identity, (2) evidence that strains are not excluded by any of the qualifications imposed for the particular taxonomic unit, (3) product-specific safety data.

Finally, EFSA adopted QPS status for species from three groups of microorganisms: gram-positive non-sporulating bacteria (e.g. *Bifidobacterium* spp., *Lactobacillus* spp., *Leuconostoc* spp., *Pediococcus* spp., etc.), *Bacillus* species from the Bacillus subtilis group (if absence of emetic food poisoning toxins with surfactant activity and absence of enterotoxic activity is demonstrated), and yeasts (e.g. *Kluyveromyces* spp., *Pichia* spp., *Saccharomyces* spp.,etc.).

Source of information: A. Fjelsted (2006) QPS – Qualified Presumption of Safety. Presentation at the 1st REBECA conference, 18–22 September 2006, Kiel; and A. Fjelsted (2007) Introduction of a Qualified Presumption of Safety (QPS) approach for assessment of selected microorganisms referred to EFSA, The EFSA Journal (2007) 587, 1–16.

Conclusions: QPS might in the future be a useful tool during the risk assessment and registration process of plant protection products based on well-known microorganisms. However, it is not applicable if the micro-organisms belong to novel groups.

### *2.6.8 Low Risk and Commodity Chemicals*

The new Regulation (EC) 1107/2009 contains separate paragraphs relating to "low-risk" and "basic" substances. The period of approval is extended to 15 years for low risk active substances. A definition for basic substances is provided and the period of their approval is extended to an unlimited time. Plant protection products containing basic substances exclusively need not be authorized. Timelines for the authorization of plant protection products based on low risk substances are defined.

The regulation does not contain a precise definition for low risk substances. However, such a definition is needed. However, it is expected that most microbials, some botanicals and/or many SCLPs will fall into this category. The category of basic substances is relevant for some botanicals, e.g. lecithine, and for some other substances such as kaolin. These are often also low-risk substances.

### *2.6.9 History of Safe Use*

During the REBECA meetings, it was repeatedly stressed that many BCAs have a long history of safe use in food or feed, in pharmacopoeia or cosmetics, or as PRI or plant protection product. Other substances have a long history of safe co-existence in the environment (e.g. pheromones). It was stressed that such "history of safe use" should be accepted as an argument instead of a study in the data requirements on human health and environmental safety. Approaches to facilitate registration of biocontrol agents are summarized in Table 2.6.

**Table 2.6** Overview of different approaches which may facilitate registration of BCAs

| Approach | • Examples |
|---|---|
| Scientific argumentation to justify non-submission of data | • For baculoviruses: OECD Consensus document No 20 on information used in the assessment of environmental applications involving baculoviruses<br>• For semiochemicals, particularly for SCLPs: OECD Consensus document No 12 on Guidance for registration requirements for pheromones and other semiochemicals used for arthropod pest control |
| Reduced data requirements | • For semiochemicals, particularly for SCLPs: OECD Consensus document No 12 on Guidance for registration requirements for pheromones and other semiochemicals used for arthropod pest control<br>• For plant extracts: SANCO/10472 (draft proposal)<br>• For low-risk substances: RUB (no longer applicable after the end of EU review program)<br>• For plant resistance improvers (PRI): German PRI regulation<br>• For micro-organisms: QPS (under development, not yet to be applied to plant protection products, but to other uses) |
| Maximum timelines for registration | • Germany, for plant resistance improvers (4 months; but additional requirements reset the clock to 0)<br>• USA and Canada for biopesticides<br>• National reg. of plant protection products containing low-risk AI, according to SANCO proposal for new Regulation |
| Reduced registration fees (in some cases only for SMEs) | Fee structure is variable across the EU. Examples are:<br>• Cancelling of fees for new active substances (e.g. Germany, Denmark)<br>• Reduced fees for certain categories of products (e.g. microbials in Belgium and Sweden, biopesticides in the UK, Austria, USA and Canada)<br>• Lower fees related to separate legal categories (plant resistance improvers in Germany) |
| Financial support for dossier preparation and/or registration fees | • GENOEG |
| Increased time, before re-evaluation is required | • Low-risk substances, according to SANCO proposal for new Regulation (15 years)<br>• Basic substances, according to SANCO proposal for new regulation (unlimited) |
| Application can be submitted by any interested party or by a member state. | • Basic substances, according to SANCO proposal for new regulation (unlimited) |
| No national registration of plant protection products required. | • Plant protection products containing basic substances, according to SANCO proposal for new Regulation |
| Pre-submission meetings or other support with dossier preparation | Occurs in a number of EU member states and other countries, for example<br>• GENOEG in The Netherlands. Here, the support comes in part from an institution which is independent of the regulatory agencies.<br>• UK for biopesticides<br>• USA for biopesticides |

## 2.7 Major Changes due to the New Regulation 1107/2009

Regulation (EC) 1107/2009 will update the EC's regulatory framework and replace Council Directive 91/414/EEC. It refers not only to active substances, but also to safeners and synergists, co-formulants and adjuvants and came into force 14.12.2009. The regulation will apply and be binding in its entirety and directly applicable in all Member States from 14 June 2011 onwards. No adoption to national law is required.

The two-step procedure with assessment of the active ingredient at EU-level and national authorisation of products is basically unchanged. The data requirements according to 91/414 as described above continue to apply, but subsequent amendments are possible. As a new data requirement, efficacy data for the representative use also need to be submitted. The exact amount of these data is not yet determined.

One major change is the introduction of fixed timelines for the evaluation of active substances and products. If no additional data are requested, inclusion into the "list of approved active substances" (corresponding to Annex I of Directive 91/414) takes 27 months from the application. Additional time is allocated if additional data are requested at different stages. However, the possibility of applying for provisional national registrations before the active substance is approved does not exist any more, except in cases where the approval is delayed by factors beyond the influence of the applicant.

Approval for active substances is valid for 10 years in general. Different times apply for "low-risk substances" (15 years) and "basic substances" (unlimited). Active substances that do not fulfil certain approval criteria may be approved under special conditions, but only for 7 years or max. 5 years with risk mitigation measures. These criteria are expected to apply for some chemicals only, and biocontrol agents most probably fulfil all approval criteria.

Low-risk substances and basic substances are new categories introduced into PPP regulation. So far, a substance can be classified as "low-risk" if certain hazard criteria are not met, but an exact definition is still missing. Criteria for low-risk may be reviewed and if necessary specified. "Basic substances" are defined as substances which are predominantly used outside plant protection (essentially commodity chemicals) or fulfilling criteria of a foodstuff. Most of the biocontrol agents currently on the market would fulfil the criteria to obtain "low-risk" status.

Plant protection products are still authorised at Member state level, but simultaneous application to several member states ("zones") is introduced and mutual recognition is facilitated. The EU is divided into 3 zones:

- **Zone A North**: Denmark, Estonia, Latvia, Lithuania, Finland and Sweden
- **Zone B Central**: Belgium, Czech Republic, Germany, Ireland, Luxembourg, Hungary, The Netherlands, Austria, Poland, Romania, Slovenia, Slovakia and UK
- **Zone C South**: Bulgaria, Spain, Greece, France, Italy, Cyprus, Malta and Portugal

For uses in greenhouses, as post-harvest treatment, for treatment of empty storage rooms, and for seed treatment, the whole EU is considered as a single zone.

The dossier for national registrations is submitted to a "Lead Member State " that evaluates the dossier on behalf of the others MS within one zone. All MS may grant authorizations with the same conditions as the lead MS, unless their specific national conditions justify alternative conditions of use (mitigation measures) or refusal of authorization. Furthermore, application in more than one zone for outdoor uses is possible: A lead MS should do the evaluation of data not related to environmental and agricultural conditions. Timelines for the evaluation of Plant Protection Products are defined as well: An application should be evaluated within 12 months. A maximum of 6 months "extra time" is given to the applicant to submit additional data requested by MS (data gap). If these data are not submitted in time the application is refused. For Plant Protection Products containing a non-approved active the MS should start the evaluation after the DAR is received. The evaluation of applications for PPP by MS should be done within 6 months after approval of active substance.

In addition to the zonal registration procedure, mutual recognition can be applied for after authorisation of the product in a first MS. If the MS where authorisation was granted belongs to the same zone, mutual recognition shall be granted within 120 days. In case authorisation was granted by a MS (lead MS) which belongs to a different zone, the authorisation can be recognized by a single MS, but not for the whole zone (e.g. France to Germany, but not France to Zone B or France to Germany and then whole Zone B). Again, no zones are applied for use in greenhouses, seed treatment and "closed systems". Data requirements, data protection and confidentiality aspects of Directive 91/414/EEC shall continue to apply with respect to active substances included in Annex I and those covered under transitional measures.

## 2.8 Overall Conclusions

In conclusion, this study shows that

- The formal data requirements are similar in the EU, Canada, Australia and the USA. The individual micro-organisms and plant extracts used in plant protection products are very heterogenous and data requirements have to cover all cases. Therefore, data are formally required even if the required information is not applicable to a particular active substance or micro-organism, a particular product or its intended uses. However, formal data requirements in all regulatory systems do not necessarily mean that this information has to be provided by a study, but may also be derived from published literature or unpublished, existing data. In the USA and Canada, this requires a formal waiver. In the EU, a scientific argument in the M Documents – without a formal waiver – serves the same purpose.
- Summaries as provided in the OECD format dossiers are required in the EU and recently also in Canada. These summaries are considered to be very useful to scientific evaluators for the preparation of reports and monographs, especially

from a time-saving perspective. However, the summaries make up a significant proportion of the applicants' efforts for dossier preparation.

- In the USA (and to a lesser extent also in Canada and Australia) the use pattern of the product and the nature of the micro-organism or substance greatly influence the data requirements. In the EU, only one set of data requirements exists, and studies are "waived" case-by-case. The flexibility of the data requirements in the EU creates uncertainty regarding the data requirements for specific cases, whereas the data requirements are more clearly defined in the USA.
- A common interpretation of data requirements has not yet evolved in the EU (e.g.: what is a "relevant metabolite"?). In this situation, many applicants choose to submit "minimum data packages", in order to save costs for studies that might not necessarily be required. This may lead to demands for further studies later on the registration process, thus considerably lengthening the process.
- The registration process in the EU has a different structure from all other systems. Registration is divided in two parts (Annex I inclusion of active substance, registration of plant protection product), while such a division does not exist outside the EU. Annex I inclusion of the active substance is evaluated at EU level. The dossier containing all information on the active substance and on at least one representative product is submitted to a member state, the designated Rapporteur Member State (RMS). Authorities of the RMS evaluate the dossier and send the Draft Assessment Report (DAR) to the European Food Safety Authority (EFSA), which then distributes it to the applicant and the other member states. Further evaluation is done by the member states and EFSA. Following this evaluation, the member states, and the European Commission decide on inclusion or non-inclusion of the active ingredient into Annex I of Directive 91/414. Plant protection products are regulated at the national level of member states. By contrast, the registration is processed mainly by one authority in the USA, Canada and Australia, as opposed to many authorities involved in Europe. In the USA, however, a state may have more stringent requirements for registering pesticides for use in that state or may also register an additional use of a federally registered pesticide product or a new end-use product to meet special local needs.
- The process has a guaranteed maximum duration in the USA, Canada and Australia (missing information "stops the clock", or even resets it, and arrival of the information starts it again). In the EU, timelines are also defined for the first step of the evaluation process (check of completeness, DAR), but may be extended, if additional information is required. In the later steps of the evaluation strict timelines are missing in the EU. Experience with registration in the EU shows that the time effectively needed for registration is much longer than in the USA.

**Acknowledgements** We warmly thank A. Fjelsted (Danish EPA), William R. Schneider (US EPA), M. Busschers and J. Meeussen (Board for the Authorization of Pesticides; The Netherlands) and Herbert Köpp (BVL, Germany) for their comments.

# Chapter 3
# An International Comparison of Invertebrate Biological Control Agent Regulation: What Can Europe Learn?

**Emma J. Hunt, Antoon J.M. Loomans, and Ulrich Kuhlmann**

**Abstract** Few European countries possess an active regulatory process for the import and release of invertebrate biological control agents (IBCAs). A number of standards, documents and guidelines have been produced over recent years in an attempt to implement a harmonised regulatory system for IBCA introduction and release in Europe. Many industries, biological practitioners and regulators, however, fear that a regulatory system would render the process of approval for IBCA introduction into a country costly and time consuming. Countries such as Australia, New Zealand, Canada and the USA, however, are far ahead in terms of regulating the import and release of exotic IBCAs, each possessing effective legislative and administrative procedures governing the process. In this paper, we revisit two analyses of the regulatory systems in place (i) in Europe and (ii) in Australia, New Zealand, Canada and the USA, and summarise and compare their findings. We then proceed to amalgamate ideas in order to offer pragmatic and effective solutions for a balanced and workable pan-European regulatory system that will minimise the costs imposed on industry without compromising risks to human health or the environment.

## Contents



U. Kuhlmann (✉)
CABI Europe-Switzerland, 2800 Delémont, Switzerland
e-mail: u.kuhlmann@cabi.org

R.-U. Ehlers (ed.), *Regulation of Biological Control Agents*,
DOI 10.1007/978-90-481-3664-3_3, © Springer Science+Business Media B.V. 2011

## 3.1 Introduction

### *3.1.1 Regulation Across Europe*

Europe as a continent has traditionally been the source rather than the recipient of a large number of alien invasive pest problems (Greathead 1976; Waage 1997; Kuhlmann et al. 2005). It is in its infancy in terms of experience in classical biological control implementation relative to other parts of the world. Thus, despite all countries having national legislations in place, very few possess an active regulatory process for the import and release of phytophagous or entomophagous invertebrate (or macrobial) biological control agents (IBCAs). Over recent years, however, Europe has witnessed an increase in the establishment and spread of exotic plant and invertebrate pest species, their introduction resulting largely from the escalation in tourism and international trade (Bigler et al. 2005a). In response to the associated costs of these invasive species to human activity and biodiversity, interest in the implementation of classical biological control programmes has grown (Waage 1997; Sheppard et al. 2006).

The acreage of protected crops, grown in glasshouses, tunnels and screen houses, has also developed rapidly in many European and Mediterranean countries over the last four decades. These environments have proven ideal for the proliferation of exotic pests, many of which have established, either temporarily or permanently. With increasing social concern about food safety and pesticide residues, and food quality regulations becoming more stringent in countries where most of the products are marketed, numerous opportunities have been created for the application of new non-chemical pest control methods. Biological control by augmentation or inundation is now a major component of (exotic) pest control in protected crops in Europe (Bigler et al. 2005a). The number of exotic IBCA species introduced, as well as the numbers released (van Lenteren 1997), has greatly increased within a few decades. Approximately 90 IBCA species are currently widely used and commercialised across Europe (EPPO 2008) and many more are under investigation for future release. Europe leads the world in this activity.

However, since Howarth's (1991) publication, attention has been drawn to the risks involved in the import and introduction of exotic species into new natural environments (Simberloff 1996; Williamson 1996; Simberloff and Alexander 1998; Bigler et al. 2006). There is a recognised risk that an increasing number of projects will be executed by persons not trained in identification, evaluation and release of biological control agents, an increasing number of agents and products will become available for the control of pest organisms and the internet will continue to increase access, sales and demands for IBCAs for public use (Loomans and van Lenteren 2005). Almost all European countries are signatories of the Convention on Biological Diversity (CBD) and are therefore obliged to "*prevent the introduction of alien species and, when prevention fails, to control as far as possible those exotic species that threaten indigenous ecosystems, habitats or species*" (CBD 1992). Regulatory procedures for the import and release of exotic IBCAs are therefore an absolute requirement across Europe, a fact that is accepted by the biological control industry (Blum et al. 2003). National governments, as the responsible authority, have an obligation to regulate and facilitate these regulatory procedures, and thus also IBCA application, in an efficient and appropriate way (Bigler et al. 2005a).

The implementation of regulatory procedures across Europe to date has been sparse (Waage 1997; Bigler et al. 2005a). Moreover, the regulation of import and release of IBCAs is not yet harmonised across Europe (Bigler et al. 2005a), thus giving rise to situations where IBCAs could be released in one country lacking regulation and migrate to a neighbouring country where its release may have been prohibited. While discussing the history of legislation and regulatory initiatives in Europe, Bigler et al. (2005a) and Loomans and Sütterlin (2005) both recognised and stressed the need for harmonisation of regulatory procedures for IBCA import and release. The first discussions regarding harmonisation took place at a joint workshop in 1997 between the European and Mediterranean Plant Protection Organisation (EPPO) and CABI (EPPO 1997). The outcome was that the workshop endorsed the Food and Agriculture Organisation of the United Nations (FAO) "Code of Conduct for the Import and Release of Exotic Biological Control Agents", which had been published the previous year as the International Standard for Phytosanitary Measures No. 3 (IPPC 1996). However, the workshop recommended that guidelines be developed to meet European needs with respect to the different legislations and regulations. There have since been a number of further initiatives and associated publications, both on a European and global scale, providing national authorities across Europe with guidelines on how to implement a regulatory system for exotic IBCA introductions, as well as providing information on how dossiers should be compiled and assessed. Those that are specific to Europe or include European countries in their scope are displayed in the timeline shown in Table 3.1. As part of

**Table 3.1** Summary of initiatives and publications, specific to Europe or including Europe in their scope, relating to the implementation and/or harmonisation of an IBCA regulation system

| Year | Initiative/ Publication | Outcome |
|---|---|---|
| 1996 | FAO Conduct for the Import and Release of Exotic Biological Control Agents, ISPM No. 3, IPPC (IPPC 1996). | A standard for countries lacking adequate legislation and procedures to regulate import and to analyse risks related to biological control agents. The document lists the responsibilities of the authorities and importers and exporters of biological control agents. |
| 1997 | EPPO/CABI Workshop on Safety and Efficacy of Biological Control in Europe (EPPO 1997). | Endorsed the FAO Code of Conduct with recommendations that guidelines be drawn to meet European needs with respect to the different legislations and regulations. Recommended a certification system be implemented instead of a registration procedure, to reduce stringency of the regulatory system. An expert panel was established to draw up more specific guidance documents and prepare a 'positive list' of invertebrate biological control agents (see next three events). |
| 1999 | EPPO Guidelines for the first import of exotic biological control agents for research under contained conditions (EPPO 1999). | Guidance stressing the importance of a two-step system for importation and release, i.e. EU countries should first establish a regulatory process for the import of exotic organisms for research under containment, the data from which can be used later for decision to approve importation of organism for release. Information to be included in an applicant's dossier is provided. |
| 2000 | EPPO Guidelines for import and release of exotic biological control agents (EPPO 2000). | As above but also provides information on how the authority should examine a dossier. |
| 2002 | List of Biological Control Agents Widely Used in the EPPO Region (EPPO 2002). | A 'positive list' of invertebrate biological control agents that are widely used in the EPPO region without any reports on adverse effects. The aim of this list was to facilitate and speed up the use of invertebrate biological control agents in the EPPO region and to regularly adapt the list depending on new information. |
| 1998–2002 | EU-funded ERBIC (Evaluating Environmental Risks of Biological Control Introductions into Europe) research project. | A proposal for the environmental risk assessment of exotic natural enemies in inundative biological control (van Lenteren et al. 2003). This paper was the first to present detailed criteria for risk assessment as well as a system for ranking biological control agents in terms of their environmental safety. |
| 2003 | OECD Guidance for Information Requirements for Regulations of Invertebrates as Biological Control Agents (OECD 2004). | Document proposing guidance to member countries on information requirements for the characterisation and identification of the organism, the assessment of safety and effects on human health, the assessment of environmental risks and the assessment of efficacy of the organism. The decision of whether and how these organisms are regulated is left to the member countries. |
| 2003 | IOBC/WPRS Commission for the Harmonisation of Regulation of Invertebrate Biological Control Agents. | Document on information requirements for import and release of invertebrate biological control agents in European countries (Bigler et al. 2005). This document provides more specific advice to applicants and national authorities on information required for risk assessment compared with the EPPO and OECD documents. It reduces data requirements for facilitating regulation but still respects concerns related to human and environmental safety. |
| 2005 | FAO Guidelines for the Export, Shipment, Import and Release of Biological Control Agents and Other Beneficial Organisms, ISPM No. 3, IPPC (IPPC 2005). | A revised version of the original FAO Code of Conduct (1996), which extends its range from classical biological control to inundative biological control, native natural enemies, microorganisims and other beneficial organisms and also includes evaluation of environmental impacts. |
| 2006 | Environmental Impact of Invertebrates for Biological Control of Arthropods. Methods of Risk Assessment (Bigler et al. 2006). | This book was compiled by 25 scientific experts at a workshop in Switzerland in 2004 to address the issue that required information and data for the submission of a dossier to the national authority were often not available to the European community. The book therefore presents a framework of environmental risk assessment for the preparation of the dossiers by the applicants and for their evaluation by national authorities. |
| 2006–2008 | EU Policy Support Action REBECA "Regulation of Biocontrol Agents" (REBECA 2007). | The aim of this project is to develop a balanced system for regulation of biological control agents (micro-and macro-organisms), semiochemicals and botanicals. It is expected therefore in a few years that EU members and other EU countries may regulate invertebrate biological control agents under uniform principles. |

the most recent European initiative to make progress towards a unified regulatory system, the REBECA project (REBECA 2006) decided to look towards countries outside of Europe that have been implementing IBCA regulation for several years, and use their experiences to inform recommendations for a pan-European regulatory scheme.

### 3.1.2 Regulation on an International Scale

Countries such as Australia, New Zealand, Canada and the USA are far ahead of Europe in terms of regulating the import and release of exotic phytophagous and entomophagous IBCAs (phytophagous IBCAs in the USA). These countries boast many years of experience with the implementation of classical biological control programmes, having long been recipients of invasive alien pest species (Coulson et al. 2000; Sheppard et al. 2003). The importance of IBCA specificity for the safety of biological control programmes was recognised during the relatively early years of biological control implementation in these countries (Waage 1997). Furthermore, as the practice of exotic IBCA import and release became more widely adopted, assessments to ensure specificity of exotic IBCAs began to be developed and implemented. Australia was one of the first countries to implement some form of legislation and risk assessment for exotic IBCAs when it introduced its Quarantine Act of 1908.

Not all IBCAs have historically been subject to the same degree of regulation. Specificity testing for weed IBCAs was first to be developed because of the more obvious threat that introduced phytophagous insects posed to economically valuable crops (Waage 2001; Sheppard et al. 2003). It thus followed that legislation and administration for IBCA regulation usually fell under the national plant quarantine service and focussed mainly on plant protection and the need to prevent introduced IBCAs from becoming agricultural pests (Waage 1997; Harrison et al. 2005). Concerns about the additional risk of introduced IBCAs to biodiversity in non-agricultural ecosystems arose much more recently (Delfosse 2005; Harrison et al. 2005). The departments responsible for the environment in New Zealand and Australia became involved with the regulatory process in the late 1990s and pre-release studies then required the incorporation of environmental impact assessments.

Specificity testing for exotic IBCAs of invertebrate pests has lagged behind that of weed IBCAs because of the traditional lack of concern for non-target effects on invertebrates (Waage 2001; Sheppard et al. 2003; Van Driesche 2004). Concern and criticism regarding the absence of data on the potential threat of exotic entomophagous IBCAs, especially to native beneficial and endangered invertebrate species and to biodiversity, has however been growing over the last 20 years. Acknowledging this fact, Australia, New Zealand and Canada have all implemented legislation for the purpose of regulating entomophagous IBCAs within the last 10 years, under the same legislation and procedures as for weed IBCAs. To date, only New Zealand imposes regulations for movement and release of native IBCAs but only for cases where the IBCA in question is a protected species.

## 3.2 Comparative Analyses of International Regulation

In separate analyses, also as part of the REBECA project, Loomans (2007) and Hunt et al. (2008) reviewed the regulatory procedures currently in place for the introduction and release of IBCAs of invertebrates in (i) Europe and in (ii) New Zealand, Australia, Canada and the USA, respectively. Loomans (2007) provided recommendations for a harmonised European regulatory system, building upon the current situation in Europe, whereas Hunt et al. (2008) focussed on the four countries analysed, determining the best components from these regulatory systems as recommendation for adoption and incorporation into a workable regulatory framework to suit the needs of Europe. Here, we employ the same criteria and sub-headings used in these previous analyses in order to summarise and compare the current regulatory status in Europe to that of New Zealand, Australia, Canada and the USA. We then combine the ideas presented and formulate recommendations for pragmatic and effective Europe-wide regulatory solutions.

A variety of sources was used to collect the information presented here and in the previous analyses. The main source of information for European national legislation, provisions and regulations are two surveys performed in 2004 and an update from 2006. The 2004 survey was developed to gather information on regulatory measures in European countries, more specifically the requirements of national authorities (NAs) in Europe. Subsequent direct consultations with employees of those national governments and with scientists directly involved with the regulatory process allowed a fine-tuning of the process. Occasionally websites of the governmental administrative bodies in each country in Europe could be consulted. However, little of the required information was readily available through internet sites or was obtainable from documents accessed. In addition, further information was obtained from already published papers and documents. Some of the results have already been presented by Bigler et al. (2005a). For gathering information about regulatory systems in Australia, New Zealand, Canada and the USA, several sources were used. The main sources were the websites of the governmental administrative bodies of each country, from which much of the required information was either readily available or easily obtainable from documents accessed via the links provided. In addition, further information was gleaned from published papers and documents as well as from consultations with government employees and scientists directly involved with the regulatory processes.

### *3.2.1 Legislation and Administration for Regulation*

#### 3.2.1.1 Europe

Except for phytosanitary and veterinary directives and decisions, there is no specific legislation in any jurisdiction within Europe controlling the import and release of non-native IBCAs for the purpose of biological control. On an EU-level, for IBCAs,

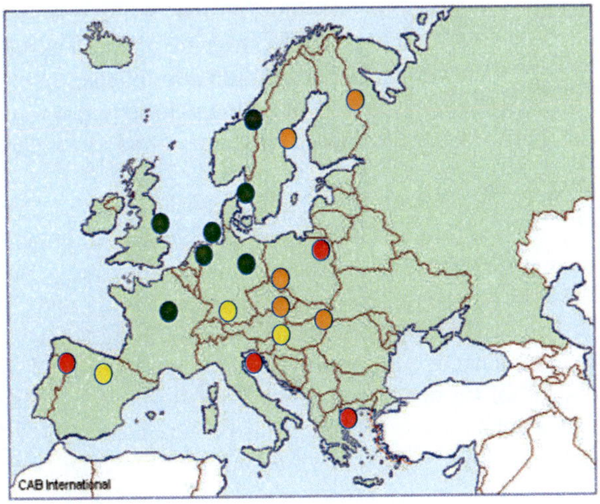

**Fig. 3.1** *Legislation*: plant protection products (pesticides) = *orange*; plant health = *yellow*; nature conservation (environment) = *green*; no/other = *red* (August 2006)

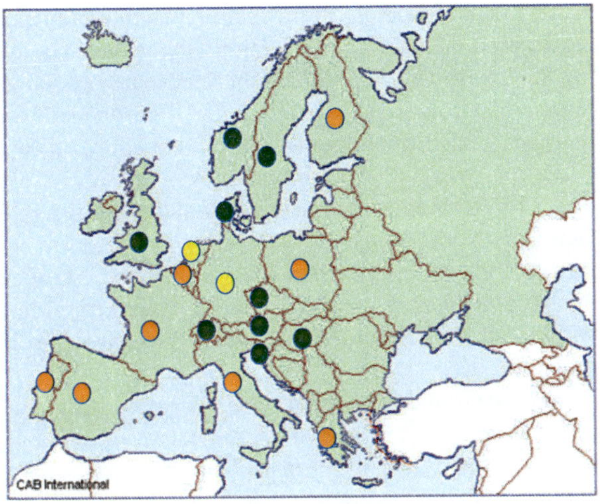

**Fig. 3.2** *Regulation:* implemented (*green*), in preparation (*yellow*), or no regulation (*orange*) (August 2006)

there is no specific EU directive available and the EU does not intend to develop such a directive. However, depending on a country's national constitution, certain types of legislation and regulation prevail when dealing with IBCAs (Fig. 3.1). Large differences exist in the degree of implementation of active regulatory measures of IBCAs in European countries. The present status of regulation in twenty European countries investigated can be divided into three different categories (Fig. 3.2):

(a) nine countries (Austria, Czech Republic, Denmark, Hungary, Norway, Slovenia, Sweden, Switzerland, UK) have regulation implemented to some degree,

(b) five countries are working on the design and implementation of a regulation system (Finland, Germany, Ireland, Netherlands, Spain), and
(c) six countries have no regulation developed or implemented yet and will not have a regulatory system in place in the foreseeable future (Belgium, France, Greece, Italy, Poland, Portugal). No contact was established in other countries.

In countries with IBCA regulation in place, legislation and regulation are often approached from a different perspective regarding the risks of IBCAs; first, in managing the risks for agriculture and facilitating pest control and, second, in managing the risks for the (native) environment and thus controlling the import and release of an IBCA. However, few countries have a regulatory system in place that suits the requirements of a proper IBCA risk-assessment.

Within European countries, three types of legislation determine the regulation framework for the protection of plants, under which IBCA regulation falls: plant health acts, pesticide acts and/or environmental acts (Fig. 3.1). In a number of European countries where regulation of IBCAs is in place, generally two types of legislation interact, in particular those pertaining to plant protection (i.e. plant health) and nature conservation (i.e. environmental). A competent authority or NA is assigned accordingly to different types of institutes; either plant health, pesticide registration or nature conservation authorities. The NA is responsible for approving the import and/or commercial release in a country, regulating the import and/or release under national legislation and also evaluating the applications. They do so, however, with different perspectives depending on the legislation.

In line with the CBD, nature conservation acts often include an article stating that is its "forbidden" to release non-native species in the wild (Belgium, Denmark, Germany, Netherlands, Norway, UK). However, in some countries (e.g. Germany, Poland (before 2004, when they joined the European Union (EU))), biological methods of plant protection may be exempted from regulatory measures, in line with nature conservation, when authorised by a specific permit based on the plant protection act.

In Switzerland, the import and release of beneficial organisms intended for use as biological control agents is subject to different legislation, depending on the nature of the agent, its form and the purpose for which it is used. For example, commercially produced agents fall under different legislation to agents for classical biological control programmes. If a biological control agent (of weed or invertebrate pests) is intended for commercial production on the market, it is considered a plant protection product and falls under the Ordinance on Plant Protection Products (Pflanzenschutzmittel – Verordnung), within the Federal Law on Agriculture. The Federal Office for Agriculture (FOAG) (*Bundesamt für Landwirtschaft, BLW*) is the competent authority for regulation in this case. For agents intended for use in a classical biological control programme, regulation for the import into containment for research, field tests and full environmental release of classical weed IBCAs also fall under the Ordinance on Plant Protection Products. For IBCAs of invertebrates, however, which are not pests of plants and where establishment is intended, the import and release is regulated under the Federal Law on the Protection of the Environment (LPE). This law is implemented by the Federal Office for the Environment (FOEN)

(Bundesamt für Umwelt, BAFU). The LPE has several (revised) ordinances associated with it, two of which deal specifically with the containment and release of non-commercial entomophagous IBCAs: the Ordinance on the Contained Use of Organisms of 1999 and the recently revised Ordinance on the Release of Organisms into the Environment of 2008.

In Germany, legislative conditions have caused a conflict of interest that has led to a standstill in the development and implementation of regulatory procedures for IBCAs. According to the federal Nature Conservation Act (Bundesnaturschutgesetz 2002), the release of exotic species is forbidden. However, the introduction and use of specimens of a native fauna species and a non-native fauna species are exempt from permit requirement if their introduction and use requires authorisation under plant protection legislation for biological methods of plant protection. The Plant Protection Act (Pflanzenschutzgesetz) does not adequately foresee such an authorisation and needs to be adapted to allow such an ordinance to regulate IBCAs. Currently, no non-native IBCA can be imported and sold on the market, but the grower, in theory, needs permission from a federal state (in German called "Bundesland") agency to release a non-native IBCA on his farm. However, this has rarely been applied for.

In Poland, prior to joining the EU in 2004, procedures were in accordance with previous acts on plant protection and relevant regulations based on these. Regulations clearly specified all documentation requirements needed for registration and implementation of plant protection products containing living beneficial (macro-) organisms. After May 1st 2004, however, when Poland joined the European Union, the process of registering macro-organisms was stopped. The Minister for Agriculture and Rural Development did not issue any regulations under the new Plant Protection Act, and it is not clear what procedures will be taken, or even if any registration of beneficial (macro-) organisms will be required in Poland.

In Greece, Italy and Portugal there is some general knowledge on IBCA regulatory issues, but no legal documents concerning the regulation of IBCAs are in force at present.

#### 3.2.1.2 International

Australia, New Zealand, Canada and the USA all have a legislative system in place for the introduction and release of IBCAs, with at least one governmental body administering the process. Depending on the country, regulation of IBCAs is covered under different acts including those pertaining to plant protection, biodiversity conservation, endangered species and environmental protection.

Australia governs the import and/or release of biological control agents under their Quarantine Act of 1908 and the Environment Protection and Biodiversity Conservation Act of 1999. Australia is unique in that it has additional legislation specific to biological control (the Biological Control Act of 1984). This act, however, generally only comes into consideration when there is controversy over the release of a biological control agent. In New Zealand, the introduction of all organisms not already present in the environment, including biological control agents of pests and weeds, falls under the Hazardous Substances and New Organisms

(HSNO) Act, which came into full effect in 1998. In Canada, "beneficial" exotic biological control agents of weeds and invertebrates are considered to be potentially injurious to plants and thus fall under the Canadian Plant Protection Act (1990). This act was enforced to prevent the importation, exportation and spread of pests injurious to plants and was preceded by a series of acts and regulations going back to the Destructive Insect and Pest Act of 1910. In the USA, the new Plant Protection Act (PPA) came into operation in 2000, allowing APHIS-PPQ authority to regulate organisms that may directly or indirectly harm plants or plant products. Thus, the import and release of exotic IBCAs fall under this act, except entomophagous IBCAs for which there is currently no comprehensive regulatory framework (Messing 2005). Current regulations for movement and release of entomophagous IBCAs are still those that were developed under the older Federal Plant Pest Act of 1957. However, no changes to these procedures are anticipated when new regulations are imposed under the Plant Protection Act (Mason et al. 2005).

In terms of administering regulatory measures and issuing approvals for IBCA import and release, it varies between countries as to whether the department of agriculture and/or the department for the environment is responsible. In Australia and New Zealand (at least in terms of IBCA import approval), both departments play a role. In Australia, this is the Department of Agriculture, Fisheries and Forestry (DAFF) and the Department of the Environment, Water, Heritage and the Arts (DEWHA) (formerly the Department of the Environment and Water Resources (DEW)). Both departments have different perspectives regarding the risks of biological control agents. DAFF has broad responsibilities for managing potential risks to primary industries, agriculture and environment, whereas DEWHA focusses on managing potential risks to the environment. In New Zealand, although the Environmental Risk Management Authority (ERMA New Zealand) (an autonomous Crown Entity, independent from government influence) implements all processes covered by the HSNO Act, it is overseen by the Ministry for the Environment and enforced by the Ministry of Agriculture and Forestry (MAF). In Canada and the USA, the responsibility currently lies only in the department of agriculture. The Plant Health Division of the Canadian Food Inspection Agency (CFIA-PHD) administers the Plant Protection Act in Canada, a process that is overseen by the Minister of Agriculture and Agri-Food (AAFC). In the USA, the Plant Protection and Quarantine of the United States Department of Agriculture's (USDA) Animal and Plant Health Inspection Service (APHIS-PPQ) administers their Plant Protection Act, overseen by the Secretary of Agriculture.

## *3.2.2 Application Procedures*

### 3.2.2.1 Europe

In countries where a regulatory system is in place or in preparation (regulation system: Fig. 3.2, NA: Fig. 3.3, dossier requirement: Fig. 3.4), the application process

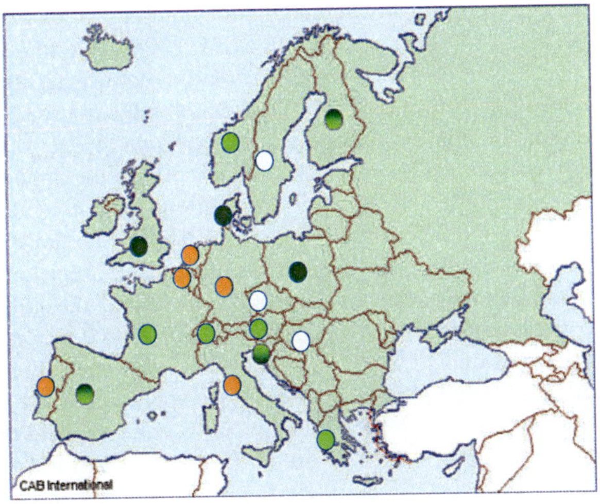


**Fig. 3.3** *Competent/National Authority:* assigned (14): environment (*dark green*), plant health (*light green*), pesticides (*white*); not yet assigned (5 = *orange*) (August 2006)

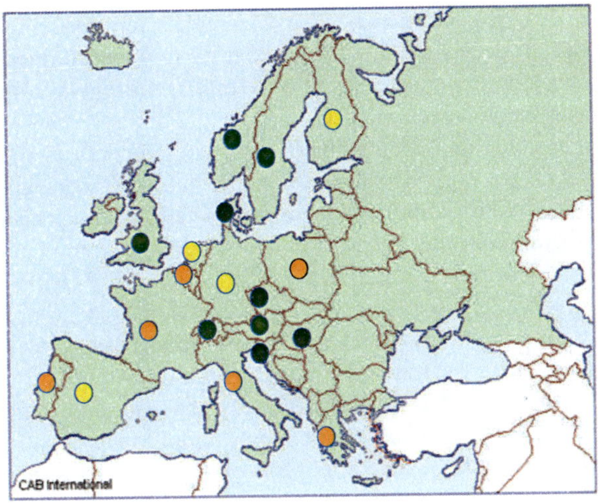


**Fig. 3.4** *Dossier for application:* required (*green*), in preparation (*yellow*), no forms required (*orange*) (August 2006)

for the (import and) release of IBCAs is hierarchically structured according to the authorisation procedure already in place for plant protection products (91/414/EEC: Austria, Czech Republic, Hungary, Sweden), plant health (Norway, Slovenia, Switzerland) or nature conservation acts (UK, Denmark, also The Netherlands). In The Netherlands, every distributor or retailer is required to apply for a permit to release a specific organism. In Switzerland, also every distributor of a specific product must apply for a permit.

Where regulation is in place or in preparation, application forms for (import and) release and dossier guidelines are available, mostly upon request from the competent or national authority and sometimes online, for example, in The Netherlands (Het LNV-Loket 2009) and the UK (CSL 2008). For an application to release an IBCA species, one (for the organism) or two application forms (for the product, where relevant) are necessary, i.e. a system of authorisation, registration, regulation and/or evaluation applies per species and per product (per distributor). The applicant usually should reside within the country where the application is submitted and, thus, foreign industries/companies/institutions can only submit through a national representative/retailer in that country. In The Netherlands, the applicant should be the person who is legally responsible and must be registered at a chamber of commerce in the EU; the applicant is the one who "owns" the authorisation or licence and mandates responsibility to the grower. Application forms, including paper-copy dossiers, are submitted to the competent authority, where they are checked for completeness, accepted and registered. When application forms or dossiers are incomplete, the applicant must resubmit it once corrections and/or additions have been made. However, information given on the application form or in the dossier is often minimal and/or not specific. Responses supporting unclear dossier issues are usually resolved directly between the applicant and the co-operator/advisor in, for example, The Netherlands.

In some countries, such as Denmark, Germany, Slovenia and the UK, only exotic IBCA species require an approval for import and release. Native species, or species that are already present, are exempt from regulation by law. In most countries in Europe, however, all IBCA species, both native and exotic, need approval or registration before they may be (imported and/or) released, and an application (including a dossier) must be submitted to the NA. In Switzerland, approval of exotic IBCAs intended for commercial use follows a two-step procedure: (1) to obtain an import permit (from the Plant Protection and Quarantine (PPQ) Service) and (2) to register the organism with the registration authority (FOAG), including submission of a dossier. Both services are part of the Ministry of Agriculture.

Import and release of IBCAs are not necessarily confluent. Imports can be made solely for research and education by universities and private or governmental institutions as well as industry. Industry can also import and mass-breed large quantities of IBCAs and subsequently export them to the country of destiny without a release in the country where it is produced. Production facilities are contained, however, they are not quarantine facilities and do not ensure prevention of an escape into the wild. In countries such as Belgium, Denmark, Germany, France, Italy, Spain and The Netherlands, where large or small commercial production facilities occur, import and mass-production of exotic species is not arranged very well.

#### 3.2.2.2 International

As in Europe, a system of authorisation, registration, regulation and/or evaluation applies per species and per product. Application forms for import and release of IBCAs may be downloaded from the websites of the administering body in each of the four countries. Together with the forms are an explanation of the application process and the specifics of the information and documentation required with each application. None of the four countries analysed applies restrictions to the use of native IBCAs, except in New Zealand when the IBCA is a protected native species.

In all four countries, approval to import into contained facilities must be sought if further experiments are to be conducted on the IBCA within the country into which it will potentially be released. In Australia, this stage constitutes the nomination and justification of the potential target species before approval to import an IBCA is granted. Once DEWHA has received sufficient information and granted approval for import, they will amend a Live Import List (DEWHA 2009a) to include the proposed biological control agent. In New Zealand, approval of containment applications is largely based on how the applicant proposes to contain the organism.

Prior to conducting any risk assessments in Australia and the USA (for phytophagous IBCAs only in the USA), it is obligatory for the applicant to seek approval of their host specificity test list. At this early stage of the approval process, recommendations on the target choice and the proposed non-target test list for host specificity testing will be made. The testing protocols against which the potential agent will be evaluated for specificity must also be provided at this stage in Australia. Initial approval of the non-target test list is not implemented in Canada. In New Zealand, a slightly different approach is taken whereby ERMA New Zealand encourages the applicant to liaise with its staff at an early stage of a biological control project so that the host specificity test list, among other issues to be addressed in the risk assessment, can be discussed. This first contact between ERMA New Zealand and the applicant is considered an essential part of the application process, ensuring that key scientific, technical and risk management issues that should be incorporated into the final application are discussed. The potential risks, cost and benefits of the introduction can also be highlighted at this time such that the necessary analyses can be carried out effectively.

When applying for the environmental release of an IBCA, applications, together with the necessary data requirements, are sent to the relevant authority (or two authorities in the case of Australia and the USA, both of which must grant their approval before a release may be made). Normally, the dossier is checked for completeness as soon as it is received and will be returned to the applicant for amendment if there is any missing information e.g. Canada and New Zealand. In New Zealand, a public hearing must be held during the application process if the submitter(s) (someone outside the application process, e.g. a member of the public or industry, who submits a comment or lodges a complaint about the proposal) asks to be heard. Various ERMA New Zealand staff, co-opted experts, the applicants and their witnesses will be present at the hearing, together with any stakeholders and members of the public who have asked to be heard or wish to attend.

## *3.2.3 Decision-Making Process and Decision Maker*

### 3.2.3.1 Europe

Except for phytosanitary, veterinary and pesticide measures and requirements, the decision making process in Europe for IBCA (and other invasive species) import and release has not been centralised. Instead, it has been drawn up according to historical lines of national legislation. NAs, to whom applications are submitted and permits are issued, are assigned for plant protection, plant protection products (pesticides) and occasionally for nature conservation or the environment.

The administrative part of the process for the approval of (import) and release of an IBCA is similar in most countries. In countries where both native and non-native species need approval, the decision-making process is different, with native species being dealt with in a more flexible way. For example, in Spain, exotics require a permit for import whereas natives only need to be registered. In Norway, Switzerland and The Netherlands, both groups require approval and submission of a dossier before release is granted; however, the evaluation process is different, with less data being required for native species.

Once a complete application has been registered, the process for approval is again different between most countries. In countries where regulation is in place or in preparation, decisions on whether to issue an import permit, a permit for release or to register the proposed IBCA are based on the quality and quantity of information and data sets provided by the applicant. To support such a decision, most countries include consultation by co-operators, reviewers and advisors, who are selected based on their expert knowledge in entomology, biological control or other relevant disciplines. For example, in Austria, an agronomist, an entomologist and an ecotoxicologist are consulted. In the UK and Norway, a national advisory committee has been established to evaluate the application and provide advice to the NA. The advice of the co-operator or committee to either accept or reject an application is mostly based on "expert opinion/knowledge".

Evaluation of the application is based on the data requirements provided in the dossier, including information on the identity and biology, the effects on human, animal and plant health, efficacy and, increasingly nowadays, environmental effects, through "expert knowledge". Objective evaluation criteria, however, have not been defined *a priori* (in contrast to, for example, the pest risk analysis standards and procedures for risk-evaluation of phytosanitary pests) and are thus not harmonised between European countries. Criteria evaluated by "expert knowledge" are very different as a result of the regulation and underpinning legislation that has been put in place in a country. In countries where nature conservation legislation and regulation is in place (e.g. Norway, Netherlands, Switzerland and UK), environmental characteristics are part of the data requirements and support the advice to approve or reject an application. In countries where legislation is based on plant protection or plant protection products, regulation is based on common principles of risks posed to animal, human and plant health, but these requirements are not tailored to suit an environmental risk analysis. During recent years, guidelines produced by EPPO, the Organisation for Economic Co-operation and Development (OECD) and the International Organisation for Biological Control of Noxious Animals and Plants (IOBC) (see Table 3.1) strongly support the use of "expert knowledge", but the expert's advice need not always have a legal basis.

The decision of whether to permit (import and) release of an IBCA is usually made by the director of the NA or by a public servant assigned by the minister in charge. Examples of licences granted, for example, in the UK or by derogation for The Netherlands, can be found on the internet (CSL 2008; Het LNV-Loket 2009, respectively).

#### 3.2.3.2 International

In terms of the dossier review process, this broadly operates in much the same way in Australia, Canada and the USA in that the dossiers are distributed to scientific experts or co-operators, usually a combination of university and government-affiliated scientists representing a broad range of expertise, within the country for independent review. In New Zealand, scientific experts are often consulted or co-opted onto the review panel and become significantly involved in the decision-making process. The main difference between countries is that Canada and the USA have both established a committee for the pure purpose of conducting these reviews. Canada's Biological Control Review Committee (BCRC) and the USA's Technical Advisory Group for Biological Control Agents of Weeds (TAG) provide science-based reviews on biological control projects, and CFIA-PHD and APHIS-PPQ, respectively, are highly influenced by their recommendations. Phytophagous agent petitions in Canada are also circulated to TAG in the USA as well as Mexico's responsible authority for IBCA releases, Servicio Nacional de Sanidad, Inocuidad y Calidad Agroalimentara (SENASICA) – Sanidad Vegetal. Entomophagous petitions are only sent to SENASICA-Sanidad Vegetal, since USDA does not formally review release petitions for entomophagous agents under their legislation. CFIA-PHD does not approve releases without comment from TAG, though it is not obliged to follow their recommendations. Similarly, comments from Canada and Mexico are taken into consideration when APHIS-PPQ makes decisions on release permits. This agreement between Canada, the USA and Mexico was established on the basis that biological control could have impacts on the entire continent and it is important to have a consensus for releases amongst all three regulatory bodies (SENASICA-Sanidad Vegetal, APHIS-PPQ and CFIA-PHD).

The ultimate decision of whether or not a release application will be approved lies in the hands of an authority figure within the governmental body administering the regulatory process, except in New Zealand where ERMA New Zealand authorises approval on behalf of the Minister for the Environment. In each country, the final decision is heavily influenced by the opinions of the co-operators/scientific experts who review the applications. Thus, although one person may ultimately make the final ruling on the application in each country, such as the Director of CFIA-PHD in Canada, the decision embodies the views and opinions of a number of different scientists representing a broad range of expertise.

### *3.2.4 Data Requirements*

#### 3.2.4.1 Europe

Although several countries (Austria, The Netherlands, Norway, Switzerland and UK) require an application with a dossier included for authorisation of IBCA species, data requirements upon which the evaluation are based vary greatly between countries and depend largely on the type of regulation underpinning the legislation (Table 3.2). In the case of approval as plant protection or plant protection products,

**Table 3.2** Overview of regulation requirements in 8 European countries (situation August 2006)

| Criteria | Norway | Sweden | Danmark | Czech Republic |
|---|---|---|---|---|
| Data requirements | species, product, efficacy | A (species) B (product) | yes | Annex 13 to Decree 329/2004 |
| Administrative procedures | Norwegian FSA | KEMI | Danish Forest & Nature Agency | State Phytosanitary Administration |
| Admin. fees / costs (in €) | 489 € | 1100 (first), 400 (renewal) € | 0 € | ? |
| Admin. time frame | 0.5 – 3 years (efficacy) | 1–2 m > 3–4 m expert | 3 month | 3 month |
| Length permit | 5 years | max 10 years | > 1 yr - -unltd | unlimited |
| Dossier: organism/product | organism / producer | product / producer | organism / producer | organism / producer |
| Opinion science/industry | scientist expert committee | scientists (ent., tox., agron.) | scientist experts | scientist experts |
| Conditions? | yes label | yes, label | yes | yes |
| Information available | http://www.mattilsynet.no/ | http://www.kemi.se | http://www.skovognatur.dk | http://www.pan-germany.org/ |
| Public scrutiny | no | no | no | no |
| Safe list? | producer: 31 species | 58 products: 19 species | no | 37 products: 23 species |
| **Criteria** | **Netherlands** | **UK** | **Austria** | **Hungary** |
| Data requirements | dossier native / exotic | yes (estabilishment, etc.) | yes, dossier | Annexes 9 –10 to Decree 89/2004 |
| Administrative procedures | Ministry Agriculture NF | DEFRA | AGES | Central Service PPSC |
| Admininstration fees/costs (in €) | 60 € (1 yr) – 100 € (5 yr) | 0 € | 1660 € (first), 1234 € (renewal) | 1000 € (applic.), >2000 (efficacy) |
| Administration time frame | 2 – 6 months | 5 weeks | 5 weeks | 12 months |
| Length permit | 1 – 15 years | case-to-case:new < renewals | case-to-case:new < renewals | 10 years |
| Dossier: organism/product | organism / producer | organism / producer | product / producer | organism / producer |
| Opinion science/industry | PPS | expertpanel ACRE | tox, ecotox, agronomist | authority |
| Conditions? | National, application, expt. | always: disease free, nondiap | yes: pest/crop | yes |
| Information available | http://www.Invloket.nl | http://www.defra.co.uk | http://www.ages.at | http://www.fvm.hu, www.ontsz.hu |
| Public scrutiny | no | no | no | no |
| Safe list? | yes, 134 species exempt | no (licensed species) | 44 products: 25 species | no BCAs exempt, 17 species |

most requirements stress human and plant health and not specific environmental criteria and characteristics. In countries where nature conservation legislation must be taken into account (Norway, The Netherlands, UK, Switzerland), specific environmental criteria, such as information on the establishment in the wild, host specificity and non-target effects need to be met. Data requirements in Norway are derived from the draft OECD requirements (OECD 2004), whereas in The Netherlands, Bigler et al. (2005b) and the application form and guidance document developed during the REBECA project (Loomans et al. 2007a, b) are used as a basis for compiling the dossier. In the UK, an extensive amount of information is required to satisfy data requirements for licence approval to release non-native animals or plants into the wild. One key requirement is information about the establishment potential in the UK. For a non-native species, the Central Science Laboratory (CSL), which is responsible for the licencing of non-native IBCA releases in association with the Plant Health Division (both part of the UK's Department for Environment, Food and Rural Affairs), requires data to be generated, when not already available, in order to properly assess the survival in the environment (CSL 2008). Information on efficacy is included as a requirement in most countries where regulation is based on plant protection or products. In Norway and Hungary, specific tests are needed before a permit is given. Host range testing is not yet a requirement in Europe, although a few countries, such as Switzerland, recommend doing so when compiling the data set.

Regarding the use of native IBCAs, when an application is required, fewer data are required to support the application for release than for an exotic agent. Sometimes native species only need registration, as is the case in Spain. Evaluation usually follows a "short track" risk assessment, whereas exotic species are assessed more thoroughly.

#### 3.2.4.2 International

For applications to import an exotic IBCA into containment for research purposes, Australia, New Zealand, Canada and the USA are fairly uniform in terms of the information required. For example, details of the applicant, the purpose of the application, the identity of the organism to be imported, information on the biology and ecology of the organism and a description of the proposed containment system (physical and operational) must be provided. In some cases, for example in Australia and New Zealand, the risks, costs and benefits of importing the agent into containment must also be analysed and supplied. Australia also specifies that possible interactions, including conflicts-of-interest, with existing biological control programmes should be considered. For example, if the target species is in the same genus as an introduced agent in an existing biological control programme, the potential agent must be tested against the existing biological control agent. A summary of the proposed activity also needs to be provided together with details of host specificity.

For an application to release an IBCA, an extensive amount of information is required in all countries, including comprehensive host range information. For example, in Australia, besides the requirements mentioned above for applications to import an IBCA, the main additional requirement for approval to release a biological control agent is a non-target risk assessment built around the results of host-specificity testing. In New Zealand, applicants are required to develop a full environmental risk, cost and benefit assessment by identifying and analysing all possible hazards, risks, costs and benefits associated with the release of the organism. Host range assessment is usually central to the risk analysis. Organisms that present a greater potential risk will require more detailed information and assessment. In Canada and the USA, petitions for the release of an IBCA must contain host specificity and other biological data on the agent to be imported and released in a format and substance that conforms to the recently updated North American Plant Protection Organisation (NAPPO) standards for the release of exotic entomophagous and phytophagous biological control agents (NAPPO 2008a, b). This includes the proposed action (specifying the need for release, the reasons for IBCA choice as well as quarantine and release procedures), target pest/weed information, biological control agent information (including host range data), host specificity data, environmental and economic impacts of the proposed release (where the benefits, risks and costs of a release are weighed against the benefits, risks and costs of other pest control choices) and plans for post release monitoring (where researchers and practitioners must demonstrate that a plan is in place to study economic and environmental impacts of programmes after the release of an agent to assist in assessing programme impacts and to validate and improve methods of release or host-specificity testing).

## 3.2.5 *Fees*

### 3.2.5.1 Europe

Fees for administration can vary largely between countries in Europe, from 0 € (Denmark), 60–100 € (The Netherlands, for a 1-year and 5-year permit, respectively), 500 € (Norway) to 1000 € (Hungary), 1100 € (Sweden) and 1660 € (Austria). The same amount is normally requested for a renewal, although Sweden (400€) and Austria (1250 €) charge less. Some countries require efficacy trials (Hungary, Norway) and risk-assessments (UK – winter survival) for which extra monetary and time costs are incurred. The costs for drawing up a dossier by the applicant, or for generating specific data requirements through experimentation, are hidden costs. Although some costs appear to be high, they are still relatively low compared to those charged for compilation and evaluation of a pesticide or microbial dossier.

### 3.2.5.2 International

Administrative costs in Australia, Canada and the USA are covered with public money via the national governmental bodies. Furthermore, the review process in these countries operates on a voluntary basis, so that scientists are not paid for the reviews they conduct. This leaves the applicant with minimal fees to pay upon dossier submission. Currently, USDA does not even charge a fee for plant pest permits. Fees charged in Canada are as follows:

Applications/permits for scientific research purposes: Ca $15.00 (approx. 10 €)
Applications/permits for purposes other than research: Ca $35.00 (approx. 24 €)
Amendment to a permit: Ca $10.00 (approx. 7 €)

In Australia, a fee of AU $180 (approximately 108 €) is charged by AQIS for issuing the permit to import biological control agents into containment. No fees are required for the applications for host-specificity test lists or for release of a biological control agent. However, when a permit has expired after the 2 years and renewal is sought, AQIS will charge the same for issuing a new import permit. New Zealand is the only country out of the four where applying to import and release an IBCA comes at a relatively high cost. This is mainly due to the fact that ERMA New Zealand operates a full ecological risk, cost and benefit analysis of biological control releases. In some cases, application fees may be reduced, for example, if a public hearing is not necessary. Application fees in New Zealand are as follows:

Notified (full release): NZ $33,750 (approx. 17,000 €)
Notified (conditional release): Negotiated
Notified (containment): NZ $11,250 (approx. 5,700 €)
Non-notified (containment): NZ $2,250 (approx. 1,115 €)
Statutory determination on grounds of reassessment: NZ $562.50 (approx. 285 €)

## *3.2.6 Administrative Time Frame*

### 3.2.6.1 Europe

The administrative timeframe for evaluating the release application varies greatly between European Member States. The NAs in the UK and Austria will approve an application within 5 weeks and in Denmark, within 3 months. In Sweden, the length of this period varies between 1 and 2 months and 3–4 months, depending on whether expert opinion is required. In The Netherlands, where the regulatory system is not fully operational, the time frame could vary between 2 and 6 months, but has been brought down to 8 weeks. The NA in the Czech Republic decides within a 3 month period from commencement of the proceedings. In Hungary, the competent authority decides on the authorisation and issues a document to the applicant within 12 months of full submission of data. In Norway, an approval can be expected after 6 months but can sometimes take up to 3 years when efficacy testing is required.

#### 3.2.6.2 International

In Australia, approval for import into containment will be given within 30 business days if all required information is provided in the application. Approval of the host specificity test plant list requires a minimum of 40 business days and for release permit applications, a response from DEWHA takes the longest; between 6 and 9 months from submission of the release application. When an application is submitted in New Zealand, according to the HSNO Act, ERMA New Zealand has up to 100 working days (if a time waiver has not been agreed) to process a publicly notified application and to inform the applicant of the decision made. The timing depends largely on the quality of the application. For an application that does not need to be publicly notified (i.e., approval to import into containment) ERMA New Zealand has up to 60 working days to process the application and inform the applicant of the decision. In Canada, once all the information has been received (petition, recommendations of BCRC and TAG etc.) and the CFIA-PHD has completed a review of the permit application form, the CFIA endeavours to issue a decision on permit to import within five to ten working days. For a new introduction (release), the total time from receipt of petition to issuance of a permit may take up to 6 months. To import a potential weed biological control organism into the USA for host specificity testing takes from 4 to 6 weeks from submission of the application to receiving a permit. In terms of release applications, the more complete the documents are upon submission to APHIS-PPQ, the faster the review process is likely to be. In practice, the full approval process takes approximately 18 months. Currently no time limits are given when applications for release of entomophagous IBCAs are sent for external review but, as with weed IBCAs, approvals for release can take approximately 18 months.

### *3.2.7 Availability of Information to Aid Applicants*

#### 3.2.7.1 Europe

Information to aid applicants with IBCA import and release is very scarce, and with few exceptions, such as in the UK (CSL 2008), little information is published online. Any information provided is very limited. Application forms for IBCAs, in the native language, can be downloaded from the websites of the respective NAs in Hungary, The Netherlands, Norway, Sweden, Switzerland and the UK.

#### 3.2.7.2 International

In contrast to the European situation, information regarding the regulatory processes in Australia, New Zealand, Canada and the USA can be found easily by performing a quick search on the government websites. The Australian DAFF website has a set of web pages providing a thorough explanation of the biological control agent import process (DAFF 2007). DEWHA also provides comprehensive information on their website for applicants regarding the Live Import List and how to amend it (DEWHA 2009b). ERMA New Zealand has also developed an extremely detailed and informative website containing comprehensive information about the roles of their organisation and the process of applying for IBCA release approval (ERMA New Zealand 2009). The site is easy to navigate around and there are numerous downloadable guidance documents for dossier preparation. Full text of all previous applications, evaluation and review reports and decisions are also accessible.

In Canada, the permit application form for import, together with information regarding import requirements are displayed on the CFIA website (CFIA 2008). The NAPPO Standards are also available on the internet (NAPPO 2008a, b). Although there are no specific guidelines for the whole process, AAFC and the CFIA have produced a comprehensive guide to provide petitioners, reviewers of petitions and interested Canadian citizens with information about the procedure (De Clerck-Floate et al. 2006). Not only does this document explain in detail the application and decision-making process for IBCA introduction and release in Canada, but it also provides examples of completed application forms so that applicants are able to clearly see the information requirements as well as methodologies and systems that can be used to gather such data. For the USA, there is some helpful information on the APHIS website (APHIS 2008) regarding the process of importing and releasing weed IBCAs. The PPQ form 526 (application form), required by APHIS-PPQ, may be also downloaded from this site. USDA has also compiled a manual detailing guidelines for evaluating the safety of candidate phytophagous IBCAs. The purpose of the manual is primarily to provide comprehensive information and guidelines to TAG reviewers but it also serves as a source of information to practitioners and researchers (USDA 2000). The only scarce information is that concerning the import and release of entomophagous agents. There is some information on the APHIS website (APHIS 2008) although the content is presently rather limited.

## 3.2.8 *Public Participation*

### 3.2.8.1 Europe

In Europe, public participation is not included in the decision-making process for IBCA import and release applications. In most cases, licensed species are published as a species register. In a few cases, including Austria, an online database can be consulted, but no European country offers the option for members of the public or stakeholders to submit their comments. Hence, public opinion is not incorporated into the evaluation of IBCA release applications.

### 3.2.8.2 International

Australia, New Zealand and the USA all offer opportunities for the public to learn about and comment on IBCA import and release applications. Canada is the only country not yet offering public participation in the review process. In Australia, there are two phases for public comment through DEWHA. The first is prior to IBCA importation when terms of reference for the assessment of likely impacts of the agent on the environment are given. The second is with respect to the draft release application. In both cases, the applications are posted on the DEWHA Public Notice website and members of the public are invited to submit their comments via the internet. The current DAFF protocol does not include public consultation; however, it is likely that a public consultation process will be introduced soon. In New Zealand, a cornerstone of the HSNO Act is the public's right to know and be heard with regard to notified applications, i.e. those that may affect the environment in some way and are categorised as being of significant public interest. Applications to import for release, or release from containment, any new organism (including an IBCA) are considered as notified applications. Receipt of such applications must be "publicly notified", in which case a 30 working day submission or comment period is open to all members of the public. Public notification involves a summary statement being advertised (i) through an alert in the major daily newspapers, (ii) on the ERMA New Zealand website and (iii) in "The Bulletin", as well as by directly notifying people who have indicated that they wish to be advised of particular types of applications. Through this procedure, New Zealanders are able to have their say and talk directly to ERMA New Zealand. In addition, public hearings of applications are held if the applicant or any of the submitters request it, or if ERMA New Zealand considers it necessary. In the USA, APHIS-PPQ publishes a 30-day notice of availability of phytophagous IBCA applications in the federal register to allow the public to comment on the proposed action. Public notification or participation has not yet been integrated into the process of entomophagous IBCA import and release evaluation.

## 3.2.9 *Length of Validity of Permit*

### 3.2.9.1 Europe

The length of the issued permits varies between countries, for example, from 1 to 5 years in The Netherlands and the UK to a maximum of 10 years in Austria, Hungary and Sweden. Within a country, the validity period may be shorter, longer or conditional according to the applications of the IBCA or crops. The length of a renewed permit is usually similar to the first authorised period.

### 3.2.9.2 International

Whereas Canada and the USA assign a validity period to full release approvals, New Zealand and Australia's release approvals are indefinite, unless for example the status of the agent changes. Once an organism is fully released into the environment in New Zealand, it is no longer considered a new organism and is thus no longer subject to HSNO Act regulation. The approval, therefore, has no validity time limit. In Canada, both permits to import IBCAs (for scientific research and for release) are valid for a period of 3 years unless otherwise stated. All permits are renewable after expiry and are valid for multiple shipments and unlimited quantities unless otherwise stated. In the USA, USDA issues permits for up to 3 years. However, under certain circumstances, the validity period may be different.

### 3.2.10 "Safe List" of IBCAs Exempt from Regulation

#### 3.2.10.1 Europe

With the exception of The Netherlands, there are no "safe lists" available in any European country where a proper environmental risk analysis has been performed on the species listed. The Netherlands published a list of 134 IBCAs (native as well as exotic species) that are exempt from regulation and permitted for release (Het LNV-Loket 2009) based on a quick scan environmental risk analysis using available information *sensu* OECD (2004) (Loomans and van Lenteren 2005). Most countries with regulation in place have a register of licensed species, for example, Austria (AGES 2009), Sweden (KEMI 2008) or commercially available species, for example, Germany (Bathon 2005) and Denmark (Wang et al. 2003). The Czech Republic currently has 32 biological plant protection products/preparations based on 23 macro-organisms in their List of Registered Plant Protection Products.

A number of countries use the recently updated EPPO list of commercially available species as a basis (EPPO 2008). This list was originally compiled and published in 2002 to facilitate decisions on the import and release of biological control agents within EPPO countries. Since the listing of agents is based on the expert judgement of available information, it enables other EPPO countries to conclude with some confidence that these agents can be introduced and used safely. Indigenous, introduced and established biological control agents are all specified and divided into two sections on the list: (1) commercially used biological control agents and (2) successfully introduced classical biological control agents. Despite the list's utility in informing sound regulatory decision in countries using it, these species are not exempt from regulation and new applications must therefore be submitted by other applicants.

#### 3.2.10.2 International

None of the four countries analysed in this review have a "safe list" as such, although there are lists available in each country documenting IBCAs that have previously been approved for release. For example, ERMA New Zealand maintains a statutory register of organisms it has approved for importation for release or release from containment, which is available on their website. The only circumstance under which the complete application process is not necessary is if the organism to be imported is not a "new organism", i.e. any species that was not present in New Zealand immediately before the date the HSNO Act came into effect. Thus, if the organism of interest features on the register of approved organisms, then HSNO Act requirements are satisfied. If the organism does not appear on the register, but is already in New Zealand, then it is possible to obtain a determination from the Authority under the HSNO Act that it was indeed present in New Zealand when the HSNO Act commenced. Otherwise a complete application process much be initiated. In Canada, there are currently about 60 arthropod biological control agents that have been historically used in commercial situations. The names of these agents can be obtained on request. Due to their safe record of use, and if imported from CFIA-approved sources, they do not have to undergo the petition process prior to importation. All other non-indigenous organisms for entomophagous and phytophagous classical biological control must be reviewed through the petition process. In the USA, there is a "safe list" of "APHIS permitted beneficials imported into the USA from other countries", which includes weed IBCAs. The list is available on the Association of Natural Biocontrol Producers (ANBP) website (ANBP 2004). In Australia, the species on DEWHA's Live Import List Part 1 includes the biological control agents permitted by DEWHA to be imported without prior approval (DEWHA 2009a). However, this is not a comprehensive list for biological control agents that have already been approved for release previously. DAFF does not maintain a published list of the released agents but any previously released agents may be imported and released again without further approval, but different strains or biotypes may require further assessment.

## 3.3 Conclusions and Recommendations

### *3.3.1 Legislation and Administration for Regulation*

Except for phytosanitary and veterinary legislation and regulation upon import of exotic species, there is no specific regulatory system for IBCAs. However, a regulatory system within Europe for the import and release of IBCAs is necessary and unavoidable. Uncoordinated regulation of biological control organisms bears the risk that approval for release in one country may have impacts for others if the organism crosses borders and establishes in other countries. Thus, a regulatory system should be harmonised across all European countries. It can be foreseen that a harmonised regulatory system will be complicated to establish in Europe due to the fact that it comprises 46 countries, all with their own governments, legislative systems and border controls. In some respects, the situation can be likened to Canada and the USA, both comprised of provinces and states, respectively, with their own provincial or state legislation and administration. In Canada, and for the most part in the USA, the regulatory process is administered at federal level, and NAPPO harmonises the regulatory needs of the three contiguous North American countries (Canada, Mexico and the USA). The establishment of a similar central governmental body within Europe may therefore be the most effective way to administer a harmonised regulatory system for IBCAs across Europe. However, past experience in Europe has also shown that over-regulation, i.e. rigid legislation with stringent data requirements, may keep biological control products off the market for a long time or even prevent industry from submitting applications in some countries. This situation has been experienced in the EU since 1992 with the registration of microbial biocontrol agents. These agents are regulated under the Directive 91/414/EEC, which largely follows requirements developed for synthetic pesticides. Thus, an EU-centralised regulatory system, whereby IBCA import and release is governed by EU legislation and administered by an EU governmental body, would most likely not be the most suitable approach.

A more pragmatic approach would be to avoid reinventing the wheel and instead adjust existing instruments to make national legislation and regulation already in place work. One competent or national authority in each country should be assigned to administer IBCA regulation. New instruments could be designed or developed using IPSM #3 as a basis in cases where specific regulatory systems for import, production, release and export of IBCAs do not already exist. Since biological control projects nowadays spark concerns about potential non-target effects in non-agricultural as well as agricultural ecosystems, Australia, New Zealand and the USA have seen the involvement of both agricultural and environmental government bodies in the regulatory processes for IBCA introductions. Provisions for IBCAs within European countries have been arranged under either nature protection, plant protection, and/or pesticide acts, depending of the historical nature of the act. Regardless of the legislation governing IBCAs, in order for a European regulatory system to remain streamlined, it is recommended that each assigned authority incorporate agricultural and environmental issues into their regulatory procedures. It is also recommended that there are no restrictions on the use of native IBCAs and that regulation applies only to exotic IBCAs, which can be described as "not native to a particular country, ecosystem or ecoarea (applies to organisms intentionally or accidentally introduced as a result of human activities)" (IPPC 1996).

### 3.3.2 Application Procedures

Different regulatory systems among European countries cause significant problems to the biocontrol industry as dossiers must respect national requirements and criteria can vary substantially between countries. Thus, with the implementation of a harmonised system across Europe, including a uniform dossier format, the application process will become less time consuming and costly. Market potential will increase relative to development costs and, thus, the likelihood of a company developing an organism into a product will also improve.

Following New Zealand's lead, a regulatory system that offers pre-submission advice to applicants through a helpdesk would be of high value in Europe, particularly for first time applicants. Based on a quick scan analysis of the organism, product, or application through consultation, decisions could be made on how and what type of procedure should be followed, what data requirements are necessary and how the dossier should to be prepared. Such an advisory service could significantly save time and costs invested into biological control programmes and ensure dossiers contain the necessary data upon first submission. Overall, the efficiency of the review process would be improved and release approval for a biological control agent could potentially be granted in a much shorter period of time. Furthermore, it would also give the NA a face and, with that, perhaps also increased credibility.

Application procedures should become transparent to the applicant and to the public. If there is an opportunity to incorporate public participation then this should also be encouraged. If the application has not been complied with within the given time frame, or when the approval has been negative, the applicant should have the right to appeal. Different application procedures should also be designed for native and exotic species.

### 3.3.3 Decision-Making Process and Decision-Maker

As with the application procedure, the decision-making process should be transparent to the applicant and public. Decision-making schemes and risk evaluation instruments, such as pest risk analyses and environmental risk analyses, should be developed or simply adjusted, if they already exist, to enable analysis of risks posed by IBCAs. Pest risk analysis schemes developed for unintentional introductions, such as those developed by EPPO, could also be adjusted for intentional introductions of species, including IBCAs. Regardless of the nature of the scheme, objective criteria upon which to base an evaluation of an application should be defined and harmonised across Europe.

Based on the success of the science-based peer-review processes in Australia, Canada and the USA, it is recommended that a similar system be set up in Europe. This could be done on an individual country basis if the relevant expertise is available within the country to perform such reviews. However, it is likely that this will not always be the case and that the NAs will therefore lack the necessary support to make a fair judgement on the environmental safety of proposed IBCAs. In order to overcome this potential problem, it is recommended that a pan-European "expert panel" be created to evaluate dossiers for IBCA introductions. Such a panel could be consulted in cases where a country lacks expertise to conduct the evaluations (or just certain ones) themselves. Experts on this panel would evaluate dossiers using a well-defined environmental risk analysis (ERA) and then provide non-binding advice to NAs on the environmental safety of proposed IBCA. Experts would be nominated to perform a review on a case-by-case basis depending on the nature of the dossier, but there should be the further possibility to seek an external review if additional expertise is required. Considering EPPO's high international reputation for biological control expertise, it is worth investigating the possibility of this organisation providing a platform for the expert panel. Experts could be accessed through the EPPO's extensive network with the biological control community. The ultimate decision on whether or not to approve an application would still remain with the NA. Therefore, in order for such a scheme to work, NAs would first need to have confidence in the regulatory guidelines and protocols supporting the ERA being performed by the group of experts, otherwise they would not consult such a pan-European expert panel.

With regards to recommendations for a pre-submission advisory service for applicants, it is also possible that if countries lack the required technical expertise to provide such a service themselves, they could also request this from the proposed pan-European expert panel.

In order to create short cuts and reduce the time frame for dossier evaluation, issues regarding the content of the application should be solved directly between the applicant and the reviewer(s) until the reviewer is satisfied.

### 3.3.4 Data Requirements

Applications for import, production and release of IBCAs should require legally enforceable risk assessments, including cost-benefit analyses that consider the potential loss of ecosystem goods and services. There is a definite need for harmonisation of data requirements across Europe in order to allow more uniform, science-based decisions. This includes fine tuning the data requirements for native and exotic species and ensuring that costly risk assessment studies are avoided since this may keep products off the market and result in few registered IBCAs.

Europe has already made substantial progress in terms of devising guidelines summarising the data requirements for dossier preparation (refer back to Table 3.1). At the time of the REBECA project, the most recent guidelines, resulting from the IOBC/WPRS Commission for the Harmonisation of Regulation of Invertebrate Biological Control Agents, entitled "Guidelines on information requirements for import and release of invertebrate biological control agents in European countries" (Bigler et al. 2005b), provided the most up-to-date comprehensive guidelines available in Europe including all the information to be included in a full dossier. Within the REBECA project, these guidelines were used as a basis for the drafting of a standard application form (Loomans et al. 2007a) and accompanying guidance document (Loomans et al. 2007b), formulating the data requirements necessary for all types of IBCAs and their application. Provided that industry, biological control practitioners and the selected national regulatory authorities consider them to be realistic and manageable, then it is recommended that these documents be finalised and adopted as the official European standard for information requirements for IBCA risk assessment.

Significant progress has also been made towards formulating a workable framework under which future IBCA risk assessments could be conducted (see van Lenteren et al. 2006). This framework attempts to simplify the process of conducting assessments for IBCAs intended for inundative release, but it is possible that IBCAs for classical biological control could be included under such a scheme. Such a framework would not only aid biological practitioners in their pursuit of conducting a thorough risk assessment, but by having a European framework, release applications would become more uniform in structure and content, thus lending themselves for a faster review process. It would also open up the possibility of sharing and accepting evaluations between neighbouring countries within the same ecoregion, provided that assessments have been conducted by biocontrol "experts".

### 3.3.5 Costs

The cost of applying to import and release an IBCA and for drawing up dossiers should be minimised through activities such as offering pre-submission advice via helpdesks, making information and application forms readily accessible via the internet and providing up-to-date and accessible safe lists etc. The cost of conducting research could also be reduced by using the hierarchical information evaluation and risk-assessment scheme as proposed by van Lenteren et al. (2006). Under such a scheme, the information required will differ depending on the IBCA being tested and the number of required tests may be lower than initially perceived by the applicant.

One of the main concerns in Europe is that a regulatory system would render the process of approval for IBCA introduction and release into a country both costly and time consuming. Extensive delays associated with environmental risk assessments could potentially multiply biological control programme costs, leaving industries struggling to afford to run them and research organisations unable to obtain funding to undertake such projects. It is clear from the analysis of the regulatory systems in Australia, Canada and the USA that the introduction and release of IBCAs does not necessarily have to be expensive for the biological control practitioner. Administrative costs in Australia, Canada and the USA are covered by public money via the national governmental bodies and the review process in these countries operates on a voluntary basis, so that scientists are not paid for the reviews they conduct. This leaves the applicant with minimal fees to pay upon dossier submission. It is recommended that public money also be used to cover some of the administrative costs in Europe to relieve the burden on biological researchers and industries. Evaluation of biological control agents is in the public interest, thus there is justification for support by public funds.

### 3.3.6 Time Frame

Establishing a regulatory system in Europe does not automatically imply that the IBCA release application review process should require protracted periods of time. Europe should endeavour to minimise the time taken for dossier turn-around to avoid delaying the progress of biological control projects. The current time frame for administration and evaluation of dossiers varies largely within Europe and can even take up to 2 or 3 years when efficacy tests are required. This time period needs to be shortened to a more reasonable and workable time-scale. It is recommended that fixed time frames for acceptance of the application form and evaluation of the dossier are set, depending on the type of evaluation required. This is similar to the situation in New Zealand where there is a legal requirement for ERMA New Zealand to provide a decision within 100 working days of receiving a dossier. The system in Australia could also be considered whereby co-operators are permitted a set time period within which they must review a dossier. When there is no response from co-operator(s) and/or experts(s) within a fixed time frame, it should be assumed there is no objection and the NA may proceed with the application.

### 3.3.7 Availability of Information to Aid Applicants

At present, there is very little or no information available online in any European country to aid applicants. This urgently needs improvement. By contrast, such information in Australia, New Zealand, Canada and the USA can be found by performing a quick search on their government websites. Europe should follow this example by enabling easy internet access to information regarding the helpdesk, forms, application procedures and the regulatory process for IBCA introduction and release into a new area. Application forms should also be downloadable. Europe would certainly benefit from following the lead of AAFC in Canada who published a comprehensive guide to provide petitioners, reviewers of petitions and interested Canadian citizens with information about the procedure (De Clerck-Floate et al. 2006).

### 3.3.8 *Public Participation*

A troubling issue for the practice of biological control is that in recent years there has been an abundance of criticism concerning its potential negative effects on biodiversity whereas, in comparison to some chemical and mechanical methods of control, biological control is normally regarded as an environmentally benign method of pest control. The public is also aware of the lengthy and costly procedures involved in investigating a potential biological control agent and this only decreases their confidence and support for such a pest control strategy. In order to restore public support, any European regulatory scheme that is implemented needs to be made transparent to the public, allowing for easy access to information so people can formulate their own informed opinions. Confidence and support would also be enhanced if members of the public were provided with the opportunity to participate in the decision-making process for release applications. Their involvement in the process would help to increase general awareness and knowledge of biological control practice, thus helping to raise its profile in society. Europe should follow the example of Australia and the USA, where release applications are placed on the government websites and members of the public are invited to submit their comments via the internet. The expert panel or NAs could then take comments or concerns from interested members of the public into consideration when evaluating dossiers. Licences that have been granted and permits that have been issued for specific IBCAs or derogations should also be posted on the NA website. Public hearings could also be considered, as they are in New Zealand, although this may be too costly and time-consuming depending on the level of administration required.

### 3.3.9 *Length of Validity of Permit*

Whereas Canada and the USA assign a validity period to full release approvals, New Zealand and Australia's release approvals are indefinite, unless, for example, the status of the agent changes. It would be advantageous to implement a similar system in Europe that permits IBCAs, which have been released safely for several years with no record of non-target impact, an indefinite release approval. At the very least, NAs should allocate a substantial enough period to allow industry to recover their expenses for the application, for example 5 years.

### 3.3.10 "Safe List" of IBCAs Exempt from Regulation

Every European country should compile a "safe list" of species/organisms that are considered to be safe for release and that are exempted from further regulation available. This would certainly aid applicants and regulators as well as help to avoid lengthy administrative procedures. The principle of a safe list on an EU or EPPO level, which could be readily accessed by industry, regulators and the public, should also be supported. Such as list would facilitate the application process both for industry and for applicants located in different countries and thus stimulate an increase in biological control applications. Furthermore, EU or EPPO "endorsement" would enhance credibility of the list and biological control in general. It is highly recommended therefore that the recently updated EPPO "safe list" list of 2008 be actively maintained and that specific criteria and data requirements for inclusion of a particular IBCA on the list be updated so that more comprehensive information can be provided on the list itself, as recommended by the REBECA project (REBECA 2006). For this particular task, the CABI BIOCAT database (data from 1900 to September 2006) would provide a valuable source of information regarding introduced classical IBCAs. As a result, the EPPO list would serve as an extremely valuable database of information for IBCA releases across Europe. It would also become an important tool for reviewers of applications and regulators, especially if it provided access to existing regulatory decisions, both positive and negative, together with their justifications. Precautions would have to be taken to ensure that confidentiality issues do not arise and threaten the utility of such a list.

Also, worth mentioning is the fact that European countries should also take action to promote the use of the vast reservoir of native European IBCA species and ensure, as mentioned previously, that there are no restrictions to their use.

### 3.3.11 Summary

Europe's goal is to develop and implement a harmonised regulatory system across all member countries for the import and release of IBCAs. To adapt or change legislation and regulatory measures already in place in a country is, however, a very difficult and lengthy task since each country has its own constitution and sovereign rights. A more pragmatic approach, one which encourages the adjustment of national instruments already in place, will reduce the time required to harmonise IBCA regulation. European countries should be willing to learn from experiences elsewhere in the world when developing their own regulatory requirements. Australia, New Zealand, Canada and the USA have all had several years of experience in implementing IBCA regulatory procedures and therefore there is great potential for Europe to benefit from their knowledge. Certainly, there are features of each system that work well and could potentially be adopted by Europe (see also Hunt et al. 2008).

Countries have an obligation to make policy decisions and application procedures transparent and to facilitate application procedures. Making better use of the internet will greatly enhance communication with applicants and the public, which in turn will help to raise the public profile of biological control.

Clearly, there will be challenges in introducing a unified scheme across so many different countries. A major gain can be achieved, firstly, by facilitating the process of IBCA release application by developing harmonised data requirements, application procedures and dossier formats and, secondly, by ensuring sound evaluation of proposals by developing uniform dossier requirements, scientific methods, tools for risk-assessment evaluation and legal instruments. Significant progress has already been made in this direction within the REBECA project. For specific details regarding the outcomes of the REBECA project regarding IBCA regulation, please refer to Chapters 11 and 16 within this volume. With continued efforts, Europe will certainly begin to see immediate movements towards an efficient, affordable and scientifically-sound harmonised regulatory system.

## References

AGES (2009) Austrian Agency for Health and Food Safety. http://www.ages.at. Accessed 20 January 2009

ANBP (2004) Association of natural biocontrol producers: permitted beneficials. http://www.anbp.org/permittedbeneficials.htm. Accessed 20 Jan 2009

APHIS (2008) United States Department of Agriculture, Animal and Plant Health Inspection Service, Permits http://www.aphis.usda.gov/plant_health/permits/index.shtml. Accessed 20 Jan 2009

Bathon H (2005) Nützlingsliste. http://www.bba.bund.de/nn_912578/SharedDocs/08__BI/Publikationen/nuetzlingsliste.html. Accessed 20 Jan 2009

Bigler F, Loomans A, van Lenteren J (2005a) Harmonization of the regulation of invertebrate biological control agents in Europe. In: Hoddle MS (ed) Proceedings of the 2nd international symposium on biological control of arthropods, Davos, Switzerland, 12–16 September 2005. FHTET-2005-08, United States Department of Agriculture, Forest Service, Morgantown, WV, USA, pp 692–700

Bigler F, Bale JS, Cock MJW, Dreyer H, Greatrex R, Kuhlmann U, Loomans A, van Lenteren JC (2005b) Guidelines on information requirements for import and release of invertebrate biological control agents in European countries. Biocontrol News Inform 26(4): 115 N–123 N

Bigler F, Babendreier D, Kuhlmann U (eds) (2006) Environmental impact of invertebrates for biological control of arthropods: methods and risk assessment. CABI, Wallingford, 299 p

Blum B, Ehlers R, Haukeland-Salinas S, Hokkanen H, Jung K, Kuhlmann U, Menzler-Hokkanen I, Ravensburg W, Strasser H, Warrior P, Wilson M (2003) Biological control agents: safety and regulatory policy. BioControl 48:477–484

Bundesnaturschutgesetz (2002) Gesetz über Naturschutz und Landschaftspflege, $ 41. http://www.buzer.de/gesetz/2122/a30126.htm. Accessed 20 Jan 2009

CBD (1992) The convention on biological diversity. http://www.cbd.int/. Accessed 20 Jan 2009

CFIA (2008) Canadian food inspection agency: application, procedures, issuance and use of a permit to import under the Plant Protection Act. http://www.inspection.gc.ca/english/plaveg/protect/dir/d-97-04e.shtml. Accessed 20 Jan 2009

Coulson JR, Vail PV, Dix ME, Nordlund DA, Kaufmann WC (2000) 110 Years of Biological Control Research and Development in the United States Department of Agriculture (1883–1993). United States Department of Agriculture, Agricultural Research Service

CSL 2008 (2008) Central science laboratory: licensing of non-native biocontrol agents. http://www.csl.gov.uk/nonnativebiocontrol. Accessed 4 Feb 2009

DAFF (2007) Australian Government Department of Agriculture, Fisheries and Forestry: Protocol for biological control agent applications – import, host-specificity test list and release. http://www.daffa.gov.au/ba/about/plant/protocol-biological. Accessed Jan 2009

DEWHA (2009a) Department of the environment, water, heritage and the arts: list of specimens suitable for live import (Live Import List). http://www.environment.gov.au/biodiversity/trade-use/lists/import/pubs/live-import-list.pdf. Accessed 20 Jan 2009

DEWHA (2009b) Department of the environment, water, heritage and the arts: amending the list of specimens suitable for live import. http://www.environment.gov.au/biodiversity/trade-use/lists/import/amend/index.html. Accessed 20 Jan 2009

De Clerck-Floate RA, Mason PG, Parker DJ, Gillespie DR, Broadbent AB, Boivin G (2006) Guide for the importation and release of arthropod biological control agents in Canada. Agriculture and Agri-Food Canada A42-105/2006E, 55 pp

Delfosse ES (2005) Risk and ethics in biological control. Biol Control 35:319–329

EPPO (1997) EPPO/CABI workshop on safety and efficacy of biological control agents in Europe. EPPO Bull 27:1–3

EPPO (1999) Safe use of biological control: first import of exotic biological control agents for research under contained conditions. EPPO Standard PM6/1(1). EPPO Bull 29:271–272. Available at: http://archives.eppo.org/EPPOStandards/biocontrol.htm. Accessed 20 Jan 2009

EPPO (2000) Safe use of biological control: Import and release of exotic biological control agents. EPPO Standard PM6/2(1). EPPO Bull 31:33–35. Available at: http://archives.eppo.org/EPPOStandards/biocontrol.htm. Accessed 20 Jan 2009

EPPO (2008) List of biological control agents widely used in the EPPO region. EPPO Bull 32:447–461. Available at: http://archives.eppo.org/EPPOStandards/biocontrol.htm. Accessed 20 Jan 2009

ERMA New Zealand (2009) Environmental risk management authority New Zealand. http://www.ermanz.govt.nz/. Accessed 20 Jan 2009

Greathead DJ (ed.) (1976) A review of biological control in western and southern Europe. Technical Communication No. 7, Commonwealth Institute of Biological Control. Commonwealth Agricultural Bureaux, Farnham Royal, UK, 182 p

Harrison L, Moeed A, Sheppard A (2005) Regulation of the release of biological control agents of arthropods in New Zealand and Australia. In: Hoddle MS (ed) Proceedings of the second international symposium on biological control of arthropods, Davos, Switzerland, 12–16 September 2005. FHTET-2005-08, United States Department of Agriculture, Forest Service, Morgantown, WV, pp 715–725

Het LNV-Loket (2009) Het Landbouw, natuur en voedselkwaliteit-Loket. http://www.hetlnvloket.nl. Accessed 19 Jan 2009

Howarth FG (1991) Environmental impacts of classical biological control. Annu Rev Entomol 36:485–509

Hunt E, Kuhlmann U, Sheppard A, Mason PG, Harrison L, Raps A, Dreyer H, Flanders B (2006) A review of the regulations for the import and release of macrobial biological control agents in Australia, Canada, New Zealand, Switzerland and the United States of America. Unpublished report of EU project REBECA [Regulation of Biological Control Agents]

Hunt EJ, Kuhlmann U, Sheppard A, Qin T-K, Barratt BIP, Harrison L, Mason PG, Parker D, Flanders R, Goolsby J (2008) Review of invertebrate biological control agent regulation in Australia, New Zealand, Canada and the USA: recommendations for a harmonized European system. J Appl Entomol 132:89–123

IPPC (International Plant Protection Convention) (1996) Code of conduct for the import and release of exotic biological control agents. International Standards for Phytosanitary Measures

Publication (ISPM) Publication No. 3, Food and Agriculture Organisation of the United Nations (FAO) Rome, Italy

IPPC (2005) Guidelines for the export, shipment, import and release of biological control agents and other beneficial organisms. ISPM Publication No. 3, FAO, Rome, Italy. Available at https://www.ippc.int/servlet/CDSServlet?status=ND0xMzM5OSY2PWVuJjMzPSomMzc9a29z. Accessed 20 Jan 2009

KEMI (2008) Swedish Chemicals Agency, Försålda kvantiteter av bekämpningsmedel (Sold quantities of pesticides). http://www.kemi.se/templates/Material____3548.aspx. Accessed 20 Jan 2009

Kuhlmann U, Toepfer S, Zhang F (2005) Is classical biological control against Western Corn Rootworm in Europe a potential sustainable management strategy?. In: Vidal S, Kuhlmann U, Edwards CR (eds) Western Corn rootworm: ecology and management. CABI, Wallingford, UK, pp 263–284

Loomans AJM (2007) Regulation of invertebrate biological control agents in Europe: review and recommendations in its pursuit of a harmonised regulatory system. Unpublished report of EU project REBECA [Regulation of Biological Control Agents]. Available at: http://www.rebeca-net.de/downloads/Regulation%20of%20Beneficials%20in%20Europe.pdf. Accessed 20 Jan 2009

Loomans AJM, Sütterlin S (2005) Regulation of invertebrate biological control agents: international context and situation in The Netherlands. IOBC/WPRS Bull 28(1):179–182

Loomans AJM, van Lenteren JC (2005) Tools for environmental risk assessment of invertebrate biological control agents: a full and quick scan method. In: Hoddle MS (ed) Proceedings of the 2nd international symposium on biological control of arthropods, Davos, Switzerland, 12–16 September 2005. FHTET-2005-08, United States Department of Agriculture, Forest Service, Morgantown, WV, pp 611–619

Loomans AJM, Bigler F, Sterk G, Bale JS (2007a) Report from the subgroup for the Development of Guidance and Administration Documents in support of an Application for the Import and Release of Macrobial (Invertebrate) Biological Control Agents. REBECA Work Package 5, Macrobial Biological Control Agents, Appendix 1, Application Form, 10 pp. Available at: http://www.rebeca-net.de/downloads/Brussels%20conference/REBECA%20Application%20form.pdf. Accessed 28 Jan 2009

Loomans AJM, Bigler F, Sterk G, Bale JS (2007b) Report from the subgroup for the Development of Guidance and Administration Documents in support of an Application for the Import and Release of Macrobial (Invertebrate) Biological Control Agents. REBECA Work Package 5, Macrobial Biological Control Agents, Appendix 2, Guidance Document, 10 pp. http://www.rebeca-net.de/downloads//Brussels%20conference/REBECA%20Guidance.pdf. Accessed 28 Jan 2009

Mason PG, Flanders RG, Arrendondo-Bernal HA (2005) How can legislation facilitate the use of biological control of arthropods in North America?. In: Hoddle MS (ed) Proceedings of the second international symposium on biological control of arthropods, Davos, Switzerland, 12–16 September 2005. FHTET-2005-08, United States Department of Agriculture, Forest Service, Morgantown, WV, pp 701–714

Messing RH (2005) Hawaii as a role model for comprehensive U.S. Biocontrol legislation: the best and the worst of it. In: Hoddle MS (ed) Proceedings of the second international symposium on biological control of arthropods, Davos, Switzerland, 12–16 September 2005. FHTET-2005-08, United States Department of Agriculture, Forest Service, Morgantown, WV, pp 686–691

NAPPO (2008a) Guidelines for petition for first release of non-indigenous entomophagous biological control agents. Regional Standards for Phytosanitary Measures #12. http://www.nappo.org/Standards/REVIEW/2008/RSPMNo.12-e_20081110.pdf. Accessed 20 Jan 2009

NAPPO (2008b) Guidelines for petition for first release of non-indigeous phytophagous biological control agents. Regional Standards for Phytosanitary Measures #7. http://www.nappo.org/Standards/REVIEW/2008/RSPMNo7-e_20081110.pdf. Accessed 20 Jan 2009

OECD (2004) Guidance for information requirements for regulation of invertebrates as biological control agents (IBCAs). OECD Environment, Health and Safety Publications. Series Pesticides 21:22 p

REBECA (2006) REBECA: regulation of biological control agents. http://www.rebeca-net.de/. Accessed 20 Jan 2009

Sheppard AW, Hill R, De Clerck-Floate RA, McClay A, Olckers T, Quimby PC Jr, Zimmermann HG (2003) A global review of risk-benefit-cost for the introduction of classical biological control agents against weeds: a crisis in the making?. Biocontrol News Inform 24:91 N–108 N

Sheppard AW, Shaw RH, Sforza R (2006) Top 20 environmental weeds for classical biological control in Europe: a review of opportunities, regulations and other barriers to adoption. Weed Res 46:93–117

Simberloff D (1996) Impacts of introduced species in the United States. Consequences 2(2):13–23

Simberloff D, Alexander M (1998) Assessing risks from biological introductions (excluding GMOs) for ecological systems. In: Calow P (ed) Handbook of environmental risk assessment and management. Blackwell, Oxford, pp 147–176

USDA (2000) Reviewer's manual for the technical advisory group for biological control agents of weeds: guidelines for evaluating the safety of candidate biological control agents. Manuals Unit, Plant Protection and Quarantine, Animal and Plant Health Inspection Service, United States Department of Agriculture. Available at: http://www.aphis.usda.gov/import_export/plants/manuals/ports/downloads/tag.pdf. Accessed 20 Jan 2009

Van Driesche RG (2004) Predicting host ranges of parasitoids and predcious insects – what are the issues?. In: Van Driesche RG, Reardon R (eds) Assessing host ranges for parasitoids and predators used for classical biological control: a guide to best practice. United States Department of Agriculture, Forest Service, FHTET–2004–03, Morgantown, WV

van Lenteren JC (1997) Benefits and risks of introducing exotic macro-biological control agents into Europe. Bull OEPP/EPPO Bull 27:15–27

van Lenteren JC, Cock MJW, Hoffmeister TS, Sands DPA (2006) Host specificity in arthropod biological control, methods for testing and interpretation of the data. In: Bigler F, Babendreier D, Kuhlmann U (eds) Environmental impact of invertebrates for biological control of arthropods: methods and risk assessment. CABI, Wallingford, pp 38–63

Waage J (1997) Global developments in biological control and the implications for Europe. EPPO Bull 27:5–13

Waage JK (2001) Indirect ecological effects in biological control: the challenge and the opportunity. In: Wajnberg E, Scott JK, Quimby PC (eds) Evaluating indirect ecological effects of biological control. CABI, Wallingford, pp 1–12

Wang et al (2003) Vurdering af muligheder for forebyggelse og alternative bekæmpelse i væksthusproducerede planter (prydplanter og væksthusgrønsager)http://www2.mst.dk/udgiv/publikationer/2003/87-7972-898-7/pdf/87-7972-899-5.pdf. Accessed Jan 2009

Williamson M (1996) Biological invasions. Chapman & Hall, London

# Chapter 4
# Regulation of Plant Protection in Organic Farming

**Bernhard Speiser and Lucius Tamm**

**Abstract** Organic farming is a system approach aiming at a sustainable ecosystem, safe food, good nutrition, animal welfare and social justice. Quantitatively, organic farming is still of minor importance, but it is one of the most rapidly growing agricultural sectors worldwide. The new EU 'organic regulation' consists of a framework regulation, complemented by implementation rules and guidelines. Other important regulations/standards are the National Organic Program of the USA, the guidelines of the Codex Alimentarius and the basic standards of the International Federation of Organic Agriculture Movements (IFOAM). Under all these standards, plant protection is strictly regulated. Organic plant protection follows a clear hierarchy: primarily, plant health is maintained by preventative measures. Only if these methods are insufficient, plant protection products may be used. However, only a very limited range of substances is authorized (substances of plant or animal origin, micro-organisms and a few other substances). In the EU, new substances can only be authorized if they are consistent with organic farming principles, necessary for sustained production, and if they are of plant, animal, microbial or mineral origin. Case studies for the codling moth and potato late blight illustrate that the practices of organic plant protection in Europe differ significantly from one country to another. The codling moth is mainly controlled by mating disruption, *Bacillus thuringiensis*, and *Cydia pomonella* granulosis virus (CpGV). To what extent spinosad will be used in the future is not yet clear, as it was authorized for organic farming only recently. To avoid the late blight epidemic, organic farmers use a variety of management practices. Within the constraints of the market, they also avoid susceptible varieties. For direct control of late blight, copper fungicides are the only plant protection products authorized in organic farming, but they cannot be used in all EU countries, and there are quantitative restrictions in some countries.

---

B. Speiser (✉)
Research Institute of Organic Agriculture (FiBL), 5070 Frick, Switzerland
e-mail: bernhard.speiser@fibl.org

R.-U. Ehlers (ed.), *Regulation of Biological Control Agents*,
DOI 10.1007/978-90-481-3664-3_4, 

## Contents



## 4.1 What Is Organic Farming?

Throughout this chapter, we use the term ‘organic farming’ which has established in english-speaking countries; the terms ‘ecological farming’ and ‘biological farming’ are synonyms also used in Europe (EC 1991). Many people primarily think of organic farming as ‘farming without chemicals’ (Lampkin 1990). This oversimplified view suggests that organic farming substitutes ‘agro-chemicals’ with ‘organic inputs’. However, organic farming defines itself primarily by what it is doing, and not by what it is avoiding. The IFOAM Basic Standards (see below) define organic farming as a system approach resulting in ‘a sustainable ecosystem, safe food, good nutrition, animal welfare and social justice’, which is ‘more than a system of production that includes or excludes certain inputs’ (IFOAM 2002). For a thorough introduction to organic farming see (Lampkin 1990).

### *4.1.1 How Organic Farming Evolved*

Organic farming principles and standards/regulations reflect the current state of agriculture and society and should not be seen as a final statement, but rather as a work in progress (IFOAM 2002). The evolution of organic farming must be seen in the context of the development of the food sector in general.

The roots of organic farming can be traced back to the 1920s, when a few pioneers searched for alternative methods of agricultural production. Their goal was to develop a production method which was appropriate for living systems and which could promote human well-being and harmony between humans and the cosmos. They objected to 'industrialized' agricultural production, and as a practical consequence rejected the use of mineral fertilizers. In the following decades, these ideas were further developed in practice (Vogt 2000). At that time, the guidelines were laid down in the form of general principles, which left some freedom to the farmer how to fulfil the principles. For the control of pests and diseases, preventative measures were considered as the most appropriate tools, but the use of very few pesticides available at that time (mainly copper and sulphur) was tolerated when needed. However, when synthetic pesticides became popular in the 1950s and 1960s, their use was explicitly banned from organic farming.

In 1976, the International Federation of Organic Agriculture Movements (IFOAM) decided to work on common, international standards. In 1980, the first IFOAM Basic Standards were published (still more in the form of general guidelines). Once such standards were in force, organic farms could be inspected and certified. Growing public awareness about environmental pollution, animal welfare and food scandals contributed to an increasing demand for certified organic produce by consumers.

Towards the end of the 1980s, some governments discovered that promotion of organic farming combined the efforts for reducing overproduction and conservation of the environment. Already in 1987, Denmark and the German federal state Saarland had started to pay subsidies for conversion to organic farming. Later, various countries started 'organic programmes' with financial, educational and legislative incentives for organic production and marketing. In 2004, the European Commission published an 'Organic Action Plan' (COM 2004), with the aim to facilitate the expansion of organic farming, Meanwhile, a broad range of organic products became available in larger quantities and with better quality. Large retailers began to sell organic products, and prices began to sink. Today, retailers have become important key players, who influence the development of the organic food sector. Organic products are now marketed as premium products with an 'added value' of environmental friendliness, animal welfare and high product quality and safety.

### *4.1.2 Importance of Organic Farming*

Quantitatively, organic farming is still of minor importance, but it is one of the most rapidly growing agricultural sectors. In 2006, more than 700,000 farms worldwide managed almost 30.4 million hectares organically. The countries with the greatest area of organic land were Australia (12.3 million hectares), China (2.3 million hectares), Argentina (2.2 million hectares) and the USA (1.6 million hectares).

Worldwide, 0.65% of all land is managed organically. The highest proportion of organic land is found in European countries, with peak values achieved in Liechtenstein (29%), Austria (13%) and Switzerland (12%). Globally, the organic area increased by 6% from 2005 to 2006, and has grown on all continents (Willer 2008).

Global demand for organic products remains robust, and sales increase continuously. Total organic sales for 2006 are estimated at 38.6 billion US Dollar. Consumer demand is largely concentrated in Europe (52%) and North America (45%). Asia, Latin America and Australasia are also important producers and exporters of organic food (Sahota 2008).

The importance of organic farming is not fully reflected by these figures: For many consumers, organic foods are the perfect example of quality and/or healthy food (e.g. pesticide-free, GMO-free). In addition, organic farming matches the goals of agricultural and/or environmental policies to a high degree. Organic farming serves as an example and thus influences the development of agriculture in general.

## 4.2 Regulation of Organic Farming

In this chapter, we discuss only plant protection. Nevertheless, the reader is reminded that the scope of organic regulations and standards is much broader and contains rules for all of the following topics:

- plant production (of which plant protection is only one aspect);
- livestock production;
- processing of organic foods;
- inspection and certification of organic farms and processing units;
- labelling of organic foods and feeds;
- importation of organic foods.

The rules can be laid down in regulations and/or private standards. Regulations are legal texts, enforced by the authorities, and must be fulfilled by all organic farms in a given country. Examples are the EU Organic Regulation and the US National Organic Program. Private standards are private initiatives which are not legally binding, but which are pre-requisites for the use of an organic label or logo. Some organic labels are far older than organic regulations and much better known to the consumers. In practice, a considerable proportion of organic farmers fulfils not only the regulation, but also the standards of a private label organization. An up-to-date overview of regulations and standards in different countries worldwide is given in Huber et al. (2008).

### *4.2.1 Legal Framework for Organic Farming in the EU*

In the EU, a legal definition of organic farming practices was first given in 1991 by the European Council regulation on organic farming (EEC) 2092/91 (EC 1991). This regulation has been subsequently amended many times. After a thorough revision, a new set of regulations is in force since the 1st of January, 2009. The new 'organic regulation' consists of a 'framework regulation' (EC 2007), complemented by 'implementation rules' and guidelines. For plant protection, the implementation rule 889/2008 is relevant (EC 2008b). The European organic regulations were developed in a lengthy process and represent a broad consensus in Europe. For a comprehensive overview of this regulation, see (Schmidt and Haccius 2008; Mikkelsen and Schlüter 2009).

### *4.2.2 Legal Framework for Organic Farming in the USA*

The United States' National Organic Program (hereafter called 'NOP') was first proposed in 1997, and has been amended in 2000. For a brief history, see (Baker 2004). It provides legally binding standards for organic production, processing and marketing of organic products.

### *4.2.3 Codex Alimentarius Guidelines*

The Codex Alimentarius is a joint food standards programme of FAO/WHO (United Nations' Food and Agriculture Organization and World Health Organization). The Codex Alimentarius is a collection of internationally adopted food standards. Their purpose is to protect the health of consumers and to ensure fair practices in food trade. The 'guidelines for the production, processing, marketing and labelling of organically produced foods' (hereafter called 'Codex guidelines') were first published in 1999 and have been subsequently revised several times (Codex Alimentarius Commission 1999). These guidelines represent a broad international consensus about the nature of organic production. Codex Alimentarius guidelines are themselves not legally binding, but they have a strong influence on national and international regulations. In the last years, a major activity was the revision of the criteria for admission of new inputs and of the list of allowed inputs.

### *4.2.4 IFOAM Basic Standards*

The International Federation of Organic Agriculture Movements (IFOAM) is a worldwide umbrella bringing together organizations of organic farmers and growers, traders and consumers. It represents some 700 member organizations in over 100 countries. The 'IFOAM Basic Standards for Production' were first published in 1980; the most recent issue is part of the 'IFOAM norms for organic production and processing' (IFOAM 2006). The IFOAM basic standards are 'standards for standards', which means that they can only serve as a basis for developing regional standards, which can then be used for certification of organic farms (Blake 2004). The IFOAM basic standards are a private initiative and have no legal standing, but their political and practical impact has been huge (Blake 2004). Because they are the oldest standards and are well rooted within the organic sector, they have directly or indirectly served as blueprints in the development of all other standards and regulations worldwide. Until now, innovations in organic regulation are mainly driven by IFOAM. The latest example is the development of 'principles of organic agriculture', which were first published in the 2005 version of the IFOAM norms (IFOAM 2006).

## 4.3 Plant Protection in Organic Farming

### *4.3.1 Hierarchy of Plant Protection Measures*

All organic farming standards establish a clear hierarchy of plant protection measures:

- Primarily, plant health should be maintained by preventative measures. This should be achieved by the choice of adapted species and varieties, crop rotation, cultivation techniques, thermal processes and the protection and/or release of natural enemies.
- Only if these methods are insufficient, plant protection products may be used. However, only a very limited range of substances is authorized for use (see Section 4.3.2). For some authorized substances, only selected uses are allowed.

A more detailed description of this hierarchy is given in Speiser et al. (2006b).

### *4.3.2 Authorized Active Substances for Plant Protection*

All regulations/standards contain positive lists of substances which are authorized for plant protection. Only those plant protection products which contain these substances as active ingredients are authorized; all other products are excluded. For some substances, the authorization is limited to certain conditional requirements or conditions for use. Here, the substances allowed under the European organic regulation are listed as an example. The lists in other standards are very similar. The substances and their authorized uses are listed in Annex II of Reg. 889/2008.

The following substances of plant or animal origin are authorized: azadirachtin extracted from *Azadirachta indica* (insecticide); beeswax (pruning agent); gelatine (insecticide); hydrolized proteins (attractant); lecithine (fungicide); plant oils (e.g. mint oil, pine oil, caraway oil) (insecticide, acaricide, fungicide and sprout inhibitor); pyrethrins extracted from *Chrysanthemum cinerariaefolium* (insecticide); quassia extracted from *Quassia amara* (insecticide, repellent); rotenone extracted from *Derris* spp., *Lonchocarpus* spp. and *Tephresia* spp. (insecticide). Rotenone was not defended in the re-evaluation process for old pesticides, and will therefore be withdrawn from the market, but the time of withdrawal is not yet clear.

Micro-organisms (bacteria, viruses and fungi) are generally allowed, as long as they are not genetically modified. Spinosad, a substance of microbial origin, is allowed as an insecticide.

The following other substances are authorized: diammonium phosphate (attractant, only in traps); pheromones (attractants, only in traps and dispensers); the pyrethroids deltamethrine and lambdacyhalothrine (insecticide; only against *Batrocera oleae* and *Ceratitis* spp., and only in traps with specific attractants); ferric phosphate (molluscicides); copper in the form of copper hydroxide, copper oxychloride, copper sulphate, cuprous oxide and copper octanoate (fungicide; maximum 6 kg copper per hectare per year); ethylene (only selected uses allowed); fatty acid potassium salt (insecticide); potassium aluminium (prevention of ripening of bananas); lime sulphur (fungicide, insecticide, acaricide); paraffin oil (insecticide, acaricide); mineral oils (insecticide, fungicide; only in fruit trees, vines, olive trees and tropical crops); potassium permanganate (fungicide, bactericide; only in fruit trees, olive trees and vines); quartz sand (repellent); sulphur (fungicide, acaricide, repellent); calcium hydroxide (fungicide; only in fruit trees, against *Nectria galligena*); potassium bicarbonate (fungicide).

The above list is relevant for substances which are used as plant protection products. In contrast, the use of substances as plant resistance improvers is not regulated by the EU organic regulation, and their use is therefore allowed (Reg. 834/2007, Article 16 (4) & (5)). In Germany, plant resistance improvers (Pflanzenstärkungsmittel) are a separate category of products recognized under the Plant Protection Act. These products may also be used in Austria (mutual recognition). The plant resistance improvers which are currently registered contain substances such as plant extracts, hydrolized proteins, stone meal, kieselgur, chitosan, etheric oils, micro-organisms, homeopathic preparations, humic acids, sugars, waxes, plant oils, kaolin, potassium and sodium bicarbonate.

## 4.4 Authorization of New Pesticides for Organic Farming

In line with the focus of this book, the present section is centered on the European situation; Section 4.4.4 summarizes the situation outside the EU. In the EU, all new plant protection products must be registered for use in general agriculture under Directive 91/414/EEC, before they can be used. This procedure is not further discussed here, as it is explicitly dealt with in other chapters. For use in organic farming, they must additionally be authorized under Regulation 834/2007. The following text refers only to this latter procedure.

### *4.4.1 Generically Authorized Substances*

In the EU organic regulation, plant oils, micro-organisms and pheromones are authorized in a generic way. New substances belonging to one of these three groups can be used in organic farming without any further authorization procedure, if they are allowed for use in general agriculture in the EU. These substances do not need to undergo the authorization procedure described in Section 4.4.2, and it is not verified whether they fulfil the authorization criteria described in Section 4.4.3.

The substances are listed in the EU Organic Regulation as follows:

- ‘plant oils (e.g. mint oil, pine oil, caraway oil)’. The conditions for use are limited to insecticides, acaricides, fungicides and sprout inhibitors.
- ‘micro-organisms (bacteria, viruses and fungi)’. The conditions for use are limited to the control of pests and diseases, and genetically modified strains are explicitly excluded.
- ‘pheromones’. The authorization is limited to the use in traps and dispensers.

### 4.4.2 Authorization Procedure in the EU

Normally, the Commission will only table requests which are submitted by an EU Member State (Speiser et al. 2005). The Commission has started to seek advice on such requests from an ad-hoc expert group (EC 2008a), which has facilitated decision-making within a short time. Recently, the Commission has set up an ‘expert group for technical advice on organic production’ (EC 2009a). At the time when this chapter was written, the expert group had not yet taken up its work, so practical experience is lacking. Note: requests for authorization can only be made by the Commission and the EU member states, but not by manufacturers. In the past, those member states more frequently requested authorization of new active substances, in which the manufacturers were based and/or in which country a need for the substance was declared.

Once an active substance is authorized for use, it is up to the certification authority to determine under which conditions commercial products may be used. In some countries, there are lists of authorized commercial products available (e.g. (Hozzank 2009; Mäder et al. 2009; Speiser et al. 2009)). In these cases, manufacturers or traders have to apply for inclusion in the lists. In Germany, it is mainly evaluated whether a product complies with the legal regulations. In Switzerland, products have to comply with the legal regulations and also with further restrictions, which were elaborated by the inputs list team in close collaboration with the organic producers’ association ‘Bio Suisse’. In the case of Austria, it is noted in the list whether a product complies only with the legal regulations, or also with the private standards of the organic producers’ association ‘Bio Austria’. In countries where no such ‘inputs list’ is available, certifiers have internal tools (lists, databases etc.) to determine compliance of products.

### 4.4.3 Authorization Criteria in the EU

The criteria for authorization of new substances are laid down in Article 16 of Reg. 834/2007. They are as follows:

- The authorization is subject to the objectives and principles of organic farming, which are laid down in Articles 3 to 5 of Reg. 834/2007. These include the health of soil, water, plants and animals, high levels of biodiversity, responsible use of energy and natural resources, animal welfare and exclusion of GMOs.

- The products are necessary for sustained production and essential for their intended use (i.e. biological, physical or breeding alternatives or other effective management practices are not available).
- The substances shall be of plant, animal, microbial or mineral origin (under specific circumstances, exceptions are possible, if the conditions for use of the product preclude any direct contact with the edible parts of the crop).

### *4.4.4 Authorization Outside the EU*

Formally, the USA have a different approach to the authorization of new substances: 'synthetic' products are generally prohibited, while 'non-synthetic' products are generally authorized. There are also some exceptions of authorized synthetic substances and of unauthorized non-synthetic substances. Despite this formal difference, the list of authorized substances is remarkably similar to that described above for Europe. For details on authorization in the USA, see (Baker 2004; OMRI 2007).

The Codex guidelines contain criteria for authorization of new substances, as well as reference lists of authorized substances, which are not binding for members, but provide advice on internationally agreed inputs. Requests for authorization of a new substance have to be accompanied by a documentation which shows that the substance fulfils the criteria for inclusion. Requests are discussed by the members and ultimately decided by voting.

In their current form, the IFOAM Basic Standards show the same pattern as the Codex guidelines. However, they are under revision and it is foreseen that in the future, they will only contain criteria for authorization of new substances, but no reference lists.

## 4.5 Organic Plant Protection in Practice: Regulation and Other Determinants

Although organic farming is governed by one regulation throughout the EU, the practices of plant protection differ significantly from one country to another. This is due to a complex interaction between organic legislation, national private standards, general pesticide legislation, commercial activities with respect to plant protection products and regional farming traditions. This chapter illustrates selected aspects of this pattern. The discussion is restricted to Europe. The two case studies comprise one horticultural crop and one arable crop, as well as one pest and one fungal disease.

### *4.5.1 Case Study Codling Moth*

The codling moth (*Cydia pomonella*) is one of the most severe pests of apples. In organic farming, it is mainly controlled by (i) mating disruption, (ii) *Bacillus thuringiensis*, and (iii) *Cydia pomonella* granulosis virus (CpGV).

Mating disruption with the specific pheromone (Zuber 1999; Brunner et al. 2002) is widely used in Spain, France, Switzerland, Croatia and Hungary, and to a lesser extent also in the Netherlands and Sweden (Casado et al. 2008). This technique is only efficient on large surfaces, and if less than 2% of the apples were attacked in the previous year (Speiser et al. 2006b).

The *Cydia pomonella* granulosis virus (CpGV) is another control method used in organic apple orchards. CpGV is most efficient against young larval stages, and must therefore be repeatedly applied during the entire flight period of the codling moth (Speiser et al. 2006b). CpGV was the first granulosis virus worldwide which was registered for food production (Andermatt 2008). It was first registered in Switzerland in the year of 1987. Because this technique is effective also in small orchards, it is frequently used in German and Swiss organic apple orchards. Recently, *C. pomonella* has become less susceptible to CpGV. Since 2004, ca 35 populations (mostly in Germany, but also in France, Italy, Switzerland and The Netherlands) of *C. pomonella* have been shown to be resistant against the first commercialized strain of CpGV. In October 2007, a new strain of CpGV was registered in Switzerland, against which *C. pomonella* is not resistant (Zingg and Kessler 2008). Recently, The European Commission has proposed to list baculoviruses on species level in Annex I of Dir. 91/414, and to add new isolates to a separate list (COM 2008). This procedure would ensure that new strains of baculoviruses can be commercialized rapidly. This is necessary, if the target pests become resistant. Differences between countries are mainly due to the presence or absence of a registered product. Further, CpGV treatment is rather costly. It is therefore only applied in situations where mating disruption is not effective (small orchards, high pest pressure).

Insecticides such as spinosad or pyrethrins can also be used for the control of *C. pomonella*. Until summer 2008, spinosad was not authorized in EU organic farming (EC 2008a). Therefore, it has not yet become part of organic farming practices, and we do not yet know to what extent it will be used for the control of *C. pomonella*. However, spinosad may play an important role in the future in the context of a strategy to manage resistance of the codlling moth against CpGV. In Switzerland, spinosad has been used successfully in combination with CpGV in spraying programmes. Pyrethrins are not widely used due to their negative impact on arthropods.

### 4.5.2 Case Study Potato Late Blight

Late blight (*Phytophthora infestans*) is one of the most important diseases of potatoes – organic and conventional alike. An in-depth inventory of the late blight situation in organic farming in seven European countries is given by Tamm et al. (2004). Organic farmers use a variety of management practices to avoid the late blight epidemic, e.g. early planting and pre-sprouting. Within the constraints of the market, they also avoid susceptible varieties (Speiser et al. 2006a).

For direct control of late blight, copper fungicides are the only plant protection products authorized in organic farming (Speiser et al. 2006a). Copper hydroxide, copper oxychloride, (tribasic) copper sulphate and cuprous oxide have traditionally been used in organic farming. Great efforts have been made to reduce the usage of copper, but a complete elimination is unrealistic at the moment. In 2002, a quantitative limit was set in the EU organic regulation (EC 2002). The amount of pure copper to be applied was restricted to 8 kg/ha per year, with a progressive reduction to 6 kg/ha until 2006. However, some countries have lower limits. In Scandinavian countries and The Netherlands, copper fungicides are not at all registered. In Switzerland, the maximum dose registered is 4 kg/ha, and in Germany 3 kg/ha. In addition, some private labels further restrict or prohibit the use of copper fungicides (e.g. Bioland Germany, Demeter, Bio Austria). Recently, copper octanoate has also been authorized for organic farming (EC 2008a), because the label rates for copper octanoate correspond to much lower amounts of pure copper than in other copper compounds (EC 2008a). Copper compounds were subject to the 3rd stage of re-evaluation of pesticides (EC 2009b).

### *4.5.3 Production Bottlenecks and Challenges for the Future*

Organic production methods, including the products allowed for plant protection, have evolved in practice over several decades before the lists of authorized substances were laid down in a legislative process. They therefore provide at least partial solutions for most production problems. Nevertheless, organic farming is in continuous evolution, and some new developments require progress in the range of authorized pesticides (Tamm 2000).

- In many parts of the world, field sizes tend to become larger and farms tend to specialize in a decreasing number of crops. This reduces crop diversity in the field and thus increases the pressure from pests and diseases. It also increases the farmers' dependency on each crop.
- The trade's requirements with respect to external quality are continuously elevated. As a result, cosmetic damage caused by pests and diseases becomes increasingly important from an economic point of view.
- With the requirement to use organic seed and planting material (which cannot be treated with synthetic fungicides), seed-borne diseases are likely to gain importance in the future.
- No-till farming systems have some ecological advantages, but are currently very difficult to implement with organic farming practices. It might be a challenge for the future to develop no-till organic farming practices.
- It is a declared aim of organic farming to reduce the use of controversial substances. In the EU for example, nicotine and metaldehyde were withdrawn, and copper fungicides restricted in quantity. In most cases, it will be necessary to find a replacement, before a substance can be withdrawn. For example, metaldehyde was withdrawn at the same time as iron phosphate molluscicides were authorized.
- Global climate change is likely to pose new challenges in plant protection. For example, the geographic distribution of host plants and pests will change, and earlier spring activity has been observed in some insects. Climate extremes may promote the outbreak of plant diseases and pest attacks (Easterling et al. 2007).

## References

Andermatt M (2008) 20 years for MADEX – 20 years for Andermatt biocontrol AG. AG J 2008:2–3

Baker B (2004) Plant protection products in organic farming in the USA. Current evaluation procedures for plant protection products used in organic agriculture. In: Proceedings of a workshop held on 25–26 September, 2003 in Frick, Switzerland. B. Speiser and O. Schmid. Frick, Switzerland, FiBL, pp 66–72

Blake F (2004) IFOAM policies concerning inputs evaluation. Current evaluation procedures for plant protection products used in organic agriculture. In: Proceedings of o workshop held on 25–26 September, 2003 in Frick, Switzerland. B. Speiser and O. Schmid. Frick, Switzerland, FiBL, pp 74–79

Brunner J, Welter S, Calkins C, Hilton R, Beers E, Dunley J, Unruh T, Knight A, Van Steenwyk R, Van Buskirk P (2002) Mating disruption of codling moth: a perspective from Western United States. IOBC wprs Bull 25:11–19

Casado D, Avilla J, Patocchi A, Samietz J, Paaske K, Lavigne C, Sauphanor B, Heijne B (2008) State of the art of control strategies for codling moth, apple scab and brown spot in Europe. In: ENDURE International Conference 2008 'Diversifying Crop Protection'. La Grande Motte, France, http://www.endure-network.eu/international_conference_2008/abstracts_online/oral/o_35_state_of_the_art_of_control_strategies_for_codling_moth_apple_scab_and_brown_spot_in_europe

Codex Alimentarius Commission (1999) Guidelines for the production, processing, labelling and marketing of organically produced foods. GL 32-1999. Rev 4 – 2007. Rome

COM (2004) European Action Plan for Organic Food and Farming. document COM (2004) 415 final. http://ec.europa.eu/agriculture/organic/eu-policy/actin-plan_en (accessed on 17 Nov 2010)

COM (2008) Guidance Document on the assessment of new isolates of baculovirus species already included in Annex I of Council Directive 91/414/EEC, European Commission, Health and Consumer Protection Directorate-General. SANCO/0253/2008 rev. 2

Easterling WE, Aggarwal PK, Batima P, Brander KM, Erda L, Howden SM, Kirilenko A, Morton J, Soussana J-F, Schmidhuber J, Tubiello FN (2007) Food, fibre and forest products. Climate change 2007: impacts, adaptation and vulnerability. Contribution of Working Group II to the Fourth Assessment Report of the Intergovernmental Panel on Climate Change. M. L. Parry, O. F. Canziani, J. P. Palutikof, P. J. van der Linden and C. E. Hanson. Cambridge, UK, Cambridge University Press, pp 273–313. http://www.ipcc.ch/ipccreports/ar4-wg2.htm

EC (1991) Council regulation No 2092/91 of June 24 1991 on organic production of agricultural products and indications referring thereto on agricultural products and foodstuffs. Off J Eur Commun L 198:22/07, 1991:0001–0015

EC (2002) Commission Regulation (EC) No 473/2002 of 15 March 2002, amending Annexes I, II and VI to Council Regulation (EEC) No 2092/91 on organic production of agricultural products and indications referring thereto on agricultural products and foodstuffs, and laying down the detailed rules as regards the transmission of information on the use of copper compounds. Off J Eur Commun L 75/21 (16.3.2002):21–24

EC (2007) Council regulation (EC) No 834/2007 of 28 June 2007 on organic production and labelling of organic products and repealing Regulation (EEC) No 2092/91. Off J Eur Commun L 189/1 (20.7.2007):1–23

EC (2008a) Commission regulation (EC) No 404/2008 of 6 May 2008 amending Annex II to Council Regulation (EEC) No 2092/91 on organic production of agricultural products as concerns the authorization of spinosad, potassium bicarbonate and copper octanoate and the use of ethylene. Off J Eur Commun L 120 (7.5.2008):8–10

EC (2008b) Commission regulation (EC) No 889/2008 of 5 September 2008, laying down detailed rules for the implementation of Council Regulation (EC) No 834/2007 on organic production and labelling of organic products with regard to organic production, labelling and control. Off J Eur Commun L 250/1 (18.9.2008):1–84

EC (2009a) Commission Decision of 3 June 2009 establishing the expert group for technical advice on organic production (2009/427/EC). Off J Eur Union L139:29–31
EC (2009b) Commission Directive 2009/37/EC of 23 April 2009 amending Council Directive 91/414/EEC to include chlormequat, copper compounds, propaquizafop, quizalofop-P, teflubenzuron and zeta-cypermethrin as active substances. Off J Eur Union L 104:23–32
Hozzank A (2010) Betriebsmittel und Verarbeitungsrichtlinien 2010. Enzersfeld, InfoXgen
Huber B, Schmid O, Kilcher L (2008) Standards and regulations. The world of organic agriculture. Statistics & emerging trends 2008. H. Willer, M. Yussefi-Menzler and N. Sorensen. Bonn, IFOAM & FiBL, pp 59–70
IFOAM (2002) Basic standards for organic production and processing. IFOAM, Tholey-Theley, Germany
IFOAM (2006) IFOAM norms for organic production and processing, IFOAM
Lampkin N (1990) Organic farming. Farming Press, Ypswich
Mäder R, Liebl B, Wilbois K (2010) Betriebsmittelliste 2010. Frankfurt am Main, FiBL Deutschland e.V
Mikkelsen C, Schlüter M (2009) The new EU regulation for organic food and farming: (EC) No 834/2007. IFOAM, Brussels
OMRI (2007) OMRI standards manual. Organic Materials Review Institute, Eugene, OR
Sahota A (2008) The global market for organic food & drink. In: Willer H, Yussefi-Menzler M, Sorensen N (ed) The world of organic agriculture. Statistics & emerging trends 2008. IFOAM & FiBL, Bonn, pp 53–57
Schmidt H, Haccius M (2008) EG-Verordnung "Ökologischer Landbau". hpslex publishing, Freiburg im Breisgau
Speiser B, Blake F, Micheloni C, Stopes C, Andersen M, Baker B, Bywater S, Canali S, Eriksen RØ, Gonzalvez V, Hozzank A, Jespersen LM, Jørgensen L, Kelderer M, Kleeberg H, Koopmans C, Kromp B, Manstretta M, Maurer da Costa A, Monnier MC, Pauler A, Schmid O, Schreyer A, Slabe A, Tamm L, Trapman M, von Fragstein P (2005) Evaluating inputs for organic farming – a new system. FiBL, Frick
Speiser B, Tamm L, Amsler T, Lambiom J, Bertrand C, Hermansen A, Ruissen MA, Haaland P, Zarb J, Santos J, Shotton P, Wilcockson SJ, Juntharathep P, Ghorbani R, Leifert C (2006a) Improvement of late blight management in organic potato production systems in Europe: field tests with more resistant potato varieties and copper based fungicides. Biol Agric Horticul 23(4):393–412
Speiser B, Tamm L, Berner A, Maurer V, Chevillat V, Schneider C, Walkenhorst M (2010) Betriebsmittelliste 2010. Frick, FiBL
Speiser B, Wyss E, Maurer V (2006b) Biological control in organic production: first choice or last option?. In: Eilenberg J, Hokkanen H (eds) An ecological and societal approach to biological control. Springer, Dordrecht, pp 27–46
Tamm L (2000) The impact of pests and diseases in organic agriculture. BCPC Conference – Pests Dis 2000:159–165
Tamm L, Smit AB, Hospers M, Janssens SRM, Buurma JS, Mølgaard J-P, Lærke PE, Hansen HH, Hermans A, Bødker L, Bertrand C, Lambion J, Finckh MR, Schüler C, Lammerts van Bueren E, Ruissen T, Nielsen BJ, Solberg S, Speiser B, Wolfe MS, Phillips S, Wilcockson SJ, Leifert C (2004) Assessment of the socio-economic impact of late blight and state-of-the-art management in European organic potato production systems. Frick, Switzerland. www.orgprints.org/2936, FiBL
Vogt G (2000) Entstehung und Entwicklung des ökologischen Landbaus. Stiftung Ökologie & Landbau, Bad Dürkheim
Willer H (2008) Organic agriculture worldwide: current statistics. In: Willer H, Yussefi-Menzler M, Sorensen N (eds) The world of organic agriculture. Statistics & emerging trends 2008 IFOAM & FiBL, Bonn, pp 23–46
Zingg D, Kessler P (2008) Madex Plus und Madex I12 break virus resistance of codling moth. AG J 2008:10
Zuber M (1999) On the increase of mating disruption in arboriculture and viticulture in Switzerland, 1996–1998. IOBC wprs Bull 22:125–127

# Chapter 5
# Policy Aspects of Regulation

**Wyn Grant**

**Abstract** The phenomenon of regulation can best be understood through the concept of the regulatory state. This is an ideal typical formation that supplants the earlier laissez faire and Keynesian welfare states. Regulation as a mode of governance can have dysfunctional aspects, although some of these, e.g., capture theory, can be exaggerated. Current regulatory philosophy has emphasised a move away from 'command and control' regulation and one consequence is an imperative for the engagement of relevant stakeholders in the regulatory process. Policy network analysis is relevant here and it is important to pay attention to problems arising from incomplete or fragmented policy networks. Solution oriented forms of stakeholder engagement in relation to biological control agents in the UK and the Netherlands are discussed. Regulatory innovation is needed to facilitate the wider availability of biological control agents, but is not easy to achieve.

## Contents



W. Grant (✉)
Department of Politics and International Studies, University of Warwick, Coventry CV4 7AL, UK
e-mail: w.p.grant@warwick.ac.uk

R.-U. Ehlers (ed.), *Regulation of Biological Control Agents*,
DOI 10.1007/978-90-481-3664-3_5, 

## 5.1 The Regulatory State

From a political science and policy analysis perspective, regulation is best understood through the concept of the regulatory state. Although biological control agents present their own very particular regulatory challenges, they can best be understood within the context of a broader framework that seeks to understand the role of regulation in contemporary governance. There has been a shift away from 'top down', hierarchical forms of government to more horizontal forms of governance that involves negotiation between various actors. Transparency and stakeholder engagement are key components of this new mode of governing.

The concept of the 'regulatory state', as developed in the work of Majone (1996) and Moran (2003) is seen as a successor to earlier state formations in European countries. These are ideal typical models so regulation is not the only form that governing takes in the 'regulatory state', it just becomes more predominant than it was in the past. Originally the state took a 'laissez faire' or 'night watchman' form in which government performed the limited functions of the defence of persons and property.

The Keynesian welfare state or the 'command state' as it is sometimes called, emerged in its most developed form after the Second World War. The state took responsibility for seeking to manage the key macroeconomic aggregates (employment, inflation and output) and to a varying extent between countries was involved in the management of production itself. The state also provided a wide range of welfare services for its citizens from public health services to housing and state pensions. However, citizens were still seen very much as 'subjects', carrying over their status from earlier periods of monarchical or authoritarian rule. They were the (presumably grateful) receipts of services provided for them by a benevolent state.

### *5.1.1 Drivers of the Regulatory State*

The failure of Keynesian techniques of demand management, the pressures of economic internationalisation and the need to control public expenditure all led to the decline of the command state, although the welfare state remained largely intact. However, although in many respects the actual level of risks that citizens encountered declined, their awareness of risk tended to heighten. Analysts such as Beck and Giddens provided an account of modern industrial society that saw it creating 'a historically novel set of risks of the sort typified by those emanating from the nuclear power industry: these are collective in character in the sense that individuals cannot separately secure protection from them; they are unknowable; and their potential magnitude means that the harm they can inflict is catastrophic' (Moran 2002).

This heightened sense of a new and threatening set of risks led in turn led to demands for more regulation of products and processes, not least in the food chain. A Eurobarometer survey conducted in 2005 found that 'Worry is an important determinant of people's attitudes to food safety and, in particular, how they respond to information about food safety' (European Commission 2006). Forty percent of those interviewed believed that the food they eat would damage their health. An index of issues that respondents were worried about in relation to food gave the highest score to pesticide residues (European Commission 2006).

More generally, the widening and deepening of the regulatory state was driven by a number of scandals in areas such as the environment, financial services and health provision which were amplified by the media. Self-regulatory arrangements were widely discredited and there were demands for more stringent and comprehensive systems of state regulation, not least in relation to pesticides. This trend was reinforced by the development of the European Union (EU) which was predominantly a regulatory state as it had limited fiscal policy instruments at its disposal and much of the budget it did have available to it was taken up by the Common Agricultural Policy (CAP).

### 5.1.2 *Operation of the Regulatory State*

Although the precise picture differs from one member state to another, regulatory agencies generally enjoy substantial operational autonomy. In order to carry out their functions effectively, they have to develop specialist expertise and this gives them an advantage over more generalist civil servants. Governments may also wish to distance themselves as far as possible from potentially controversial topics such as pesticides, so that if anything does go wrong, the blame can be absorbed by the regulatory agency.

The operational staff of a pesticides agency requires relevant scientific expertise. The question then arises, what sort of scientific expertise is relevant? In practice, a range of disciplines and expertise will be required in order to properly carry out the approvals process. However, given that agencies were set up to register synthetic pesticides, they generally have understandably concentrated on developing a range of expertise that is relevant to chemical products. Thus, in 1997, before a number of subsequent improvements took place, Waage (1997) was able to write:

> [The] entire pesticide regulatory process... has not adapted itself to the new opportunities which biopesticides provide. In their emphasis on high efficacy standards typical of fast-acting potent chemical products, registration procedures make little allowance for new products whose effect is a combination of direct kill and the conservation of natural enemies.

## 5.2 Dysfunctional Aspects of Regulation

The regulatory state is characterised by a continual tension between the regulators and the regulated. A recurring theme in the literature has been the ability of the regulated to outwit the regulators, partly as a consequence of asymmetries of information. This is why it is important for regulators to develop the relevant knowledge base and expertise to undertake the regulatory task. Some of the literature has, however, over emphasised the danger of regulators being 'captured' by the regulated. 'Capture theory' has American origins and arose in an environment where there was a particular relationship between the regulatory agency, the relevant Congressional committee (that approved its budget) and related interests (who often funded the election campaigns of the legislators).

There has been an increasing recognition that 'command and control' used as the principal mechanism of regulation can incur high transaction costs and can be difficult to enforce. Hence, there has been growing interest in alternative policy instruments, particularly those that make use of the price mechanism through taxes or various forms of pollution or emission trading. Pesticide taxes are used as policy instruments in Denmark and Norway, but remain controversial. What is generally agreed is that there needs to be more stakeholder involvement in policy design and implementation.

### 5.2.1 The Creation of Barriers to Market Entry

Regulation can create substantial barriers to entry to an industry or market which can serve the interests of existing market participants. A classic example would be the airline industry where substantial regulation protected the interests of 'flag' or 'legacy' airlines. Once regulation was removed or reduced, new budget airlines were able to enter the market and reduce the cost of travel for consumers. Regulation can have the effect of raising mark ups and reducing innovation, investment and productivity growth. Allocative efficiency losses seem to be generally greater than compliance costs, although it is the latter that are often emphasised in political discussions of regulation.

Societal and political processes produce systems of regulation that may have the unintended effect of benefiting existing players in the market and their modes of production. Thus, synthetic chemical producers did not create the system of pesticides regulation which was driven by concerns about impacts on the environment, consumers and workers handling pesticides. Nevertheless, it created a barrier for manufacturers of biological control agents because of the prevalence of a chemical pesticides model. This was exacerbated by the fact that biological controls were generally produced by small enterprises who could not afford the regulatory affairs divisions maintained by large multinational chemical companies. Hence, the orientation and complexity of the regulatory system acted as a barrier to the wider availability of biological controls because the registration process was perceived to be too difficult and expensive. One consequence was the marketing of products outside the regulatory system as 'plant strengtheners' or 'leaf conditioners'.

Pesticides are toxic substances and there needs to be a through and informed evaluation of their impact. The issue is therefore not about removing or reducing regulation, but reconfiguring it so that it enables the benefits of more environmentally friendly products to be realised. This involves a learning process for the regulators and a willingness to engage effectively with a range of stakeholders.

It is here that understandings of governance and policy network analysis become important.

## 5.3 Governance

Governance has been a key organising concept in political science in the last decade. Rhodes has defined governance in terms of 'self-organizing, inter-organisational networks' (Rhodes 2000). In his view it has four key characteristics. First, 'Interdependence between organisations. Governance is broader than government, covering non-state actors.' An underlying theme here is the enhanced importance of various non-governmental organisations, given that traditional representative institutions are seen not have to capacity to represent. Second, there are 'Continuing interactions between network members, caused by the need to exchange resources and negotiate shared purposes.' In other words, governance requires the existence of policy networks that operate effectively. Third, there are 'Game-like interactions, rooted in trust and regulated by rules of the game negotiated and agreed by network participants.' Fourth, there is 'A significant degree of autonomy from the state. Networks are not accountable to the state: they are self-organizing' (Rhodes 2000).

### *5.3.1 Policy Networks*

As the Rhodes formulation makes clear, policy networks facilitate negotiation and the development of shared understandings among participants. In order to function properly, policy networks must be constitutive of all relevant stakeholders. 'Networks can be characterised in terms of their internal modes of governance, and their modes of coping with a wide range of problems integral to the network process. Of particular importance is the question of coordination – the problem of how an array of often very different organisations and individuals with clearly divergent strategic intentions and motivations, criteria of success and failure, time-horizons and strategic resources at their disposal can be cemented and drawn together around' (Hay 1998).

#### 5.3.1.1 Policy Networks in Biological Control

Biological control agents are typically characterised by relatively incomplete or fragmented policy networks. This reflects the relative youth of the policy arena and the lack of resources possessed by some of the participants. In broad terms one can identify the following potential participants in a biological control policy network at member state level (for a discussion of the EU level, see below):

1. The regulatory agency (which may often form the hub of the policy network)
2. The growers (and their representative organisations)
3. The biocontrol manufacturers (and their representative organisation)
4. Consultants (who can be important intermediaries)

5. Environmental non-governmental organisations
6. Retailers
7. Consumer organisations
8. Academic researchers

The relevant national government department is excluded from this list as under a governance arrangement, its role should be one of 'steering'. This task may be imperfectly performed, but in any event it is expected to be softer, less intrusive and less hierarchical than under traditional systems of government.

#### 5.3.1.2 Network Components

Let's consider each of the components in the self-steering policy network. The national regulatory agency can play a key role in both creating and sustaining a policy network. Consider, for example, the case of the Pesticides Safety Directorate (PSD) in Britain (Now re-named as The Chemicals Regulation Directorate, CRD) which has devoted considerable resources to stakeholder engagement. It has, for example, set up regular joint liaison arrangements with the International Biocontrol Manufacturers Association (IBMA). Its Availability Action Plan Implementation Group comprises a range of stakeholders. Of course, one cannot assume that the national regulatory authority either thinks it is appropriate or is able to take on this role and even in the UK case, the network is incomplete.

The growers are a key group because their ability to grow the crops depends on the availability of appropriate plant protection products. National farmers' unions are usually well resourced and effective organisations, but in some countries, the horticulture sector may not be well represented in such organisations or may have fragmented representation organised on a crop basis.

Biocontrol manufacturers are afflicted by the fact that they are generally small scale firms who can spare very limited resources of money or time to support collective organisation. Nevertheless, the IBMA has become an increasingly effective organisation and its annual conference in Switzerland has become a 'one stop shop' for networking in relation to biological control issues. However, the IBMA would itself admit that its technical knowledge has not always been matched by political sophistication.

In a loosely coupled network, consultants can play a key role as facilitators and intermediaries that can assist the integration of the network and its effective functioning. They invariably have a relevant science background and in some cases have worked in the national regulatory agency. They are, however, sometimes regarded with a measure of suspicion by national regulatory agencies who think that they charge their clients for information that is freely available or which they obtain themselves from the regulatory agency.

Environmental non-governmental organisations generally have a wider remit than pesticides, the exception being the Pesticides Action Network (PAN). Although environmental non-governmental organisations have generally been critical of pesticides and called for greater restrictions on their manufacture and use, they have not generally been very supportive of biological control agents. This may in part be because of a suspicion that they are 'still pesticides' and because of incomplete network links between the environmental organisations and bodies like IBMA.

Retailers can have an important influence on the use of biological control agents, particularly in countries like the UK where they have substantial market power. They then use this to impose restrictions on pesticide use which go beyond regulatory requirements. However, this restriction of the use of synthetics has not been generally matched by an active promotion of biological control agents. This has started to change recently in the UK where retailers such as Marks and Spencer and Sainsbury's have started to promote the use of biologicals to their growers. In 2008 Sainsbury's held a conference to discuss advances in the use of biopesticides with their suppliers.

Nevertheless, retailers are not effectively integrated into the policy network in the UK where they are well placed to participate. It was evident from our interviews that the links between retailers and PSD were relatively weakly developed. Often they were confined to attendance at open meetings of the Advisory Committee on Pesticides (ACP) or perhaps membership of the Pesticides Forum. One retailer commented, 'Only interact with PSD if they want specific information from us.'

Retailers see themselves as proxies for the consumer and consumer organisations themselves are not generally all that involved in the discussion of biological control agents. Consumers are generally information takers rather than policy makers. One of the problems here is that while consumers generally have a relatively clear if sometimes ill-informed image of organic produce, they have relatively little understanding of the potential contribution of biocontrol agents to a more environmentally sustainable agriculture. One retailer referred in interview to the relative ignorance of consumers about organic produce: 'You get some daft responses. If you ask them about organics, they say no pesticides are applied, when you explain there are pesticides applied, they get very upset.'

Academic researchers can play a key role in linking various participants in a more effective network. Investments made in work on biological control agents in the UK by the Rural Economy and Land Use (RELU) programme have facilitated information exchange, a heightening of awareness of the contribution of biopesticides and workshops at which different network participants can work together. Whilst academics must be careful not to cross the dividing line between being analysts and advocates, they should be able to win the trust of participants whose interests or perspectives do not always coincide and hence facilitate constructive dialogue and the identification of policy solutions.

### *5.3.2 Policy Networks at the EU Level*

Pesticide regulation offers a characteristic case of multi-level governance with a complex and changing division of responsibilities between the EU and member states.

It also fulfils the criteria for multi-level governance in the sense that there is a considerable horizontal as well as vertical dimension to policy formation and implementation.

However, if policy networks are incomplete or fragmented at the member state level, this is even more evident at the EU level. What is immediately apparent is the absence of any functional equivalent at the EU level of the national regulatory authorities which serve as a hub around which a policy network can cluster, for example by providing a location for meetings. They also have authority resources so the chance of influencing the way in which those resources are used creates an incentive for meeting them. Farmers' organisations, the IBMA and environmental organisations are all present in Brussels, but the greater complexity of the decision-making process resulting from co-decision makes it even more difficult to focus representative efforts efficiently than at a national level.

In a sense REBECA itself filled a significant vacuum by creating a neutral yet informed policy space in which various actors could interact. The attendances at REBECA conferences showed the considerable level of interest in the subject of biological control agents, but also the relatively lack of opportunities to interact on a systematic basis. Regulators can, of course, meet in the Biopesticides Steering Group of the OECD or in various EU level committees. Informal, bilateral links between regulators are also continually developing. The IBMA annual conference provides one meeting point. However, with the end of REBECA, there is no general umbrella framework that can facilitate the discussion of issues at a European level among a wide range of actors. One consequence has been that debates in the European Parliament have not always been as well informed as one might wish.

## 5.4 Solution Oriented Stakeholder Engagement

There is an increasing recognition that contributions to environmental sustainability such as biological control agents cannot be advanced by 'top down' arrangements but require new cooperative arrangements that engage a range of stakeholders. Such arrangements are likely to enjoy great legitimacy among stakeholders and hence become more effective when it comes to implementation. Fortunately, two working examples of such arrangements exist in member states. Although the two schemes are somewhat different in terms of the way in which they operate, their fundamental objective is the same: to facilitate more registrations of biological control agents.

### *5.4.1 The UK Biopesticides Scheme*

One of the disadvantages of policy networks is that they tend to foster only relatively incremental forms of change and some kind of exogenous shock is often required to bring about change. In the case of the UK, this came about through the Better Regulation Executive in the Cabinet Office encouraging the PSD to think about ways in which the registration rate of biopesticides could be improved. This led to the introduction of the Pilot Project in 2003 which was converted into a permanent Biopesticide Scheme in 2006.

The Biopesticide Scheme had three key elements. Substantially reduced fees are charged for biological control agents and these lower fees have been maintained while charges for the registration of synthetics have increased. Applicants were invited to pre-submission meetings with PSD staff. These provided an opportunity to identify gaps in the application dossier, leading to the identification of ways in which these gaps might be filled, for example by the use of published data. As part of our research we observed a number of these meetings and it was evident that they were highly informative for both applicants and PSD staff and enhanced mutual understanding. Finally, the Scheme involved the appointment of a 'Biopesticides Champion' within PSD to look after the needs of applicants.

One of the consequences of the Scheme has been the development of an informal internal network of staff within PSD with interest and expertise in issues related to biological control. PSD staff members are used to team working as they work in groups on approval processes and the agency has a relatively horizontal structure and informal working style in which relationships are based on collegiality and mutual respect. They have received training to help them to develop their skills and have been very receptive to this career development opportunity. The Scheme has not overcome all problems, as the level of applications is still relatively low. This reflects in part the fragmented character of the policy network and the fact that not all developers are IBMA members.

### *5.4.2 The Genoeg Scheme in the Netherlands*

Genoeg is the acronym for 'Gewasbeschermingsmiddelen van Natuurlikje Oorsprong Effictief Gebruiken' which can be translated as using plant protection products of natural origin more effectively, or, more colloquially, as the effective use of natural pesticides. Exploratory work began in 2001 and the first phase of the scheme ran from 2003 to 2005 with a second phase from 2004 to 2007. Registration fees and some extra studies for applicants were funded up to a level that was not allowed to exceed 50% of registration costs and to a maximum of €100,000.

One of the striking aspects of this project is the way in which policy networks have been utilised to facilitate consensus and coalition building. In part this was a consequence of the importance of the protected crops sector in the Netherlands and the pre-existence of policy networks which could be used to promote biological control agents. Hence, a less 'top down' stimulus was necessary than in the UK. The project was also facilitated by its management by a consultancy called the Centre for Agriculture and Environment (CLM) which has a long standing track record in sustainable agriculture. Hence, it was possible to obtain political and financial support from the Ministry of Agriculture, as well as the active involvement of producer organisations in the form of the Dutch Organization for Agriculture and Horticulture (LTO Glastuinbouw) and the Product Board for Horticulture (Productschap Tuinbouw).

The Board for the Authorisation of Pesticides (CTB) has been actively involved in the project as part of a broader strategy of trying to promote low risk profile pesticides. Within CTB what was described in interview as a 'great team open minded to new approaches' was built in a process similar to that of PSD. Apart from the team leader, there were five team members with two concerned with residues; one concerned with environment; one concerned with characterisation; and one responsible for links with the EU, OECD and REBECA. An important element of the whole process has been a Genoeg meeting two or three times a year with stakeholders, including organic farmers.

## 5.5 Regulatory Innovation

Innovation remains a central challenge for systems of regulation. Although growing, the number of biological control agents registered in the EU and available on the market is still behind the figure for the United States. Biological control agents have the potential to make a substantial contribution to environmental sustainability and to the rural economy. Although improvements have been made in the regulatory process, much remains to be done both in some member states and at the EU level.

As Greaves (2009) notes, 'Bureaucrats and regulators are typically risk averse. The desire to avoid things going wrong means they are not natural innovators. Risk averseness does not create an encouraging environment for regulatory innovation (indeed, the term is almost a contradiction).' Greaves develops a model in relation to biopesticides which illustrates the interaction of endogenous and exogenous factors in regulatory innovation. These can come together to create a window of opportunity in which regulatory innovation can occur.

It has been observed that 'The regulatory state is becoming a risk management state' (Power 2004). Pesticides regulation is a clear example of this trend. There is 'an increasing emphasis on communication with different publics as a basis for managing reputation' (Power 2004) and this might be regarded as a positive trend. However, 'the risk management of everything is characterised by the growth of risk management strategies that displace valuable – but vulnerable – professional judgement in favour of defendable process' (Power 2004). In a regulatory bureaucracy like any bureaucracy process may displace goals. That is why it is important to continue to focus on the goal of environmental sustainability to which the wider availability and use of biological control agents can contribute. If environmental sustainability is emphasised, then hopefully regulatory sustainability should follow, provided that fragmented and incomplete policy networks undergo continuing development.

## 5.6 Note

This chapter draws on research funded by the UK Research Councils as part of the Rural Economy and Land Use (RELU) programme, RES 224-25-008.

## References

European Commission (2006) Special Eurobarometer 238: risk issues. European Commission, Brussels

Greaves J (2009) Biopesticides, regulatory innovation and the regulatory state. Pub Policy Admin 24:245–264

Hay C (1998) The tangled webs we weave: the discourse, strategy and practice of networking. In: Marsh D (ed) Comparing policy networks, Open University Press, Buckingham, pp 33–51

Majone G (1996) Regulating Europe. Routledge, London

Moran M (2002) Review article: understanding the regulatory state. Br J Pol Sci 32:39–413

Moran M (2003) The British regulatory state: high modernism and hyper innovation. Oxford University Press, Oxford

Power M (2004) The risk management of everything. Demos, London

Rhodes R (2000) The governance narrative: key findings and lessons from the ESRC's Whitehall programme. Public Management and Policy Association, London

Waage JK (1997) Biopesticides at the crossroads: IPM products or chemical clones?. In: Evans H (ed) Microbial insecticides. Novelty or necessity?, British Crop Protection Council, Farnham, pp 11–19

# Chapter 6
# Cost-Benefit, Risk and Trade-Off Analysis of Regulation

**Ralf-Udo Ehlers**

**Abstract** Several reasons have made registration of biological control agents (BCAs) in Europe a time-consuming and costly effort, resulting in these environmentally friendly products being kept off the market. To be cost effective, regulatory states must become cost-benefit states. To avoid over-regulation with all its negative consequences, government regulation should ask whether the benefits of regulation justify the costs of regulation. Such an analysis includes a cost-benefit-analysis, which is supported by a risk-trade-off-analysis. A survey among biocontrol companies indicated that average costs for registration are 1.9 million €. Benefits of regulation are put relative to costs for the different groups of BCAs. Trade-off effects of regulation on biocontrol companies, plant protection practice, consumer safety and the environment are described. Consequences resulting from the joint regulation of BCAs with chemical compounds and the organisation of registration in Europe are specified and proposals for overcoming problems are presented. Contradictions between the objectives of the European agriculture policy and the limited support of biological control are discussed.

## Contents



R.-U. Ehlers (✉)
Department for Biotechnology and Biological Control, Institute for Phytopathology, Christian-Albrechts-University Kiel, 24118 Kiel, Germany
e-mail: ehlers@biotec.uni-kiel.de

R.-U. Ehlers (ed.), *Regulation of Biological Control Agents*,
DOI 10.1007/978-90-481-3664-3_6, 

## 6.1 The Risk Society and Regulation of Biological Control

Our modern societies in industrialized countries are risk societies. Risks are well publicised, benefits often ignored. Any activity of mankind causes benefits or hazards to human beings or the environment. Of course this applies also to biological control agents. But the perception of risks and benefits plays a major role in political decisions on risk management.

The general public considers biological control to be an environmentally safe method of plant protection. One symbol of biological control is the ladybird beetle (*Coccinella septempunctata*), an effective antagonist of aphids. In some rare occasions these beetles bite without causing major pain or damage and consequently they are considered as safe. With the wide-spread presence of the exotic harlequin lady beetle *Harmonia axyridis* the perception of risks of biological control is changing (see Chapter 11) and the introduction of *H. axyridis* was the major motivation to include regulation of invertebrate biological control agents into the REBECA Action. Although benefits and damage caused by the beetle have not been assessed and the beetle is now found in the out-door environment, biocontrol companies refrain from marketing this insect because of its negative image. Compared with the invasive exotic maize pest Western Corn Rootworm (*Diabrotica virgifera virgifera*), which causes millions of Euros damage to agriculture, the harlequin beetle is publicised in the media to a much greater extent (Roy et al. 2006). The magnitude of risks and damage is not so important; more important to the media is the perception of risks and the negative story behind it.

When one explains to the general public that microbial biological control agents (MBCAs) like bacteria, fungi and viruses are used in biological control, the first reactions are concerns about public health as the general public recognizes these organisms as human pathogens. A particular problem is that people think that products or activities are either “safe” or “unsafe”. Risk perception depends on knowledge and less knowledge often results in exaggerating risks. Public perception of risks thus differs from the real magnitude of risks.

When damage is difficult to predict and risk assessment lacks scientific backup, the precautionary principle comes into play (see Chapter 1). One might think: Better safe than sorry. For example, if we do not release an exotic invertebrate BCA into the environment, it cannot cause any harm; but are we aware of the tradeoffs of regulation? Risks are on both sides of the equation and the precautionary principle therefore guards against one set of risks while ignoring the others. Regulation might keep older, riskier technology like pesticides in use. Hazards related to the use of chemical compounds by far outweigh the risks related to the use of biocontrol. Environmental damage caused by biological control is of much less magnitude than hazards associated with alternative control measures (chemical pesticides), particularly when damage to humans is taken into consideration. By over-regulating BCAs we allow potential risks to the environment from more risky technology. If, as a consequence of exaggerating regulation, chemical pesticides continue in the market, farmers, consumers and the environment can be harmed. Thus, we can conclude that regulation of BCA will produce a contradictory situation: while policy and society demand a reduction of chemical control, as laid down in the EU Common Agriculture Policy or Directive 2009/128/EC (EU 2009), over-regulation of BCAs can result in a more widespread use of chemical control measures or at least keep older and more damaging technology in the market.

The ideal situation is not a risk-free existence. Risks are part of our daily life. Everyone analyses risks many times during the day and decides which risks can be accepted and which need to be avoided. As with everything else, biological control can cause hazards. Hence it is generally accepted that an evaluation of the risks of BCAs is necessary and according to potential risks they should be subjected to some kind of registration process. The purpose of the authorisation is to develop a reasonable risk management in order to minimise potential risks. The question is whether the existing approaches (Dir. 91/414) are reasonable or whether the data requirements are exaggerated, not only in Europe, but around the world. If we exaggerate the risks of biological agents, and as a consequence governments implement major regulation on biological control, regulation can easily be a risk to the society as well.

We are dealing with a mismatch of the regulatory strategy and social goals if risks of biological control are exaggerated. This problem has produced bad policy. The existing laws have failed to protect health, safety and the environment. This was not because anybody exaggerated the risks of biological control agents but because the risk analysis for biological control agents was never done before implementing regulation of BCAs. Policy just did not differentiate between risks of chemical and biological agents when introducing the Dir. 91/414 legislation, for whatever reasons.

## 6.2 Methods for Assessing the Cost-Benefit Relation and Economic Efficacy of Regulation

The Commissions communication on the precautionary principle (European Commission 2000) demands "a comparison between the most likely positive and negative consequences of the envisaged action and those of inaction in terms of the overall costs to the Community." This might be the major point on which the current system has failed. To avoid over-regulation with all its negative consequences, governments should first ask whether the benefits of regulation justify the costs of regulation. Such an analysis includes a cost-benefit-analysis (CBA), which is supported by a risk-trade-off- analysis (RTA). Has the magnitude of risks related with

BCAs ever been assessed prior to introduction of regulation measures? No, because we lack precise comparative information.

To be cost effective, regulatory states must become cost-benefit states. To avoid over-regulation with all its negative consequences, government regulation is increasingly assessed by asking whether the benefits of regulation justify the costs of regulation. Three steps should be taken.

1. Cost-benefit-analysis (CBA) of regulation: such an assessment should try to see if there is a human health risk or estimate the potential environmental damage. Where scientific knowledge does not allow for specific estimates, ranges should be identified. The first step is to explore the costs of regulation, the second the benefits governments and the society gain from regulation. The probability of occurrence of the hazard will be of major value to analyse the magnitude of potential risks. The result of the CBA will answer the question: do the benefits of regulation justify costs of regulation?

   We often lack quantitative data on the costs in relation to the methods for risk assessment of biological control agents as in some cases (e.g. sensitisation by microbials) we have not even decided what methods to use to assess risks of BCAs. Often the data cannot easily be produced or benefits obtained by implementation of regulations cannot easily be quantified. Eventually this analysis will have to be performed, so governments can evaluate whether the benefit gained justifies costly regulations.
2. Risk-Trade-off-Analysis (RTA): Governments should attempt to assess trade-offs, also in quantitative terms if possible. Risk trade-off occurs when, in a portfolio of risks, a countervailing risk is generated by an intervention to reduce a target risk (Graham and Wiener 1995). Once trade-offs are identified in a quantitative or qualitative way, target risks and countervailing risks must be assessed and affected populations (e.g. farmers *vs* endangered species) be estimated. RTA can help to avoid the most serious risks on either side.
3. Cost effective analysis: Governments should attempt to use effective and inexpensive tools. If we take costly steps to address all risks, however improbable they are, we will quickly impoverish ourselves. The search for cheaper and more effective tools to achieve the basic goal is of major importance and might produce creative solutions for risk assessment.

These three principles are simple but also quite powerful. If they were taken seriously and implemented in the right way, they would have an extremely important effect on risk regulation, potentially saving money and damage. The analysis ensures that policy is driven by full appreciation of relevant risks and not by hysteria and alarm (Sunstein, 2002).

The REBECA Action was not able to perform a profound CBA and RTA of regulation of biological control agents. However, one task of the Action was first to identify potential risks of BCAs before proposing risk assessment strategies. Thus, catalogues of potential risks exist and have been summarised in this book in Chapters 7 and 8 on bacterial agents by Alabouvette and Cordier (2011) and Berg et al. (2011), in Chapter 9 on fungal BCAs by Strasser et al. (2011), on botanicals and semiochemical in Chapter 10 by Regnault-Roger (2011) and on invertebrate BCAs in Chapter 11 by de Clercq and Bale (2011). The chapters also summarise the benefits of these biological control agents; however, economic data on benefits are rare. The Workpackage 6 of the Action dealt with risk trade-off and cost-benefit analysis of regulation and was organised by Heikki Hokkanen and Ingeborg Menzler-Hokkanen (University of Helsinki, Finland).

## 6.3 Costs of Regulation

Not only costs but also time is of concern when authorisation of biological control agents is approached. A direct comparison of the time necessary to get the products registered by the EPA in the USA and by SANCO on Annex 1 in Europe was possible as almost identical dossiers had been used. *Paecilomyces fumosoroseus* (PreFeral) took 85 months in the EU and 60 months in the USA, *Coniothyrium minitans* (Contans) took 63 in the EU and 23 months in the USA, *Gliocladium catenulatum* (Prestop) 67 in the EU and 13 months in the USA. The list could be enlarged, but the results are the same. The EU system takes considerably more time to register BCAs. Regulation of macrobials is usually quicker but can also take between 1 and 2 years.

Hokkanen and Hokkanen-Menzler (2008) summarized the results of a survey among 21 biological control manufacturers in Europe. Member State (MS) registrations vary widely and range from a few months up to over 100 months, averaging around 24–36 months. Companies reported average costs for Annex I inclusion of approximately 1.9 million €. Of these costs, 21% was spent on efficacy tests, 43% on toxicology, 23% on ecotoxicology studies and 13% for other studies.

Several companies reported that they would bring more products to the market should conditions be less stringent, while several others stopped investment into development of new products that need registration. In many companies products not requiring registration have priority. Many enterprises replied that they will focus on other geographical regions for marketing their products because they judged the registration environment to be more favourable. Three companies shelved products due to costs and time required for registration, although they already had spent on average 0.2 million € in development of these products (Hokkanen and Hokkanen-Menzler 2008).

A company will compare costs with the market potential of a product. The economic potential of the biocontrol markets has increased significantly during the last decade but the overall economic potential of single products is small and often does not exceed 0.2 million per year. Due to the nature of BCAs, in particular their host specificity, the market potential is limited. Their introduction needs additional advice and, due to limited shelf life, distribution logistics need more economic input. In the past, *Bacillus thuringiensis* had an 80% market share in biological control (Lisansky and Coombs 1994). In 2000, 55% of the markets were products based on invertebrate BCAs (insects, mites and nematodes) and microbials had a 26% share.

Why did the IBCAs gain such an importance? The answer is quite simple. Most OECD countries have exempted IBCAs from registration. Revenues in the microbials market are severely restricted by requirements to register new products (Frost and Sullivan 2001).

Products based on microbials often lack cost effectiveness when registration costs come into the calculation. Comparing registration cost of chemical products estimated at 200 million €, costs for biological control agents appear minor. But we need to compare the turnover. The chemical insecticide imidacloprid (Bayer Crop Science) was the second largest pesticide in the market after Roundup, with a turnover, in 2003, of 590 million € (Agrow 2004). It is obvious that the costs for risk assessments and registration are commensurate with the value of such a market. But are they for any given BCA product?

## 6.4 Benefits of Regulation and Cost-Benefit Ratio

When discussing benefits, most people think of benefits of biological control agents but a cost-benefit analysis of regulation would have to evaluate benefits of regulation of BCAs. Such an approach must consider the potential hazard and damage related to the use of BCAs because regulation tries to predict potential damage and develop risk management strategies to avoid damage. The avoidance of damage, and the safety for humans and the environment are the benefits resulting from the regulation. The most stringent risk management decision would be to prohibit use. Conditional authorisation tries to reduce damage by restricting the use, an approach to minimise risks.

Direct benefits related with the use of BCAs can come into play as well, when potential risks for damage need to be weighed against the benefits of use.

It is desirable that decisions on regulation would be based on the cost-benefit ration. If costs for regulation and data acquisition exceed the benefits of regulation, the ratio is $> 0$ and governments should review the implementation of regulation. If benefits are higher than costs, potential risks and expected damage is high, the ratio will be $< 0$ and regulation is justified. This aspect could also be taken into consideration during risk assessment or even during pre-submission meetings of applicants and authorities in order to have a parameter to hand that can help review data requirements in the context of magnitude of potential economic risks and damage.

### *6.4.1 Macrobials*

Potential risks related to the use of macrobials are low. Major risks are to the environment, like establishment of an exotic species, competition or displacement of native species, effect on non-targets, and perturbation of ecosystem functions (see Chapter 11). The damage caused by the release of *H. axyridis* is hard to estimate, because its beneficial effect on populations of aphids might outweigh its negative effects. Analysing the data provided by the EPPO website (EPO 2010) on the use of invertebrate BCAs, currently 42% of the macrobials used in biological control in Europe are exotic species and it looks as if the benefit from introduction outweighs the damage, as no damage has been reported so far (except for *H. axyridis*). Costs to produce safety data for invertebrate BCAs can easily reach 0.1 million €, mainly for non-target testing (see Chapter 16), not taking into account the workload and costs related to authorisation in each MS. Consequently, a cost-benefit analysis of regulation of macrobials will result in a extremely high ratio, meaning that the costs far exceed the benefits and, consequently, governments should decide against regulation. However, this ratio is based on the BCAs currently in the market. The history of early biological control introductions provides evidence for severe damage, e.g., the introduction of *Bufo marinus* into Australia (Easteal 1981). Thus, some regulation targeted at the control and documentation of introductions, particularly but not exceptionally of non-specific antagonists, is necessary. Europe has been far behind Australia, New Zealand, the USA and Canada in definition of requirements for regulation and the organisation for introduction of exotic IBCAs. However, the system currently in force in Europe (that authorisation needs to be granted in each MS) is a waste of money and resources. Even if, as a consequence of the REBECA activities, harmonisation in data requirements is achieved, it is still difficult to understand why a risk assessment has to be performed in every MS, particularly as IBCAs will not stop their migration at national borders.

### *6.4.2 Microbials*

For microbials regulation is in place. Microbials with biocontrol potential can belong to species that have been identified as human pathogen. Details are reported in Chapter 8 (Berg et al. 2011). Of course, human pathogens must be exempted from use in biocontrol but modern molecular methods can distinguish between pathogenic and non-pathogenic species. Environmental risks caused by microbials are economically negligible. The major problems that still need to be solved are microbial metabolites as described in Chapter 9 (Strasser et al. 2011). Damage is difficult to estimate, but potential damage can be serious enough to justify registration, e.g. should toxic metabolites enter the food chain. All current scientific information available, also summarised in Chapter 13 (Strauch et al. 2011), indicates that metabolites could not be found in the food chain (see also results of RAFBCA EU Project QLK1-CT-2001-01391). Fundamental questions on safety need further research efforts and an applicant for a registration of a plant protection product cannot be made responsible for the lack of information or be compromised by the costs of the required investigations.

In many cases the potential damage can be estimated from long-term experience of safe use (e.g., *Bacillus thuringiensis*). Data requirement could be reduced; however, is a bureaucratic registration system flexible enough to be able to reduce registration costs in order to keep the ratio $< 0$? The solution to optimising the cost-benefit ratio lies in the flexibility of a regulation system, which is increasingly more cost-effective with increasing flexibility.

### *6.4.3 Semiochemicals*

This group of compounds has never been reported to cause any damage. The cost-benefit ratio is probably even higher than for the comparable group of macrobials. If a well balanced cost-benefit ratio is the basis for a positive decision on the introduction or continuation of regulation, then semiochemicals should not be regulated at all.

### *6.4.4 Botanicals*

Compounds originating from plant extracts are not necessarily safe. Some might be highly toxic and thus are not allowed to be used in plant protection, e.g., nicotine. The cost-benefit ratio is more balanced and similar standards like those used for synthetic chemical compounds need to be used. Again, long term experience of safe use can make a difference, e.g. garlic oil or other compound also used in human alimentation, and the cost-benefit ratio will be much better should the registration system allow and include such information for the risk assessment.

In order to avoid unnecessary over-regulation the REBECA consortium proposed to

- analyse costs and benefits prior to introduction of new regulation demands
- consider cost-benefit ratio during the registration/regulation process
- take into account trade-off effects of regulation
- minimise trade-off effects and maximise efficiency of regulation
- develop cost-effective procedures and accelerate the registration process

## 6.5 Trade-Off Analysis

A trade-off analysis examines the drawbacks of regulation in terms of consequences for the different stakeholders. Interests of the different stakeholders are described in Chapter 1 (Ehlers 2011). The trade-off analysis should involve a monetary analysis and a qualitative analysis in terms of worker and consumer health, environment, etc. The qualitative analysis is also necessary when monetary data cannot be assessed easily. REBECA WP6 organised a workshop on the topic with the objective of filing possible trade-off effects resulting from registration/regulation of BCAs. The REBECA Action did not have at its disposal the resources to perform a monetary analysis.

### *6.5.1 Trade-Offs Affecting Biocontrol Industry*

Costly and lengthy registration of BCAs restricts the market entry of products of the biocontrol industry. Considerable negative impacts can be expected for small- and medium-sized enterprises (SMEs), start-ups or spin-offs. Should a product not make it to the market within a few years after a company has been founded, the small-size enterprises experience problems of liquidity. Many companies have not even been started because their business plans would not withstand a stress test, particularly because the timescale for market access is impossible to plan when registration is necessary. Calculation of costs for registration is a black-box. It is difficult to anticipate what data will be requested and when authorisation will be granted. After Annex I inclusion the national registration further prolongs market introduction. A precise calculation of the return of investment is impossible, a situation that is not particularly attractive for venture capital or investors. Evidence for this trade-off effect is that many successful enterprises in the biocontrol market started selling invertebrate BCAs, which were not regulated or regulation requirements produced minor costs. Those European start-ups, which have microbial BCAs in the market, all have larger companies in the background or, despite the unattractive business chances, were able to attract venture capital. However, the latter ones are the vast minority.

Compared with production of the safety data, fees are of less monetary magnitude. However, the fees charged for national registration have often resulted in abandoning smaller markets. If one compares national authorisation catalogues for products based on BCAs in different MS, the trade-off effect is well documented.

Another problem for biocontrol companies is that regulation keeps innovation off the markets. Several biocontrol companies have very active research co-operations with public enterprises or run their own research and development departments. Results from these activities, however, are not transferred into practice. The substitution of products already registered by, for example, strains or species with increased control potential or better economic effectiveness, is impossible due to the enormous economic investment necessary to produce a new dossier. The same is true for just improving the formulation of a product based on a BCA. The national authorisation has to be renewed and the dossier needs to be supplemented with new efficacy data. The current legislation and fees thus keep scientific and technical innovation off the markets and make the biocontrol industry less effective and less competitive than industry outside Europe.

### *6.5.2 Trade-Off Effects on Plant Protection Practice and Consumer Safety*

The consequences for plant protection are clearly visible. Fewer products based on BCAs make it to the market and the agricultural sector is increasingly reliant on chemical control agents. Today, fewer and fewer chemical products make it to the market and the chemical industry is not registering their new products. BCAs could fill gaps, particularly in the minor use markets. These markets are small crops, like vegetables, fruit and ornamental, which currently depend on exceptional authorisation of chemical products under the minor use legislation. Over 2,000 of these authorisations exist in the German market and many of the authorisations would not be necessary if more BCAs were on the market. The minor use markets are an excellent business opportunity for biocontrol companies as they, rather than the chemical industry, are prepared to serve these "smaller" customers.

The limited number of pesticides causes problems with resistance. Farmers react to the development of pesticide resistance by increasing dosage and application frequency of pesticides. The immediate consequences are problems with residues. Pesticide residues in fruits and vegetables in Germany have increased during the past 10 years. In 2004 German authorities documented pesticides residue levels exceeding the Acute Reference Dose (ARfD) in 8% of vegetable and fruit samples from conventional production (BVL 2006 summarised by Reuter 2007). As a consequence, retailers have developed their own rules on Maximum Residue Levels (MRLs) they tolerate in agricultural produce, which in all cases fall below those prescribed by governments. Policy makers lag behind retailers in the development of consumer protection. Growers are increasingly aware of this problem as whole batches of agriculture produce were refused by retailers due to problems with pesticide residues. Growers now search for residue-free plant protection products. If there were more BCAs in the market, then growers would have more viable alternatives to meet the new standards required by retailers.

Keeping BCAs off the market thus results in a continuous high exposure of users and consumers to chemical compounds and fewer alternatives to avoid development of resistance. The consequences also influence the production of organic agriculture as their production costs would be lower and their economic efficacy would increase if they have better access to biocontrol innovation.

### 6.5.3 Macro-economic Effects

As fewer products based on BCAs reach the market, this has consequences also for the environmental effects caused by the use of chemical pesticides. This chapter is not meant to summarise negative effects of chemical pesticides but it is important to mention that the use of chemical pesticides produces external costs. They include costs caused by pesticide contamination in water, soil and air, environmental costs related to biodiversity and wildlife, human health costs mainly due to irregular use of pesticides but also related to treatment of chronic diseases, and costs accruing by regulation and monitoring of residues, among others. Assessments of the external costs of chemical pesticides have indicated that external costs sometimes greatly exceed the purchase value of pesticides. For example, in Germany over 50% of the estimated external costs arise from ground water contamination (Waibel et al. 1998).

Pimentel et al. (1992) analysed the environmental and socio-economic costs of pesticide use in the USA and calculate external costs amounting 8.3 billion US$ every year, exceeding the purchase value of all pesticides, which was about 6.5 billion US$ per year in those days. The highest cost from pesticide usage was calculated to arise from bird losses ($2.1 bn/a), followed by costs of groundwater contamination ($1.8 bn/a), costs of pesticide resistance ($1.4 bn/a) and public health impacts ($0.93 billion/a). These authors concluded that if it was possible to measure the full environmental and social costs of pesticide usage, the total cost would still be significantly greater than estimated.

Chemical pesticides have developed since these studies were conduced and now are supposed to be safer and are subject to more restrictive registration practices. Whether these costs can be verified under current conditions needs further investigations. However, the external costs related to the use of chemical pesticides cannot be concealed. These external costs could at least be reduced by introducing more biological control measures into agricultural practice and society would not only benefit from a cleaner environment and an increased biodiversity in agriculture ecosystems, but also from a reduction of external effects.

### 6.5.4 Trade-Off Effects from Joint Regulation of BCAs and Synthetic Compounds

Problems for BCAs can also result from policy decisions on chemical plant protection compounds. As BCAs are covered within the same legislation together with synthetic compounds, restrictions on the use of synthetic compounds automatically also apply for BCAs. These trade-off effects are often neglected by policy-makers. For instance, the decision to exclude sensitising compounds from the list of low risk products automatically excluded microbial BCAs from this list. No test systems exist to assess sensitisation and consequently products containing microbials are always labelled as sensitising. Thus, they were automatically excluded from the possibility of being grouped with the low risk products and get easier and longer authorisation.

Another example: during the discussions on the new regulation, SANCO proposed a zonal authorisation of PPP after Annex 1 inclusion (mutual recognition), which would reduce the necessary number of national authorisations and save costs and time. Some EU Parliament members had concerns about ground water pollution with chemicals and wanted national authorisation to be maintained. Ground water pollution does not apply to microbial BCAs. The stringent rules applied for synthetic compounds would have had negative consequences for the introduction of BCAs.

As long as the legislation for synthetic compounds is not strictly separated from the legislation from biologicals, biological control will often experience the same restrictions put on the use of synthetic compounds. A separation of BCAs from the legislation for synthetic compounds would avoid bureaucratic hurdles, which apply only for synthetic pesticides. Registration would be more balanced and can be adapted to the needs of BCAs. A legislation adapted to the potential risks related to the use of BCAs would be more flexible and allow for fast track systems for compounds or organisms that are obviously of low risk. This situation seems to be ignored in European policy, otherwise there would be more attempts to avoid trade-off effects on authorisation of BCAs from decisions on chemical pesticides.

At an early stage of the REBECA Action the participants decided to concentrate on proposals that could provide a short term improvement for the conditions of

BCA regulation. The development of a completely new system specifically adapted for the needs of biocontrol was never the objective of the Action and, consequently, the reader of this book will not find any proposals for a completely new organisation of the registration of BCAs. However, the possibility should not be neglected that future activities might pick up such an idea. REBECA participants were aware that this would be a long term project and the problems for BCAs might remain unsolved in the meantime. Therefore, the proposals of the consortium to policy-makers concentrated on improvements with potential for short term implementation. REBECA participant recommended to:

- produce definitions for low risk
- acknowledge the lower risk status of BCAs in the development of new rules
- consider the possibilities to separate legislation of BCAs from synthetic compounds
- develop more flexible risk assessment procedures
- introduce fast track systems for low risk products

### 6.5.5 *Trade-Offs Caused by the Organisation of the Registration Process*

In Europe, regulation is the remit of authorities in charge of agricultural, environmental and health affairs. All three sectors have conflicting interests; finding consensus is time-consuming. Another problem is that personnel dealing with files and monographs on BCAs often are experts in reviewing information on synthetic chemistries. If BCAs are regulated by these agencies, which lack background information on the risks of BCAs, the process of consensus finding will be particularly long and expensive and can result in exaggeration of risks. Lack of knowledge is resulting in an overestimation of risks.

In addition to the described problems, the two level registration involving 25 MS is complex, complicated, time-consuming and expensive. One would wish for an agency like that existing in the USA, where one department of the Environmental Protection Agency takes care solely of dossiers on BCAs. This agency can produce continuity in expertise. Personnel with long term experience in BCA regulation will be more reluctant to give waivers. However, the idea of a centralised European authority only dealing with dossiers of BCAs probably does not match well with the subsidiary principle of the European Union.

The microbial biological control sector in the EU has suffered significantly from implementation of registration requirements following the rules developed for synthetic agrochemicals. Many potential biocontrol products are not submitted for EU registration due to costly data requirements. Sometimes unnecessary data packages have been required because applicants and regulators could not agree on waivers. In order to prevent bureaucratic hurdles and unnecessary consensus finding costs, attempts should be made to get regulation of BCAs into the hands of experts. Should European policy makers seriously want to promote the further introduction of biological control strategies, then they should take measures to equip authorities with more personnel with experience and expert knowledge. They even might want to waive fees and support the data production necessary for the risk assessment. The REBECA consortium therefore proposes to:

- reduce consensus finding costs
- equip registration authorities with skilled personnel
- consider expert knowledge in the regulation process
- not allow abuse of registration system to protect markets
- waive fees for registration of BCAs
- support production of safety data

## 6.6 Agriculture Policy and Biological Control

The policies of the EU and MS do not meet their stated objectives. Reduction programmes (EU 2009) concentrate on more efficient use of chemical pesticides but neglect the potential of biocontrol agents. In the EU project ENDURE (www-endure-network.eu) biological control plays a minor role. Policy wants to improve on food safety and have less pesticide residues in agricultural produce and the environment, but the sales of chemical pesticides has increased over the years and so do the problems with residues. Policy aims to support innovative technology, support start-ups and SMEs and produce qualified job opportunities for young scientists; however, their support for the biological control sector is decreasing. Considerable research funds support investigations of safety of transgenic crops and into molecular biology, whereas support for projects dealing with biological control has decreased over the past decade.

There is profound evidence for trade-off effects caused by the current EU regulation policy. Many more products based on microbials, pheromones and botanicals are on the market in the USA compared with the number in Europe, although the diverse structure of the European markets would fit much better for biological control strategies. Costs and fees for registration are lower in the USA and registration is less time-consuming. The survey among biocontrol companies in Europe has revealed that 3 out of 5 companies do not invest in new biocontrol products because of the high costs and the long time needed for registration. Three companies had shelved biocontrol products mainly due to registration hurdles. Several companies indicated readiness to bring new products to the market should the conditions for registration be more favourable (Hokkanen and Hokkanen-Menzler 2008).

Biocontrol has the potential to replace several chemical control agents, and be used more widely. A good example is the Spanish vegetable producer market, which switched from 500 ha with biological control to approximately 38.000 ha within 5 years (van der Blom 2009).

With the support from the REBECA Action, EU policy makers have finally tackled one of the major problems, which prevent the more wide-spread use of biological control agents – the current registration process. The Action gave valuable contributions to improve regulation of BCAs in Europe, to develop more balanced, and more economically balanced, regulation procedures without compromising safety for consumers and the environment. It is now up to policy makers to implement these proposals and to continue the dialogue on strategies for implementing a better regulation system for biological control agents and to further promote the introduction of biological control strategies into European agriculture.

## References

Agrow (2004) Aventis products boost Bayer's 2003 results. Agrow World Crop Protect News 445:5–6

Alabouvette C, Cordier C (2011) Risks of microbial biocontrol agents and regulation: are they in balance? In: Ehlers R-U (ed) Regulation of biological control agents. Springer, Dordrecht, pp 157–173

Berg G, Zachow C, Cardinale M, Müller H (2011) Bacterial ecology and human pathogenicity of plant-associated bacteria. In: Ehlers R-U (ed) Regulation of biological control agents. Springer, Dordrecht, pp 175–189

De Clercq P, Bale JS (2011) Risks of invertebrate biological control agents – Harmonia axyridis as a case study. In: Ehlers R-U (ed) Regulation of biological control agents. Springer, Dordrecht, pp 243–255

Easteal S (1981) The history of introductions of *Bufo marinus* (Amphibia: Anura); a natural experiment in evolution. Biol J Linnean Soc 16:93–113

Ehlers R-U (2011) Regulation of biological control agents and the EU Policy Support Action REBECA. In: Ehlers R-U (ed) Regulation of biological control agents. Springer, Dordrecht, pp 3–23

EPPO (2010) EPPO Standards on Safe use of Biological Control - PM 6/3 - Version 2010. List of biological control agents widely used in the EPPO region. http://archives.eppo.org/EPPOStandards/biocontrol_web/bio_list.htm

EU (2009) Directive 2009/128/EC of the European Parliament and of the Council of 21 October 2009 establishing a framework for Community action to achieve the sustainable use of pesticides. Off J Eur Union L 309:71

European Commission (2000) Communication from the Commission on the precautionary principle. Commission of the European Communities COM 2000–1. http://eur-lex.europa.eu/LexUriServ/site/en/com/2000/com2000_0001en01.pdf

Frost and Sullivan (2001) European biopesticides market. http://www.frost.com/prod/servlet/report-toc.pag?ctxixpLink=FcmCtx3&searchQuery=European+biopesticide+market+2001&repid=3905-01-00-00-00&bdata=aHR0cDovL3d3dy5mcm9zdC5jb20vc3JjaC9jYXRhbG9nLXNlYXJjaC5kbz9xdWVyeVRleHQ9RXVyb3BlYW4rYmlvcGVzdGljaWRlK21hcmtldCsyMDAxJnBhZ2VTaXplPThAfkBTZWFyY2ggUmVzdWx0c0B%2BQDEyOTEwNDAxOTcxOTE%3D&ctxixpLabel=FcmCtx4

Graham JD, Wiener JB (1995) Risk versus risk – tradeoffs in protecting health and the environment. Harvard University Press, Cambridge

Hokkanen HMT, Hokkanen-Menzler I (2008) Deliverable 24: Cost, trade-off, and benefit analysis. REBECA-Regulation of biological control agents. Specific Support Action, Project no. SSPE-CT-2005-022709. http://www.rebeca-net.de/downloads/report/deliverable%2024.pdf

Lisansky SG, Coombs J (1994) Development in the market for biopesticides. In: Proceedings of the Brighton Crop Protection Conference – Pest and Diseases, Farnham, Surrey, UK, pp 1049–1054

Pimentel D, Acquay H, Biltonen M, Rice P, Silva M, Nelson J, Lipner V, Giordano S, Horowitz A, D'Amore M (1992) Environmental and economic costs of pesticide use. BioScience 42: 750–760

Regnault-Roger C (2011) Risks of biocontrol agents containing compounds of botanical origin or semiochemicals. In: Ehlers R-U (ed) Regulation of biological control agents. Springer, Dordrecht, pp 216–242

Reuter W (2007) Pestizide am Limit II. Veränderungen von Höchstmengen für Pestizide in pflanzlichen Erzeugnissen in Deutschland 2004–2006. http://www.greenpeace.de/gpd/user_upload/themen/umweltgifte/greenpeace_pestizide_am_limit_2.pdf

Roy H, Brown P, Majerus M (2006) Harmonia axyridis: A successful biocontrol agent or an invasive threat?. In: Eilenberg J, Hokkanen HMT (eds) An ecological and societal a approach to biological control. Springer, Dordrecht, pp 295–309

Strasser H, Hutwimmer S, Burgstaller W (2011) Metabolite toxicology of fungal biocontrol agents. In: Ehlers R-U (ed) Regulation of biological control agents. Springer, Dordrecht, pp 191–213

Strauch O, Strasser H, Hauschild R, Ehlers R-U (2011) Proposals for bacterial and fungal biocontrol agents. In: Ehlers R-U (ed) Regulation of biological control agents. Springer, Dordrecht

Sunstein CR (2002) Risk and reason - safety, law, and the environment. Cambridge University Press, Cambridge, UK, pp 342

van der Blom J (2009) Microbiological insecticides against lepidopteran pests in greenhouse horticulture in Almeria, Spain. IOBC/wprs Bull 45:59–62

Waibel H, Fleischer G, Becker H, Runge-Metzger A (1998) Kosten und Nutzen des chemischen Pflanzenschutzes in der deutschen Landwirtschaft aus gesamtwirtschaftlicher Sicht. Agrarökonomische Monographien und Sammelwerke. Wissenschaftsverlag Vauk, Kiel

# Part II
# Risks and Risk Assessment

# Chapter 7
# Risks of Microbial Biocontrol Agents and Regulation: Are They in Balance?

**Claude Alabouvette and Christelle Cordier**

**Abstract** In this chapter we will review the requirements of the Annex II of the Directive 91/414 and discuss whether these requirements are justified in relation to risk assessment. It is first necessary to identify the hazards before the associated risks can be evaluated. Regarding microbials, the first requirement is an accurate identification of the micro-organism at the species and strain levels. Whether the biological control agent belongs to a species known to include pathogens for man or plant is crucial information for conducting the risk assessment. If there is no report in the medical data banks reporting cases of pathogenicity, infectivity or toxicity due to strains belonging to the same species as the biological control agent, a minimum data set in connection to risk for human health should be accepted. When original studies are needed, adapted experimental methods should be made available. Introducing micro-organisms into the natural environment requires assessment of their fitness and behaviour under various conditions. Today molecular techniques enable the population kinetics of a microbial strain to be followed after its release into the environment. Results showed that when reintroduced into the environment from which it has been previously isolated, a micro-organism will establish but not become dominant; therefore, it does not present major risks. Effects on non-target organisms have to be studied. However, most of the methods that have been designed for chemicals do not apply to microbials, and the usefulness of some studies should be questioned. In many cases waivers should be accepted. In order to address the specific problems linked to the use of micro-organisms as plant protection products, specific regulations should be established.

## Contents



---

C. Alabouvette (✉)
Agrene, 47 rue Constant Pierrot, 2100 Dijon, France
e-mail: c.ala@agrene.fr

R.-U. Ehlers (ed.), *Regulation of Biological Control Agents*,
DOI 10.1007/978-90-481-3664-3_7,  

## 7.1 Introduction

In the beginning of the twenty-first century, humanity is facing several challenges regarding food safety in relation to climatic change and the energy crisis. Indeed, in contrast with past years, the price of food is increasing drastically because of a shortage of production. At the same time the world population continues to increase and agricultural production should increase to meet the need for food. Thus, we have to increase the agricultural production for both food and energy and at the same time decrease the use of fertilizers and pesticides of chemical origin. In this context there is interest in biological control of pests, diseases and weeds. Despite efforts made during the last century to discover new biological control agents and to study their modes of action, there are still very few products on the market. Among the diverse reasons which could be evoked to explain the lack of success of biological control (Alabouvette et al. 2006), one is related to the regulatory status of biological control agents, which must be registered according to guidelines originally developed for chemical pesticides. Indeed, in the European Union, the Directive 91/414 EEC applies to any type of plant protection products, including natural products such as plant extracts, semiochemicals and micro-organisms (viruses, bacteria and fungi). Requirements of this Directive represent the main constraint of putting a biological control agent on the market. The directive was written to avoid risks linked to the use of chemical pesticides, whose hazards have been clearly demonstrated. An analysis of the Directive requirements for listing a new active substance on Annex I shows that some requirements are not justified for biological agents and that some specific hazards might not be taken into consideration by the Directive. In the present chapter we intend to analyse the requirements of Directive 91/414 EEC and discuss whether they are useful to assess the risks linked to the use of microbials to control pests and diseases in crops. Based on the conclusions of the REBECA project, we will suggest different ways for a more realistic approach of microbial risk assessment in plant protection.

## 7.2 Regulation as It Exists Today

The aim of the Directive 91/414 EEC, which will soon (14 June 2011) be replaced by Regulation (EC) 1107/2009 concerning the placing of plant protection products on the market, is firstly to ensure a high level of protection for both human and animal health and the environment, and secondly to improve the functioning of the internal market through harmonisation of the rules on placing plant protection products on the market, while improving agricultural production. We all agree with these two principles but we should question whether or not Directive 91/414 EEC has achieved them, especially in regard to availability of biological control products.

Indeed for small-and medium-sized-enterprises (SMEs) involved in the production and commercialisation of biological control products, the regulation is perceived as a bottle neck responsible for slowing down progress and innovation in the conception of new products. Regulation is time-consuming and costly, especially in comparison with the turn-over the products can achieve in niche markets targeted by most of the microbial plant protection products. Moreover, regulation requires scientific knowledge not always available in small enterprises and the justifications for some studies are not easily accepted or understood.

As stated above, the Directive 91/414 EEC has been gradually established to prevent risks associated with the use of chemical pesticides. The use of chemical pesticides preceded the regulation and, unfortunately, the hazards of chemical pesticides were documented before the regulation was put into force. For microbials, the situation is totally different. Based on the fact that Directive 91/414 applies to all plant protection products, including natural products and micro-organisms, the producers of biological control agents must respect the regulation that was established for chemicals and which is not adapted to risks related to the use of BCAs. Specific requirements for micro-organisms were set up by Directive 2001/36 EEC, which has adapted Annex II and Annex III of Directive 91/414. But these Annexes IIB and IIIB did not simplify the dossier required for registration of microbials; on the contrary, they made the dossier more complex since it includes the fact that biocontrol agents are living organisms able to grow on the treated plants and in the environment. The only difference between synthetic chemicals and microbials regarding registration is that, most often, there is little difference between the active substance and the formulated preparation; thus, data provided for Annex II can be used for Annex III, without modification. Indeed most of the preparations are made of bacteria or fungal propagules corresponding to the active substance mixed with food grade additives. Thus, as risk assessment is of concern, the requirements of Annexes II and III are almost identical and in this chapter we will review both together.

Risk is defined as the probability for a hazard to occur. Based on this definition the principle on which the directive is based consists of identifying the hazards, which are any toxic or negative effect that can occur with the use of the product in plant protection, and then to calculate the probability for the hazard to occur. This probability depends on exposure and thus for the same hazard the risk will be different for farmers applying the product, for by-standers and for consumers. The first steps consist of identifying the hazards and considerable efforts are expended identifying all the potential hazards even without any evidence for toxicity of the plant protection product. Obviously, if there is no hazard there will never be any risk, but by contrast, in the absence of exposure there will be no risk, even in the presence of a well identified hazard. Unfortunately, the directive, which is based on the precautionary principle, imposes the requirement to identify all potential hazards before assessing the risks.

## 7.3 Requirements of Annex II B of Directive 91/414

In this section we will review the requirements of Annex II of the Directive and discuss whether these requirements are necessary to assess the risks linked to the use of microbials.

### *7.3.1 Identity of the Microorganism*

The Directive clearly states that "the identification together with the characterization of the micro-organism provides the most important information and is a key point for decision making". All experts agree with this statement: a correct identification of the micro-organism is the first step to ensure the safety of the plant production product. Although there was a debate about the level of identification of the micro-organism and about methods to be used, scientists agree that identification must be at the strain level, using the most accurate method available. The necessity to identify the biological control agent at the strain level is justified by the fact that many species of micro-organisms include both pathogenic and beneficial strains. Moreover, in many cases the identification at the species level is not easy. The following example will demonstrate this. Fungal strains belonging to the genus *Trichoderma* are known for many years as beneficial organisms having antagonistic activities against many plant pathogenic fungi. The genus *Trichoderma* is ubiquitous in soil, on organic debris, on roots and other plant parts. It is mainly present as its telemorphic stage producing many unicellular microconidia. There are many different species of *Trichoderma* that are very difficult to identify based on morphological characters, as described by Rifai (1969). Only recently the use of molecular techniques targeting the sequence of the ribosomal DNA enables a clear identification of the species. Indeed these tools enable the strain to be placed in a phylogenetic tree and thus to see the relationship with other species belonging to the same genus. This enables the potential hazards to be predicted based on the proximity with known pathogenic species. Cordier et al. (2006) compared the 18 s rDNA sequence of strains of *Trichoderma* to identify strains that showed some abilities for biological control. Results showed that one strain was in the same bootstrap with several strains belonging to the species *T. longibrachiatum*. This species being known to include strains pathogenic to man, it was decided to stop the development of this strain since it would not be possible to put on the market a biological control agent potentially pathogenic to humans.

Having identified the species of a biological control agent, it is useful to develop a method to identify the biocontrol strain itself among other strains belonging to the same species. This is necessary for regulation procedures since a tool to verify the identity of the biological control agent must be provided. It is also very useful to track the strain after release in the environment and finally having a tool specifically to identify the biocontrol strain will ensure its commercial protection. Indeed, since in Europe a natural microbial strain can not be patented, registration is a way to protect the plant protection product if the strain itself is well characterised.

To return to the example of *Trichoderma*, since many strains are already on the market the question is how to be sure that the plant protection product contains the strain that has been registered and not another. The best approach today is to design a SCAR marker (specific-characterised-amplified-region). Several strategies can be used to identify a unique sequence that could be amplified (Cordier et al. 2006). We designed such a SCAR for the strain T1 of *T. atroviride* which is in the process of registration, and can be specifically identified among other strains belonging to the same species (Cordier et al. 2006).

In some cases the design of a SCAR is absolutely needed since it will be the only solution to identify the biological control strain among strains belonging to the same species that includes plant pathogens. The best example comes from the species *Fusarium oxysporum*. This species of asexual fungus is also very common in soil and in the rhizosphere of many different plant species. This fungus is well known as a very aggressive plant pathogen. It attacks the roots and can provoke either rots or wilts. Interestingly, some strains belonging to this species are not only non-pathogenic but protective on certain plant species. Applied to plants they can prevent infection of the plant by pathogenic *formae speciales* of *F. oxysporum* (Alabouvette et al. 2009). In this example, one must not only identify the biological control strain at the species level, but also describe a tool enabling its unique identification among strains of the same species (Edel et al. 2008).

To conclude with this first requirement of the directive, one must agree that a perfect identification of the biological control agent at the species and strain level is an absolute necessity for risk assessment.

This first chapter of the directive includes also other requirements such as the specification of the material used for manufacturing of formulated products. This includes the "identity and content of impurities, additives, contaminating micro-organisms".

The level and nature of acceptable contaminants pose problems in relation to risk assessment. Some methods are recommended to check for the presence of contaminants of human concern and most of the European countries put limit levels for *Listeria*, *Salmonella* etc. From the microbiologist's point of view it would be preferable to have a plant protection product without contaminants. Indeed, the presence of contaminants means that the production process has not been optimized.

Among impurities are the "relevant metabolites (i.e. if expected to be of concern to human health and/or the environment)". This point is one of the most controversial since micro-organisms are able to produce many different secondary metabolites whose properties are not known. Moreover the production of these secondary metabolites depends on many factors, such as the age of the culture, the growth medium or the plant organ on which the biological control agent is applied (Woo and Lorito 2007). Whether these secondary metabolites contribute to the modes of action of the biological control agent will be discussed below. However, it is quite impossible to predict which metabolite will be produced and in what quantity, and it is economically not possible to analyse all the metabolites present in a culture at trace levels. Thus, this question raised by the directive is a source of debate between regulators and industry and introduces distortion between member states since there is no common rule to be applied.

### 7.3.2 Biological Properties of the Microorganism

The directive requires summary of knowledge available in the literature dealing with the origin of the micro-organism: its habitat, its biological cycle, its relationships with known pathogenic or beneficial strains. Providing the strain has been correctly identified and belongs to a well known species this literature review is important. If information can be provided from the literature, regulators might be able to waive data requirements and costs related to the production of original data can be avoided.

Among biological properties, the modes of action of the biological control agent must be described. This is not an easy task even if the strain belongs to a well studied species. Indeed there are always several modes of action by which a biocontrol agent controls an arthropod or a disease. To take again the example of *Trichoderma* spp., many different modes of action have been documented: direct antagonism through competition for nutrients, competition for space, hyperparasitism, antibiosis and, more generally speaking, production of secondary metabolites and enzymes having a direct effect on the target organism and indirect antagonism through stimulation of plant defence reactions (Woo and Lorito 2007).

Is knowledge on the modes of action absolutely needed to assess the risks? It is useful to understand how a biological control agent interferes with the target pathogen and the plant. It is obvious that when the main mode of action is hyperparasitism, or competition for nutrients, there is no risk linked to these mechanisms. When antibiosis based on the production of secondary metabolites is the main mode of action there might be some concern in regard to toxicity for man. How can this aspect be addressed?

If we followed the directive we are supposed to characterise the secondary metabolites potentially of concern, and to study their toxicity as it is required for a chemical pesticide. However, there are many important differences between a chemical pesticide with an active substance possibly representing 95% of the preparation and a living micro-organism, which will locally release small amounts of a toxic metabolite. Moreover, these metabolites toxic for arthropods or for fungi are usually not toxic for vertebrates.

As stated above, strains of *Trichoderma* spp. might produce many different types of secondary metabolites. Some of them might be toxic and of concern. Indeed, among many other molecules, *Trichoderma* strains can produce trichotecenes and peptaibols (Kubicek et al. 2007). It is possible, but time and money consuming, to detect and quantify these molecules in the culture filtrate (Stoppacher et al. 2007) but it is impossible to know if these molecules are produced in situ after application of the biological control strain to soil or to the plant organ to be protected. Indeed, it has been demonstrated that the same strain does not produce the same secondary metabolites on different plants or pathogens (Marra et al. 2006). Again, it is important to state that the secondary metabolites are only produced locally and in very limited quantities. Thus, it might not be necessary to spend much effort on the characterisation and quantification of all toxic metabolites.

A recent example of the complexity of this question concerns the production of DDR by *Pseudomonas chlororaphis*. A strain of *P. chlororaphis* was in the process of registration when it was demonstrated that it produces a toxigenic compound 2,3-deepoxy-2,3-didehydrorhizoxin (DDR). This secondary metabolite was isolated, characterised and its mutagenic potential demonstrated. As a consequence discussions started about whether the strain should be authorized or not. A method of analysis was developed by the company and validated by the regulators. The quantification limit in the fermentation medium is 2 mg/l and the seed detection limit is 1 μg/kg (Directive 2004/71/EC). The highest amounts of DDR were detected at the end of the fermentation process, but since DDR decomposes rapidly no detectable quantities of the metabolite could be detected on treated seeds (Hökeberg 2006). The risk might have been dangerous for producers but not for farmers. However, producers would have had to consume several litres of the fresh bacterial culture to be harmed. Finally the decision was taken to allow this strain to be put on the market with the restriction that it has to be used in closed seed dressing apparatus (Directive 2004/71/EC).

One should understand that we know little about the mechanisms of action of biological control agents. What is known today and presented as the main mode of action might not be an essential trait of the biocontrol activity. At the present time, regarding disease control, much emphasis is given to the indirect mode of action through induction of resistance of the host plant. With adequate tools one can demonstrate that in addition to a known mode of action, a biocontrol strain is also able to induce resistance in the plant. The protective strain *F. oxysporum* Fo47, developed in our laboratory, has been shown to compete with the pathogen for nutrients in the soil and the rhizosphere (Fravel et al. 2003). It is also able to compete for the colonisation of the root tissues and a few years ago we believed that theses mechanisms of competition were the most important among the modes of action of Fo47. Today, thanks to a novel approach, we demonstrated that this non-pathogenic strain is also able to induce systemic resistance in the plant. The relative importance of these mechanisms probably depends on the plant species to which it is applied (Alabouvette et al. 2007).

Inducing resistance in the plant is commonly considered to be a relatively safe mode of action because it does not involved the production of secondary metabolites possibly toxic to man. However, recently there have been some concerns regarding the pathogenesis-related proteins accumulated by the plant in response to contact with the biological control agent. Being proteins these defence molecules might induce sensitising reactions (Niskanen and Dris 2004).

To finish with these aspects of risk assessment in relation to the mode of action of the biological control agents, we must recall that the pathogenic micro-organisms that we are aiming to control also produce secondary metabolites and toxins, such as mycotoxins, which are tolerated at low levels in feed and food. We should not be more restrictive on biological control agent than for pathogenic micro-organisms.

### 7.3.3 Further Information on the Microorganism

The Directive insists on the description of the method of production of the active substance and on quality control. This is important since in many preparations already on the market one can find microbial contaminants that could be more dangerous than the active substance itself. It would not be logical to insist on assessment of the risks due to the active substance and tolerate high levels of unknown contaminants. The production process must ensure the harvest of a clean active substance at the end of the process. There is a list of pathogenic bacteria that should not be present in the formulated product but, again, it would be preferable to harvest a pure active substance. However, acceptable contamination levels should not be lower than those acceptable for food and feed.

The Directive also asks questions about the methods to prevent loss of virulence. Virulence, or more precisely the biocontrol capacity of the strain, has to be preserved in order to deliver an effective biological control product. This aspect should be addressed in relation to quality control but has nothing to do with risk assessment. Moreover, we must admit that in most cases we do not know how to address this question. If a secondary metabolite is involved in the mode of action we can propose either a phenotypic or a molecular approach to determine that the strain still has the capacity to produce this molecule. However, as stated above, in many cases we do not have information on the main mode of action of the biological control agent. Thus, it is almost impossible to verify in vitro that it is stable. Only bio-assays can be designed to assess the biocontrol capacity of the product. These bio-assays are usually time consuming but are the only solution to determine the capacity of the active substance to control the target pest or disease; however, efficacy has nothing to do with risk assessment.

As for chemical compounds, the Directive requires information on the compatibility of packaging material and the preparation. This is not relevant in connection with risk assessment. Indeed, biological control products based on micro-organisms will be compatible with most materials used for packaging. Similarly, there should be no problem in case of an accident during transport or storage. Most biocides can be used to kill the biological control agents.

### 7.3.4 Analytical Methods

As for chemicals, accurate descriptions are required for the methods used to identify the micro-organisms and the contaminants, detect and quantify the secondary metabolites, analyse the micro-organism as manufactured, and determine and quantify the residues.

It is very important to describe accurately the methods used to characterise and identify the micro-organisms, both at the species and strain levels. Indeed, only accurate identification methods will guarantee that the product put on the market contains the strain whose properties have been assessed. According to the Directive, secondary metabolites produced by the biocontrol agent are considered as contaminants and/or residues. Therefore, methods to detect and quantify these secondary metabolites should be described but, as stated above, many different secondary meatabolites can be produced by a single strain of biocontrol agent and specific methods of detection of these metabolites are frequently not available. When they have been developed, on laboratory scale, they have to be validated by regulators and their use is often not economically possible routinely to check the quality of the products and to detect the production of secondary metabolites in the environment.

### 7.3.5 Effects on Human Health

This point, which is obviously very important in regard to risk assessment, has been the subject of controversy during discussions between regulators, scientists and industrial partners. Everybody agrees that regulation must protect human health but the question is whether the studies required by the Directive are all needed, and whether the methods designed to study the toxicity of chemicals are adapted to study microbials. First of all it is admitted that micro-organisms are very diverse and therefore this group needs to be assessed case by case. However, there are some exceptions; for example the *Baculoviruses*, for which OECD and REBECA have proposed a common approach leading to the listing of the family *Baculoviridae* or at least of a given species on Annex I. As soon as we have more data for certain groups of micro-organisms we can expect that a more general approach will be possible.

The Directive clearly states that evaluation should be carried out in a tier-wise manner. The first Tier includes all available information and basic studies which have to be performed for all micro-organisms. Tier II studies are required if tests under Tier I have shown adverse effects. Regarding Tier I, the debate is whether all the studies have to be performed or whether waivers can be accepted based on knowledge already acquired on strains belonging to the same species as the biological control agent under review. Regarding the first paragraph dealing with "basic information", it must be said that a review of the literature and of medical data banks will provide much useful information which is not always correctly used during the evaluation process.

It must be clearly stated that humans are regularly exposed to a wide range of micro-organisms and the probability that man has never been exposed to a biological control agent isolated from the environment is very low. Since the medical data banks report all cases of human infection, including those of immuno-depressed patients, related to opportunistic micro-organisms, the survey of these data banks will give useful information on the infectivity and pathogenicity of a given species of micro-organism. If there is no indication of symptoms linked to this microbial species there is a strong probability that a micro-organism belonging to that species is neither infectious nor pathogenic to humans.

The second paragraph entitled "basic studies" is the most discussed aspect of the Directive. It concerns (i) sensitisation, (ii) acute toxicity, pathogenicity and infectiveness, and (iii) genotoxicity. Indeed, some regulators will ask for original data for all the studies recommended by the Directive; others will accept waivers for some studies. In fact one of the most critical problems is that the recommended methods that have been set-up to study the toxicity of chemical are not adapted to the study of microbials. Why should we ask for time- and money-consuming studies when we know that the results will not be relevant to identify the hazards and assess the risks? This is perfectly illustrated with the sensitisation studies. The Directive itself recognised that "as a consequence of the absence of proper test methods, micro-organisms will be labelled as potential sensitisers, unless the applicant wants to demonstrate the non-sensitising potential by submitting data". Thus, biological control products will have the bad image associated with sensitisers just because there is no method available to check their sensitisation potential. In addition, when an applicant provides results they will be evaluated differently by different experts because the method is not valid for micro-organisms. There is an urgent need to develop adapted protocols to assess the sensitising capacity of biological preparations.

Regarding "acute toxicity, pathogenicity and infectiveness" the Directive lists a series of studies: acute oral toxicity, acute inhalation toxicity, intraperitoneal/subcutaneous single dose, all of which are required. Moreover, since micro-organisms are able to grow and multiply in many different environments, clearance (elimination/excretion) of the micro-organisms has to be studied, making the studies more difficult and costly. The question is whether all these studies are required for risk assessment.

The toxicity tests to be conducted should take into account the main route of exposure. Considering that inhalation is the most probable route of contamination, intratracheal acute toxicity seems appropriate. There was also a consensus among REBECA participants to consider the oral toxicity test, because ingestion of the active substance enables to test both the direct pathogenicity/infectivity and also the indirect toxicity linked to the presence of secondary metabolites that may be potentially toxic. In our opinion there is no need to perform the intraperitoneal/subcutaneous test since this represents the "worst case", which is not realistic according to the use of the biological control product.

Similar to sensitisation, genotoxicity testing poses methodological problems. The first level of test (Ames test) is based on the detection of mutations induced by the pesticide when growing strains of *Salmonella typhimurium* on a nutrient medium enriched with the active substance. This test cannot be used when the active substance is a living micro-organism. It has been proposed that a sonicated preparation of the micro-organism should be used to perform the test and some Rapporteur Member States (RMS) asked for these data. What is the significance of such a test since sonication liberates many molecules that are normally not released in the environment? This test might be required when secondary metabolites are produced; in such circumstances the culture filtrate, containing the excreted metabolites, can be used to perform the test. Indeed, the Directive clearly states that "relevant metabolites" must be purified and their toxicity studied as for a chemical. But, what metabolite is relevant in relation to toxicity studies? Admittedly, when secondary metabolites are essential for the mode of action of the biocontrol agent they must be studied per se. This has been the case for the DDR produced by *Pseudomonas chlororaphis*. However, as stated above, most of the micro-organisms are able to produce many secondary metabolites at different times of their life cycle and depending on the target organism.

The actual requirements of the Directive regarding concerns for human health are not adapted to the study of microbials. They are based on the principle of precaution and on the "worst case" approach, neglecting the fact that man has already been in contact with these naturally occurring micro-organisms during his evolution. Moreover, the available methods designed for chemicals are not adapted to produce the required data. Should we continue to ask for data knowing that these data will not be valid and thus will not improve risk assessment? There is an urgent need to promote research aiming at setting up methods better adapted to the study of microbials and at addressing important questions in relation to human health.

### 7.3.6 Residues in or on Treated Products, Food and Feed

The next requirement of the Directive concerns the residues, which are defined as both the living micro-organisms and the residual traces and metabolites (toxins) remaining in or on plant products. In general, when the plant production product is based on a living micro-organism the residues on treated products and food is the micro-organism itself. Thus, if no toxicity has been demonstrated previously, there should be no problem of toxic residues. Again the question arises about when the micro-organism produces secondary metabolites that are suspected to be toxic.

Consumer concerns regarding mycotoxins entering the food chain has prompted closer scrutiny of the secondary metabolites of all fungal biocontrol agents. Regulatory authorities often require detailed information on "relevant metabolites". It is not clear what constitutes a relevant metabolite when most fungi secrete a disparate array of bioactive compounds with different ones produced under different conditions (Magan et al. 2002). Thus, it is almost impossible to detect and quantify all the secondary metabolites potentially produced on the treated products. Moreover, one must remember that in contrast to chemicals, these secondary metabolites produced by naturally occurring micro-organisms have always been present in the environment. Thus, these biological molecules are usually quickly metabolised or degraded under normal environmental conditions.

In conclusion, in relation to risks for human health, it appears that micro-organisms originating from the environment can be considered safe as soon as they have been correctly identified and are not known to be related to a pathogenic species.

### 7.3.7 Fate and Behaviour in the Environment

Study of fate and behaviour of the plant protection product in the environment poses quite different questions, depending on whether the product is a chemical or a living micro-organism. The chemical product can accumulate with repeated applications and its degradation might take time since it is a synthetic molecule, which has not been present in the natural environment. By contrast, the biological control product is based either on a naturally occurring biological molecule or a living micro-organism. There is an unjustified fear that an introduced micro-organism can multiply in the environment and become a pest. This fear is not justified by facts. Several reports in the literature show that pathogenic arthropods or fungi have expanded over the world by adapting to new environmental conditions (Graniti 1998) but in these examples the micro-organisms or arthropods were favoured by the presence of the crop on which they feed. They had a competitive advantage compared with other micro-organisms or insects.

In relation to biological control agents, we do not yet have a clear example of an organism becoming dominant in the environment where it has been introduced. In the absence of any selection pressure, introduced bacteria or fungus originating from the natural environment will not become dominant when reintroduced in the same environment (Alabouvette and Steinberg 1998). However, since it is not possible to rule out any associated hazard, it is necessary to study the fate and behaviour in the environment.

Again the methodology recommended for the study of chemical pesticides is not adapted to microbials. Behaviour of the biological control agent is first studied in laboratory experiments then under natural environmental conditions. In the latter case, it is necessary to develop a tool enabling the introduced micro-organism to be traced and separated from other strains belonging to the same species that occur naturally in the environment. Indeed, it is important to realise that often the biological control agent already exists but at a density too low to be effective in controlling the pest or disease. To trace an introduced strain, the easiest approach is to use a mutant, resistant to an antibiotic or a fungicide. For example, we used a UV irradiated mutant of Fo47 resistant to Benomyl to study the population dynamics of this biological control strain in two soils of different physico-chemical properties over 1 year. Results are summarized in Figs. 7.1 and 7.2 taken from Edel-Herman et al. (2009). In the disinfested soils, this strain grew and established itself at a high population density whatever the dose of inoculation and the soil type (Fig. 7.1). By contrast, in the non-disinfested soil, i.e. in the presence of a native community of micro-organisms, the biological control strain was not able to proliferate. It did not disappear but established at a population density lower than that at which it was introduced. These results presented Fig. 7.2 illustrate the fact that a naturally occurring micro-organism re-introduced in the environment from which it has been isolated neither disappears nor proliferates more than the native microbial community (Edel-Herman et al. 2009). The biological control organism becomes part of the native microbial communities. However, this approach using antibiotic or fungicide resistant mutants can only be used in a confined environment since it is not safe to release mutants in the environment. Moreover, the mutation might have modified the dispersal or survival behaviour of the micro-organism. To address the fate of an introduced micro-organism in the environment the most elegant approach consists in designing a SCAR marker that will enable the natural organism to be traced among other strains of the same species in the environment. We followed this approach

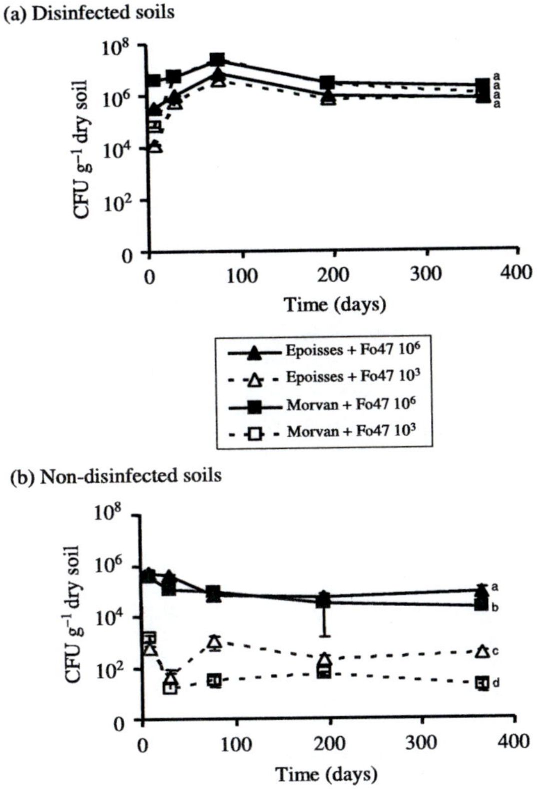


**Fig. 7.1** Population dynamics of *Fusarium oyxsporum* Fo47b10 (mutant of Fo47 resistant to Benomyl) introduced at $10^3$ and $10^6$ conidia $g^{-1}$ into the disinfected soils (**a**) and the non disinfected soils (**b**) of Epoisses and Morvan. The two soils showing different physico-chemical properties were regulated at the same water potential corresponding to 18 and 15% of water content respectively and incubated at 25°C. Values with similar letters are not significantly different ($P = 0.005$) (Edel-Hermann et al. 2009)

for the strain T1 of *Trichoderma atroviride*. The results of the population dynamics study in two soils of different physico-chemical properties were analogous to that obtained for Fo47. Indeed the strain T1 neither disappeared nor proliferated in the non disinfested soils (Fig. 7.3 taken from Cordier et al. 2007). Based on these results and on many others from the literature, we can conclude that a soil-borne micro-organism re-introduced into a soil will survive but will not proliferate; it will become part of the native populations of the same species.

There will be no need in the future to ask for time-consuming studies to state that a naturally occurring biological control agent is not posing any risk of proliferation when re-introduced in the milieu from which it has been isolated.

### *7.3.8 Effects on Non-target Organisms*

Again, the Directive is asking for the same requirements for microbials as for chemical, and again, the problems are not similar since most of the non-target organisms have already been exposed to the natural micro-organisms developed as biocontrol agents. Moreover, many animals including mammals, birds, fish and crustaceans

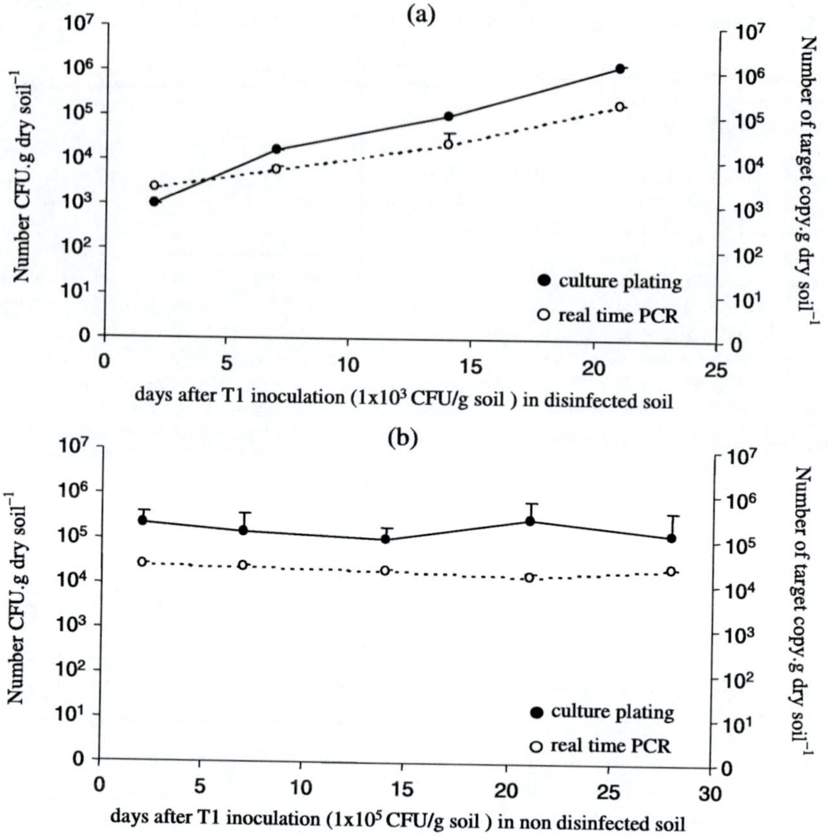


**Fig. 7.2** Population dynamics of a strain T1 of *Trichoderma atroviride* introduced in Epoisses soil, disinfested (**a**) or not (**b**), followed by the traditional soil dilution technique or by real time PCR using a SCAR marker (Cordier et al. 2007)

are under intensive animal husbandry. Therefore, the pathogens of these animals are surveyed. Possible non-target effects could be identified by a literature survey and the absence of reported hazards despite regular exposure of the non-target organisms would indicate a negligible risk. As stated above for effects on human health, most of the methods proposed to assess the risks towards aquatic organisms, birds and beneficial arthropods are not adapted to the study of microbials. In assessing the hazard to bees, microbials must not be considered as chemicals. Potential pathogens of bees are known and there is no reference in the literature showing an adverse effect of biological control agents. On the contrary, bees can be used as vectors to deliver biological control agents (Fravel 2005). This is the case for strains of *Trichoderma* used to control *Botrytis cinerea.* The bees will deliver the conidia of the biocontrol agents on the flower where they have to be present to protect the fruit against grey mold (Shafir et al. 2007).

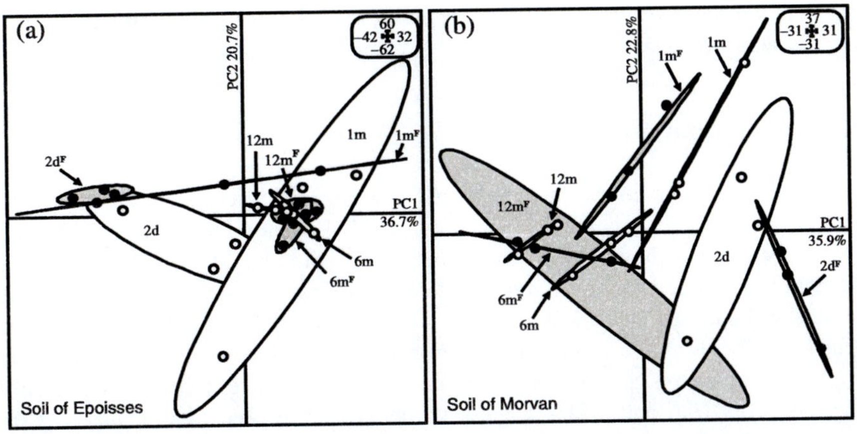


**Fig. 7.3** Fungal community structures: principal component analysis (PCA) of the 18S terminal restriction fragment length polymorphism (T-RFLP) data sets from the soils of Epoisses (**a**) and Morvan (**b**), non-inoculated (o) or inoculated (•) with *Fusarium oxysporum* Fo47b10 at four sampling times: 2 days (2d, 2d$^F$), 1 month (1m, 1m$^F$), 6 months (6m, 6m$^F$) and 12 months (12m, 12m$^F$). The soils of Epoisses and Morvan were regulated at 18 and 15% of water content, respectively and incubated at 25°C. Ellipses represent 90% confidence limits (Edel-Herman et al. 2008)

Data requirements for adverse effects on earthworms are not necessary since there is no pathogen of earthworm described in the literature and because earthworms have probably already been exposed to naturally occurring biological control agents.

Finally, modern technology helped us to demonstrate that there is no need to worry about non-target effects on soil micro-organisms. Obviously the soil microbial communities play very important roles in the ecosystem, but the soil microbiota are characterised by a redundancy of the functions. Thus, the functional characteristics of component species are as important as the number of species for maintenance of essential processes, such as nitrogen or carbon cycling. The use of molecular tools enabled the presence of genes encoding for important functions to be traced, and showed that release of a relatively small quantity of a biological control agent did not modify the soil functioning (Sessitsch et al. 2002).

Another family of methods enables a global assessment of the impact of the introduction of a biocontrol agent on the structure of the microbial communities. Results of such studies (Figures 4 and 5 taken from Edel-Herman et al. 2009) showed that even when an impact is detected shortly after introduction of the biological control agent, the structure of the microbial communities tended to revert rapidly to their initial stage. After a few weeks there was no difference in the structures of the microbial communities between the infested soil and the non-infested control. Moreover, similar studies have shown that traditional agricultural practices have much more impact on the soil microbiota and the soil functions than release of a biological control agent. This is especially the case with manure or compost amendments that release millions of unknown micro-organisms.

Based on these results REBECA proposed to waive data requirements on non target effects on soil micro-organisms.

## 7.4 Conclusion

Microbial biological control agents (viruses, bacteria and fungi) used in plant protection in the EU are regulated according to the EU council Directive 91/414 EEC. This directive was amended by the Commission Directive 2001/36/EC regarding the data requirements for the Annex I inclusion of micro-organisms as active substances and national authorisation of products (Annex IIB and IIIB). The Uniform Principles for evaluation and authorisation of plant protection products containing micro-organisms are laid down in the council Directive 2005/25 EC. Even if Directives 2001/36 and 2005/25 were written with the objective of adapting the Directive 91/414 to the case of living micro-organisms, one must say that this corpus of regulation does not permit a rapid and satisfactory evaluation of risk linked to application of biological control products.

Indeed, the Directive 2001/36 follows exactly the same scheme as the Directive 91/414, which has been written to prevent risks due to application of chemical pesticides with demonstrated negative effects on man and the environment. By contrast, based on a survey of the literature one can assume that most of the biological control agents are safe; there is a lack of evidence of proven deleterious effects. The micro-organisms used in plant protection products are very diverse and data requirements have to cover all cases. Therefore, data are formally required even if the required information does not apply to a particular active substance. Thus, following the Directive requirements, applicants are obliged to undertake expensive and time consuming studies to demonstrate that micro-organisms do not present risks, which are normally linked to the use of chemicals. Moreover, the Directive imposes some methods of investigation that are not adapted to living micro-organisms. Thus, why should we set up experiments when we know from the beginning that the results will be questionable because the methodology is not adapted to the question asked?

In our opinion, regulation and risks are not in balance; the actual regulation is too strict. It is not our opinion that all biological products are safe and, obviously, there is a need for risk assessment. However, we should take into account knowledge from the literature and accept waivers either when the question is not pertinent or the methodology is not adequate to address the question. In our opinion, there is a need for a regulation that differs from the regulation applied to chemical and is appropriately adapted to the situation with living naturally occurring micro-organisms.

## References

Alabouvette C, Olivain C, Migheli Q, Steinberg C (2009) Microbial control of soil-borne phytopathogenic fungi with special emphasis on wilt inducing *Fusarium oxysporum*. New Phytol 184:529–544

Alabouvette C, Olivain C, Steinberg C (2006) Biological control of plant pathogens: the European situation. Eur J Plant Pathol 114:329–341

Alabouvette C, Olivain C, L'Haridon F, Aimé S, Steinberg C (2007) Using strains of *Fusarium oxysporum* to control Fusarium wilts: dream or reality? In: Vurro M, Gressel J (eds) Novel biotechnologies for biocontrol agent enhancement and management, Springer, Dordrecht, NL, pp 157–178

Alabouvette C, Steinberg C (1998) Fate of microorganisms introduced into soils, effects on autochthonous communities and activities. Workshop on microbiological plant protection products: scientific basis for risk assessment, KEMI, Stockholm, 26–28/10/1998

Cordier C, Edel-Hermann V, Martin-Laurent F, Blal B, Steinberg C, Alabouvette C (2007) SCAR-based real time PCR to identify a biocontrol strain (T1) of *Trichoderma atroviride* and study its population dynamics in soils. J Microbiol Meth 68:60–68

Edel-Hermann V, Brenot S, Gautheron N, Aimé S, Alabouvette C, Steinberg C (2009) Ecological fitness of the biocontrol agent *Fusarium oxysporum* Fo47 in soil and its impact on the soil microbial communities. FEMS Microbiol Ecol 68:37–45

Edel-Hermann V, Steinberg C, Gautheron N, Brenot S, Aimé S, Alabouvette C (2008) Molecular tools to monitor a biocontrol agent and to assess its environmental impacts. In: 10th international *Fusarium* workshop and genomics workshop, August 30th September 2nd 2008. Alghero, Sardinia

Fravel D (2005) Commercialization and implementation of biocontrol. Annu Rev Phytopathol 43:337–349

Fravel D, Olivain C, Alabouvette C (2003) *Fusarium oxysporum* and its biocontrol. New Phytol 157:493–502

Graniti A (1998) Cypress canker: a pandemic in progress. Annu Rev Phytopathol 36:91–114

Hökeberg M (2006) *Pseudomonas chlororaphis* MA342 active organism in the Lantmännen BioAgri products Cedomon and Cerall. In: REBECA workshop 18–22 September 2006 Salzau, Germany

Kubicek C, Komon-Zelazowska M, Sandor E, Druzhinina IS (2007) Facts and challenges in the understanding of the biosynthesis of peptaibols by *Trichoderma*. Chem Biodivers 4:1068–1082

Magan N, Hope R, Colleate A, Baxter ES (2002) Relationship between growth and mycotoxin production by Fusarium species, biocides and environment. Eur J Plant Pathol 108:685–690

Marra R, Ambrosino P, Carbone V, Vinale F, Woo SL, Ruocco M, Ciliento R, Lanzuise S, Ferraioli S, Soriente I, Gigante S, Turrà D, Fogliano V, Scala F, Lorito M (2006) Study of the three-way interaction between *Trichoderma atroviride*, plant and fungal pathogens by using a proteomic approach. Curr Genet 50:307–321

Niskanen R, Dris R (2004) Pathogenesis-related proteins- an important group of plant-derived allergens. J Food Agr Environ 2:14–24

Sessitsch A, Howieson JG, Perret X, Antoun H, Martinez-Romero E (2002) Advances in *Rhizobium* research. Crit Rev Plant Sci 21:323–378

Shafir S, Dag A, Bilu A, Abu-Toamy M, Elad Y (2007) Honey bee dispersal of the biocontrol agent *Trichoderma harzianum T39*: effectiveness in suppressing *Botrytis cinerea* on strawberry under field conditions. Eur J Plant Pathol 116:119–128

Stoppacher N, Reithner B, Omann M, Zeilinger S, Krska R, Schuhmacher R (2007) Profiling of trichorzianines in culture samples of *Trichoderma atroviride* by liquid chromatography. Rapid Commun Mass Spectrom 21:3963–3970

Woo SL, Lorito M (2007) Exploiting the interactions between fungal antagonists, pathogens and the plant for biocontrol. In: Vurro M, Gressel J (eds) Novel biotechnologies for biocontrol agent enhancement and management. Springer, Dordrecht, NL, pp 107–130

# Chapter 8
# Ecology and Human Pathogenicity of Plant-Associated Bacteria

**Gabriele Berg, Christin Zachow, Massimiliano Cardinale, and Henry Müller**

**Abstract** Plant species and organs are colonised by diverse bacterial communities, which fulfil important functions for their host. Plant-associated bacteria have a great potential in diverse areas of biotechnology, e.g. as biological control agents (BCAs) in plant protection. Although many of them have a positive interaction with their host plants, they can interact with other eukaryotic hosts like humans in a pathogenic way. This review presents an overview about these bacteria that have bivalent interactions with plant and human hosts. We discuss mechanisms of the interactions and their behaviour and ecology. Another important issue is to detect those potentially dangerous bacteria by reliable test systems, and to exclude them from biotechnological applications. The *Caenorhabditis elegans* slow killing assay is such a bioassay, which is presented and discussed with examples. Besides human health, effects on the environment, especially on structure and function of microbial communities, are discussed. Diverse studies show that BCA application resulted only in transient, short-term effects.

## Contents



## 8.1 Molecular Ecology of Plant-Associated Bacteria

One single plant can be divided into several microenvironments, in which different biotic and abiotic conditions for microorganisms occur. There exists specific terms

G. Berg (✉)
Graz University of Technology, Environmental Biotechnology, Graz 8010, Austria
e-mail: gabriele.berg@tugraz.at

R.-U. Ehlers (ed.), *Regulation of Biological Control Agents*,
DOI 10.1007/978-90-481-3664-3_8, © Springer Science+Business Media B.V. 2011

for each microenvironment: phyllosphere (leaves), carposphere (fruit), caulosphere (stem), endorhiza (root), and endosphere (inner parts of the plant). The interface between soil and plant roots – the rhizosphere – is, due to root exudates and the resulting high nutrient content, a unique microenvironment in terrestrial ecosystems (Sørensen 1997; Raaijmakers et al. 2008). Almost all parts of plants are colonised by bacteria. Although ubiquitous and cosmopolitan plant-associated bacterial genera are known, e.g. *Pseudomonas, Bacillus* and *Methylobacterium*, specific populations have been detected for each microenvironment. For example, endophytic and ectophytic potato-associated bacterial communities differ in structure and antagonistic activity against plant pathogenic fungi (Berg et al. 2005b). This was confirmed for *Sphagnum* mosses, which belong to the phylogenetically oldest land plants (Opelt et al. 2007b). The rhizosphere is influenced by diverse parameters, e.g. by soil quality, climate, grazers and animals, pesticide treatments and plant health (Siciliano et al. 2001; Graner et al. 2003; Rasche et al. 2006; reviewed in Garbeva et al. 2004). Interestingly, the plant species also showed a significant influence on the structure and function of bacterial communities (Marschner et al. 2001; Smalla et al. 2001; Berg et al. 2002; Costa et al. 2007; reviewed in Berg and Smalla 2009).

Cultivation-independent methods on the basis of DNA/RNA developed at the end of the last century, such as microbial fingerprinting techniques or fluorescence-in-situ-hybridization (FISH) (reviewed in Smalla 2004), gave interesting insights into the structure of plant-associated bacterial communities. Figure 8.1 shows two

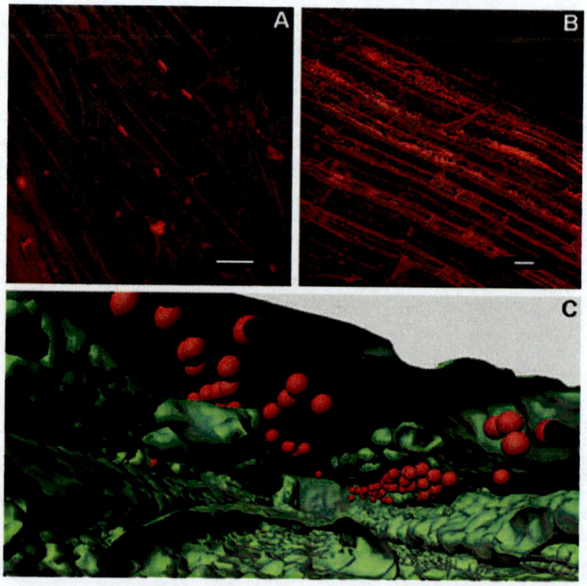


**Fig. 8.1** Confocal laser scanning microscopy of 2-week old sugar beet roots colonised by DsRed2-labeled bacteria: (**a**) *Pseudomonas fluorescens* L13-6-12 and (**b**) *Pseudomonas trivialis* RE*1-1-14. (**c**) Three-dimensional re-construction of a cross section of an area densely colonised by *Pseudomonas trivialis* RE*1-1-14 with Imaris® 6.0 (Bitplane AG, Zürich, Switzerland) clearly shows that the bacterial cells also colonised endophytic parts of the sugar beet root

examples for visualization of *DsRed2*-expressing isolates using confocal laser scanning microscopy (CLSM). While *Pseudomonas fluorescens* L13-6-12 (Grosch et al. 2005; Scherwinski et al. 2008) colonised cavities of the sugar beet rhizosphere and formed cluster (Fig. 8.1a), *P. trivialis* RE*1-1-14 (Zachow et al. 2008) colonised the surface and deeper regions of the root in lines (Fig. 8.1b). Digitalized CLSM images were converted to three-dimensional models with the Imaris® software. Surprisingly, results revealed an endophytic lifestyle of *P. trivialis* (Fig. 8.1c).

But what do we know about the functions of plant-associated bacteria? Firstly, bacteria play a role in plant growth. They can support nutrient uptake, which is well-studied for N-fixing bacteria, and they can enhance the availability of phosphorous. Furthermore, bacteria produce a broad range of phytohormones, which can have a significant influence on growth (Costacurta and Vanderleyden 1995). An interesting phenomenon is the enhancement of stress tolerance by lowering the ethylene level (Glick et al. 1998). Another important function is the involvement of plant-associated bacteria in pathogen defence. Many pathogens attack plants, especially fungi, oomycetes and nematodes. Although it is difficult to find accurate data, it is estimated that they cause yield losses of more than 30% worldwide. Whereas resistance against leaf pathogens is often encoded in the plant genome, it is difficult to find resistance genes against soil-borne pathogens. Cook et al. (1995) suggest that antagonistic rhizobacteria fulfil this function. Interestingly, besides direct antagonism, plant-associated bacteria can induce a systemic response in the plant, resulting in the activation of plant defence mechanisms (Pieterse et al. 2003). However, several studies suggest that there are many more plant-microbe interactions and resulting functions. A fascinating example is the endophytic methylobacteria, which use C1-bodies from the plant for their energy production (Zabetakis 1997). The chemical compound hydroxypropanol is given back to the plant and works as precursor of the flavour compounds mesifuran and 2,5-dimethyl-4-hydroxy-2H-furan (DMHF). The latter posses additional antifungal activity and can be responsible for pathogen defence. These bacteria also show a strong plant growth promoting effect. Interestingly, another report provided evidence that hormone-producing methylobacteria are essential for germination and development of protonema of the moss *Funaria hygrometrica* (Hornschuh et al. 2002). Another function of rhizobacteria can be the degradation of root exudates with allelopathic or even autotoxic functions (Bais et al. 2006).

To study plant-associated bacteria and their structure and functions is important not only for understanding their ecological role and the interaction with plants and plant pathogens, but also for any biotechnological application. In biotechnology, plant-associated bacteria can be applied directly for biological control of plant pathogens as biological control agents (BCAs), for growth promotion and enhancement of stress tolerance as biofertilisers and phytostimulators or as rhizoremediators (Whipps 2001; Lugtenberg et al. 2002; Berg 2009). Indirectly they can be used for production of bioactive substances, e.g. antibiotics, enzymes, volatiles, and osmoprotective substances and as new targets for drug research (Sokol et al. 2007).

## 8.2 Plants as Reservoir for Opportunistic Human Pathogenic Bacteria?

For any biotechnological application it is also important to study and assess the risk for humans and the environment. In this section, plants as a reservoir for human pathogenic bacteria will be discussed. During the last few years, it has been shown that plants, especially in the rhizosphere, can harbour not only beneficial bacteria, but also those that potentially can cause diseases in humans (Berg et al. 2005a; Opelt et al. 2007a). These pathogens are called opportunistic <Lat. = highly adaptable> or facultative human pathogens and they cause diseases only in patients with a strong predisposition to illness, particularly in those who are severely debilitated, immuno-compromised or suffering from cystic fibrosis or HIV infections (Parke and Gurian-Sherman 2001; Steinkamp et al. 2005). This group of bacteria cause the majority of bacterial infections associated with significant case/fatality ratios in susceptible patients in Europe and Northern America. A special group are those bacteria responsible for hospital-acquired diseases which are called nosocomial infections. For example, in intensive care units in Europe 45% of the patients were infected by opportunistic pathogens (Vincent et al. 1995). In the last two decades, the impact of opportunistic infections on human health has increased dramatically.

Many plant-associated genera, including *Burkholderia, Enterobacter, Herbaspirillum, Ochrobactrum, Pseudomonas, Ralstonia, Serratia, Staphylococcus* and *Stenotrophomonas* contain root-associated bacteria that enter bivalent interactions with plant and human hosts. Several members of these genera show plant growth promoting as well as excellent antagonistic properties against plant pathogens and therefore were utilised as BCAs and for the development of biological control products (Whipps 2001). However, many strains also successfully colonise human organs and tissues and thus cause diseases. The problems with biofungicides based on strains of the genus *Burkholderia* underlines the importance of thorough risk assessment studies prior to registration (Govan et al. 2000; Parke and Gurian-Sherman 2001).

Furthermore, it is necessary to understand plant-microbe interactions as well as the important question "What turns bacteria into opportunistic pathogens?". Plant-associated bacteria with antagonistic activity against eukaryotes are able to interact with their hosts using various mechanisms. These mechanisms include (i) inhibition of pathogens by antibiotics, toxins and bio-surfactants [antibiosis], (ii) competition for colonisation sites and nutrients, (iii) competition for minerals, e.g. for iron through production of siderophores or efficient siderophore-uptake systems, (iv) degradation of pathogenicity factors of the pathogen such as toxins, and (v) parasitism that may involve production of extra-cellular, cell wall-degrading enzymes such as chitinases and β-1,3 glucanases (Lugtenberg et al. 2002; Raaijmakers et al. 2008). Furthermore, the importance for all plant-associated bacteria to recognise and adhere to plant roots is underlined in many biocontrol studies (Lugtenberg and Dekkers 1999). Other factors that contribute to rhizosphere fitness include the ability to use seed and root exudates as carbon sources or, more generally, ecological and nutritional versatility. In addition, synthesis of compatible solutes by bacteria contributes to survival under changing osmolarities, which occur in the rhizosphere (Miller and Wood 1996). Steps of pathogenesis are similar and include invasion, colonisation and growth, and several strategies to establish virulence (the relative ability of a pathogen to cause disease in the host). In addition, recognition and adherence to human cells is necessary to establish pathogenicity. Many mechanisms involved in the interaction between antagonistic plant-associated bacteria and their host plants are similar to those responsible for pathogenicity of bacteria (Rahme et al. 1995). These mechanisms may also be involved in colonising the human body (Cao et al. 2001), as shown in Fig. 8.2. An additional important feature, which is necessary to survive on/in humans, is the ability to grow at 37°C. Interestingly, we found that the majority of rhizobacteria isolated from oilseed rape and strawberry in Northern Germany are able to grow at 37°C (G. Berg, unpublished results).

Several studies provided evidence that similar or even identical functions are responsible for beneficial interactions with plants and virulence in humans. For example, the involvement of siderophore-uptake systems or extra-cellular enzymes is common to both beneficial bacteria and human pathogens (Tan et al. 1999). Dörr et al. (1998) reported that type IV pili of the plant-associated *Azoarcus* sp. BH72 are responsible for the adhesion on plant and fungal cells. Furthermore, the amino acid sequence of the pilus showed a high similarity to pili of the human-associated strains of *P. aeruginosa* and *Neisseria gonorrhoeae*. While a mutant of *Pseudomonas fluorescens* deficient in a laurolyl transferase involved in lipid A biosynthesis resulted in an impaired root colonisation (Dekkers et al. 1998), a similar mutant of *Salmonella typhimurium* is limited in its ability to colonise organs of the lymphatic system of mice (Jones et al. 1997). Type III secretion systems are responsible for the introduction of effectors into eukaryotic host cells; they were found for pathogenic bacteria

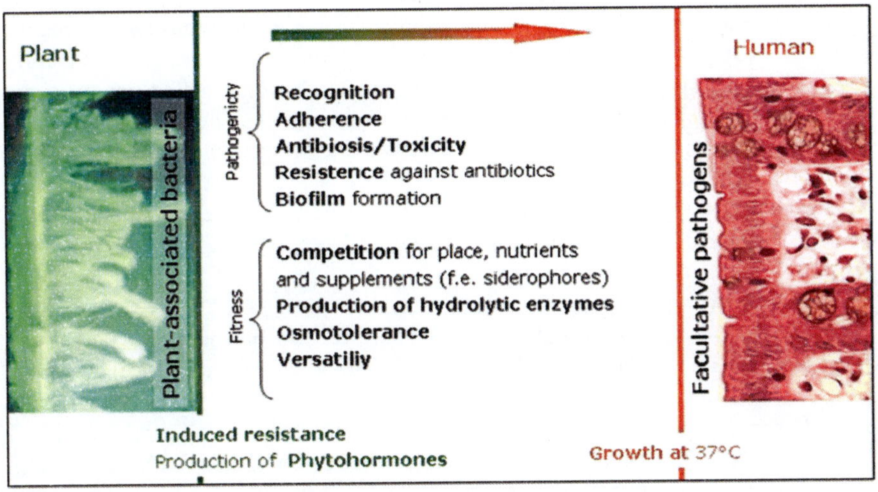


**Fig. 8.2** Potential mechanisms of plant-associated bacteria to interact with plants and humans

as well as plant-associated bacteria with beneficial effects on host plants (Preston et al. 2001).

In a study published by Alonso et al. (1999) it was shown that clinical and environmental isolates of *P. aeruginosa*, which is the major causal agent for morbidity and mortality of patients with cystic fibrosis, share several phenotypic traits with respect to both virulence and environmental properties. Several studies support the view that the environmental strains are indistinguishable from those from clinical sources in terms of genotypic, taxonomic or metabolic properties (Kiewitz and Tümmler 2000; Wolfgang et al. 2003; Finnan et al. 2004). Restriction fragment length polymorphism based on 14 single nucleotide polymorphisms (SNPs) of conserved loci in 111 *P. aeruginosa* isolates of diverse habitats allowed specific fingerprinting and a discrimination of strains (Morales et al. 2004). Interestingly, the highly virulent clinical strain CHA shared their SNP genotype with two environmental strains, which again supports the view that *P. aeruginosa* isolates thriving in non-clinical habitats possess all the functions required potentially to infect mammals. In addition, differences between environmental strains and those that cause infections may occur at the level of regulation of genes, rather than their presence or absence (Parke and Gurian-Sherman 2001). Similar studies to that on *P. aeruginosa* were published on *Stenotrophomonas maltophilia* (Berg et al. 1999; Minkwitz and Berg 2001; Ribbeck-Busch et al. 2005; Hagemann et al. 2008; reviewed in Ryan et al. 2009) and *Burkholderia cepacia* (Parke and Gurian-Sherman 2001). Nevertheless, antagonism studies and biocontrol effects were reported for all mentioned species, and one product derived from *B. cepacia* was on the market (Hebbar et al. 1998; Nakayama et al. 1999; Dunne et al. 2000; Govan et al. 2000). All species are common inhabitants of the rhizosphere; due to their medical relevance they are grouped into risk group 2 in the public databases, e.g. those by the German Collection of Microorganisms and Cell Cultures (www.dsmz.de), and should be excluded from direct biotechnological applications.

An important mechanism by which harmless bacteria can behave as pathogens is change of host or host niche, upon which their virulence potential is frequently revealed to its full extend. This mechanism is clearly relevant for opportunistic pathogens from plant-associated habitats. In addition, other mechanisms such as structural changes of the bacterial chromosome due to gene acquisition and loss, recombination and mutations can lead to bacterial pathogenicity (for a review see Hacker et al. 2003). Genes responsible for pathogenicity or fitness of bacteria often occur as genomic islands, which are blocks of DNA with signatures of mobile genetic elements (Hacker and Carniel 2001). They are called "fitness islands" or "pathogenicity islands" according to their function.

Plant-associated bacteria with a high capacity for biocontrol can be potentially dangerous for human health. Therefore, it is important to understand the mode of action and specific properties of the BCA. It is well known that antagonistic properties and underlying mechanisms are highly strain-specific (Berg et al. 2002, 2006) but identification of bacteria is based mainly on 16S rDNA sequencing. Thus, from sequencing information it is difficult to draw conclusions about potential pathogenicity: neutral bacterial strains can be dangerous due to pathogenicity islands or pathogenic bacteria can be harmless because of the absence of any pathogenicity factor. Therefore, models to assess the pathogenicity of individual BCAs are important as risk assessment studies.

## 8.3 *Caenorhabditis elegans*: A Model to Assess Pathogenicity Factors

To assess the pathogenic potential is particularly difficult in many opportunistic human pathogens as well as BCAs because of the lack of adequate animal models. Until now, this procedure for BCAs is based on rules originally developed for synthetic pesticides (EU Council Directive 91/414/EEC, see also http://www.rebeca-net.de). Methods adopted from standardized tests for chemical-based agents, including elaborate animal tests are not only time-consuming and expensive but also their results are difficult to interpret. Pathogenicity and the mode of action of facultative pathogens such as *Burkholderia* and *Stenotrophomonas* could not be analysed in traditional animal models. Therefore, alternative models using the slime mould *Dictyostelium discoideum* (Alonso et al. 2004) or the nematode *Caenorhabditis elegans* (Tan et al. 1999) were developed. The model organism *C. elegans* has valuable advantages, enabling it to be used in many bacteria-pathogen interaction analyses to evaluate the pathogenic potential of these bacteria (Aballay and Ausubel 2002; Cardona et al. 2005). *C. elegans* is a free-living terrestrial nematode that feeds on bacteria in its environment (Beale et al. 2006). Self-reproducing hermaphrodites grow to 1 mm in length and have simple growth requirements. The worm has a small and fixed number of 959 cells. Moreover, it is transparent and therefore easy to observe under a standard light microscopy. Extensive information about *C. elegans* research is available in well-resourced internet databases (www.wormbook.org, www.wormbase.org). Calculations made after the completion of the *C. elegans* genome sequence (Wilson 1999) indicated that 74% of human gene sequences had nematode matches (Consortium 1998).

In an extensive study, we applied a broad range of BCAs, pathogens and plant-associated bacteria to a rapid and inexpensive bioassay, using *C. elegans* to estimate the risk of bacteria to harm human health (Zachow et al. 2009). The nematode killing assay described as slow killing assay by Köthe et al. (2003) was used. Movement and reproduction behaviour of nematodes with BCAs were compared with those fed with the human pathogen *Pseudomonas aeruginosa* QC14-3-8 (positive control) and *Escherichia coli* OP50 (negative control). In Fig. 8.3, the kinetics of killing *C. elegans* under slow killing conditions is shown for three BCAs (*Serratia plymuthica* HRO C48 = RhizoStar©, *Bacillus subtilis* B2g = Phytovit© and *Pseudomonas trivialis* 3Re2-7 = Salavida©) and two rhizobacteria (*Serratia liquefaciens* and *Salmonella thyphimurium*). The latter were selected according to their identification and grouping into risk group 2. Indeed, bacteria from risk group 2 showed a significantly higher rate of killing. Altogether, results indicate that the nematode *C. elegans* provides a reliable model system to assess the human pathogenic potential of BCAs prior implementation of extensive studies using animal test systems.

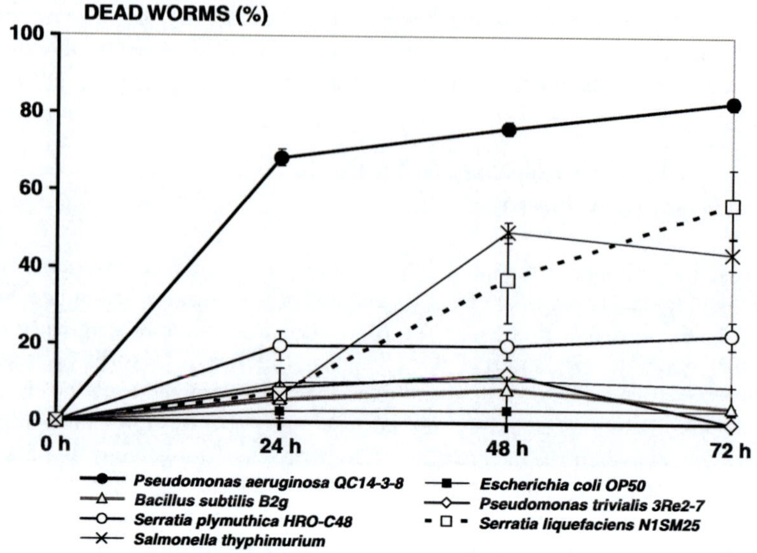


**Fig. 8.3** Kinetics of killing of *Caenorhabditis elegans* by different biological control agents under slow killing conditions. Worms grown on NGMII and feeding on *Pseudomonas aeruginosa* QC14-3-8 (positive control, *black circle*), *Escherichia coli* OP50 (negative control, *black square*). The biological control products were represented by *Bacillus subtilis* B2g = Phytovit© (white triangle), *Pseudomonas trivialis* 3Re2-7 = Salavida© (*white diamond*), and *Serratia plymuthica* HRO C48 = RhizoStar© (*white circle*). The rhizobacteria were represented by *Serratia liquefaciens* N1SM25 (*white square*) and *Salmonella typhimurium* LT2 (*black cross*). Data points represent means ± standard errors of at least five independent experiments

The *C. elegans* assay can be integrated into initial screens for BCAs and is useful to exclude pathogens in a very early stage of the product development.

There are some restrictions for the *C. elegans* assay. The model of pathogenicity is limited by the amount of bacteria infecting the worm, which was shown for *P. aeruginosa, P. fluorescens, Serratia marcescens, Burkholderia cepacia, B. pseudomallei, B. thailandensis, Salmonella* spp. and *Bacillus megaterium* (Aballay et al. 2000; Tan and Ausubel 2000). Therefore, in this study we used an overnight culture with approximately $10^7$ cells/ml, which provides an appropriate thin cell layer to evaluate the behaviour of the transparent worm on the Petri dishes. In a pilot study, this concentration of cells was found to permit detection of differences in survival among worm strains after 24 h (Schulenburg and Ewbank 2004). Furthermore, the developmental stage of the applied worms influenced the slow killing. Adult worms were more sensitive and died faster than fourth-stage larvae. Therefore, in this study we used second-stage larvae in the assay, exactly 48 h after egg preparation. Another restriction is associated with *B. thuringiensis*, a well-known BCA. Although *B. thuringiensis* is used world-wide against insect pests without reports that it has caused harm to humans, the bioactive toxin acted against *C. elegans* (Devidas and Rehberger 1992).

**Table 8.1** Risk assessment studies for biological control agents and antagonistic bacteria

| Strains | Plant/pathosystem | Results | References |
|---|---|---|---|
| *Pseudomonas putida* QC14-3-8<br>*Serratia grimesii* L16-3-3 | Potato | No differences between the inoculated and not-inoculated communities | Lottmann et al. (2000) |
| *Pseudomonas fluorescens* CHA0 and GMO | Cucumber | Differences in the composition and/or relative abundance of species in the fungal community, no effect on species diversity indices, and species abundance<br>Impact of treatments was smaller than the effect of growing cucumber repeatedly in the same soil | Girlanda et al. (2001) |
| *Pseudomonas putida* WCS358r and GMOs | Wheat | Transient change in the composition of the rhizosphere microflora<br>No influence on soil microbial activities | Glandorf et al. (2001)<br>Viehbahn et al. (2003) |
| *Pseudomonas fluorescens* F113Rif | Clover | No influence on the structure of the *Rhizobium* community<br>Small influence on the proportion of Phl-sensitive isolates | Walsh et al. (2003) |
| *Microbispora* sp. strain EN2<br>*Streptomyces* sp. strain EN27<br>*Nocardioides albus* EN46<br>Single isolates; mixture | Wheat | Treatment with mixture disrupted the natural actinobacterial endophyte population, reducing diversity and colonization levels<br>Treatment with single isolates – population was not adversely affected | Conn and Franco (2004) |
| *Pseudomonas fluorescens* strains 2-79, Q8rl-96, and a recombinant strain, Z30-97 | Wheat | Inoculation with Z30-97 resulted in several shifts in rhizosphere bacterial community structure | Blouin-Bankhead et al. (2004) |
| *Serratia plymuthica* HRO-C48<br>*Streptomyces* sp. HRO-71 | Strawberry and potato – *Verticillium dahliae* | Only negligible, short-term effects due to the bacterial treatments | Scherwinski et al. (2006) |
| *Serratia plymuthica* 3Re4-18<br>*Pseudomonas trivialis* 3Re2-7<br>*Pseudomonas fluorescens* L13-6-12 | Lettuce – *Rhizoctonia solani* | Only negligible, short-term effects due to the bacterial treatments | Scherwinski et al. (2007), (2008) |
| *Pseudomonas fluorescens* SBW25 | Wheat | Only minor impacts were found on native microflora due to bacterial (GMO or wild-type) inoculation | Jäderlund (2008) |

## 8.4 Influence of Antagonistic Bacteria on Indigenous Microbial Communities

Although originating from plant-associated microenvironments themselves, beneficial bacteria, if applied to plant roots in adequate numbers, may perturb indigenous microbial populations and their associated important ecological functions. Therefore, unwanted, unspecific actions of the introduced beneficial microbes against non-target organisms have to be assessed. To this end, sufficient knowledge about the microbial ecology of the target habitats is necessary for reasonable risk assessment studies concerning the release of beneficial microorganisms. As only a small proportion of the microorganisms can be analysed by common cultivation techniques, several DNA-based, cultivation-independent methods, have been developed to overcome the limitations of cultivation (reviewed in Smalla 2004). The use of such molecular methods is urgently needed in order to include the highest possible number of total microorganisms in risk assessment studies to determine non-target effects of introduced beneficial bacteria (Winding et al. 2004). Several studies using cultivation-independent methods exist. They focus mainly on the effects of genetically modified microorganisms (GMOs) such as *Pseudomonas* (Viebahn et al. 2003; Glandorf et al. 2001) and *Sinorhizobium* (Schwieger and Tebbe 2000) on non-target microorganisms. Examples of studies, which analysed the fate and ecosystem effects of introduced BCAs and antagonistic bacteria, are given in Table 8.1. Generally it can be concluded from these studies that the impact of bacterial inoculants is either negligible or small compared with effects of general agricultural practices, and more or less all effects are transient. However, for strains with a strong production of antifungal antibiotics or genetically modified strains with additional genes to synthesise antibiotics, effects were observed (Viehbahn et al. 2003; Walsh et al. 2003; Blouin-Bankhead et al. 2004).

## 8.5 Conclusions

Plant-associated bacteria are an interesting bio-resource for biotechnology. For example, biocontrol offers environmentally friendly and sustainable alternatives for the control of plant pathogens. On the other hand, problems with opportunistic infections, which are originally from plants, will become even more severe due to the increasing numbers of at-risk individuals in the human population. Therefore, it is important to understand the biology and ecology of plant-associated bacteria, especially the ambivalent strains. It is important to exclude potential pathogenic bacteria at an early stage of product development. Criteria are growth at 37°C, grouping in risk groups (www.dsmz.de or Dir. 2000/54 EC) or any toxic effect in the *C. elegans* assay. Otherwise, more research and toxicological data are necessary for risk assessment. In studies assessing the risk for the environment, mainly short-term effects were reported. New studies seem to be only relevant if strong antimicrobial metabolites are produced by the BCA.

**Acknowledgements** We would like to thank Jamshid Fathei (Uppsala, Sweden) for cooperation regarding DsRed2-transformation of bacteria and Anton Hartmann (Neuherberg, Germany) for help with Fig. 8.2.

## References

Aballay A, Ausubel FM (2002) *Caenorhabditis elegans* as a host for the study of host-pathogen interactions. Curr Opin Microbiol 5:97–101

Aballay A, Yorgey P, Ausubel FM (2000) *Salmonella typhimurium* proliferates and establishes a persistent infection in the intestine of *Caenorhabditis elegans*. Curr Biol 10:1539–1542

Alonso A, Morales G, Escalante R, Campanario E, Sastre L, Martinez JL (2004) Overexpression of the multidrug efflux pump SmeDEF impairs *Stenotrophomonas maltophilia* physiology. J Antimicrob Chemother 53:432–434

Alonso A, Rojo F, Martinez JL (1999) Environmental and clinical isolates of *Pseudomonas aeruginosa* show pathogenic and biodegradative properties irrespective their origin. Environ Microbiol 1:421–430

Bais HP, Weir TL, Perry LG, Gilroy S, Vivanco JM (2006) The role of root exudates in rhizosphere interactions with plants and other organisms. Ann Rev Plant Biol 57:234–266

Beale E, Li G, Tan MW, Rumbaugh KP (2006) *Caenorhabditis elegans* senses bacterial autoinducers. Appl Environ Microbiol 72:5135–5137

Berg G (2009) Plant-microbe interactions promoting plant growth and health: perspectives for controlled use of microorganisms in agriculture. J Appl Microbiol Biotechnol 84:11–18

Berg G, Eberl L, Hartmann A (2005a) The rhizosphere as a reservoir for opportunistic human pathogenic bacteria. Environ Microbiol 71:4203–4213

Berg G, Krechel A, Ditz M, Faupel A, Ulrich A, Hallmann J (2005b) Endophytic and ectophytic potato-associated bacterial communities differ in structure and antagonistic function against plant pathogenic fungi. FEMS Microbiol Ecol 51:215–229

Berg G, Opelt K, Zachow C, Lottmann J, Götz M, Costa R, Smalla K (2006) The rhizosphere effect on bacteria antagonistic towards the pathogenic fungus *Verticillium* differs depending on plant species and site. FEMS Microbiol Ecol 56:250–261

Berg G, Roskot N, Smalla K (1999) Genotypic and phenotypic relationship in clinical and environmental isolates of *Stenotrophomonas maltophilia*. J Clin Microbiol 37: 3594–3600

Berg G, Roskot N, Steidle A, Eberl L, Zock A, Smalla K (2002) Plant-dependent genotypic and phenotypic diversity of antagonistic rhizobacteria isolated from different *Verticillium* host plants. Appl Environ Microbiol 68:3328–3338

Berg G, Smalla K (2009) Plant species and soil type cooperatively shape the structure and function of microbial communities in the rhizosphere. FEMS Microbiol Ecol 68:1–13

Blouin-Bankhead S, Landa BB, Lutton E, Weller DM, McSpadden Gardener BB (2004) Minimal changes in the rhizobacterial population structure following root colonization by wild type and transgenic biocontrol strains. FEMS Microbiol Ecol 49:307–318

Cao H, Baldini RL, Rahme LG (2001) Common mechanisms for pathogens of plants and animals. Ann Rev Phytopathol 39:259–284

Cardona ST, Wopperer J, Eberl L, Valvano MA (2005) Diverse pathogenicity of *Burkholderia cepacia* complex strains in the *Caenorhabditis elegans* host model. FEMS Microbiol Lett 250:97–104

Conn VM, Franco CMM (2004) Effect of microbial inoculants on the indigenous actinobacterial endophyte population in the roots of wheat as determined by Terminal Restriction Fragment Length Polymorphism. Appl Environ Microbiol 70:6407–6413

Consortium (The *C. elegans* Sequencing Consortium) (1998) Genome sequence of the nematode *C. elegans*: a platform for investigating biology. Science 282:2012–2018

Cook RJ, Tomashow LS, Weller DM, Fujimoto D, Mazzola M, Bangera G, Kim DS (1995) Molecular mechanisms of defense by rhizobacteria against root disease. Proc Natl Acad Sci USA 92:4197–201

Costa R, Gomes NC, Krogerrecklenfort E, Opelt K, Berg G, Smalla K (2007) *Pseudomonas* community structure and antagonistic potential in the rhizosphere: insights gained by combining phylogenetic and functional gene-based analyses. Environ Microbiol 9:2260–2273

Costacurta A, Vanderleyden J (1995) Synthesis of phytohormones by plant-associated bacteria. Crit Rev Microbiol 21:1–18

Dekkers LC, van der Bij AJ, Mulders IHM, Phoelich CC, Wentwoord RAR, Glandorf DCM, Lugtenberg BJJ (1998) Role of the O-antigen of lipopolysaccharide, and possible roles of growth rate and NADH:ubiqunone oxidoreductase (nuo) in competitive tomato root-tip colonization by *Pseudomonas fluorescens* WCS365. Mol Plant-Microbe Interact 11:763–771

Devidas P, Rehberger LA (1992) The effects of exotoxin (Thuringiensin) from *Bacillus thuringiensis* on *Meloidogyne incognita* and *Caenorhabditis elegans*. Plant Soil 145:115–120

Dunne C, Moënne-Loccoz Y, de Bruijn FJ, OGara F (2000) Overproduction of an inducile extracellular serine protease improves biological control of *Pythium ultimum* by *Stenotrophomonas maltophilia* strain W81. Microbiol 146:2069–2078

Dörr J, Hurek T, Reinhold-Hurek B (1998) Type IV pili are involved in plant–microbe and fungus–microbe interactions. Mol Microbiol 30:7–17

Finnan S, Morrissey JP, O'Gara F, Boyd EF (2004) Genome diversity of *Pseudomonas aeruginosa* isolates from cystic fibrosis patients and the hospital environment. J Clin Microbiol 42:5783–5792

Garbeva P, van Veen JA, van Elsas JD (2004) Microbial diversity in soil: selection of microbial populations by plant and soil type and implications for disease suppressiveness. Annu Rev Phytopathol 42:243–270

Girlanda M, Perotto S, Moenne-Loccoz Y, Berbero R, Lazzari A, Defago G, Bonfante P, Luppi P (2001) Impact of biocontrol *Pseudomonas fluorescens* CHA0 and a genetically modified derivative on the diversity of culturable fungi in the cucumber rhizosphere. Appl Environ Microbiol 67:1851–1864

Glandorf DCM, Verheggen P, Jansen T, Jorritsma JW, Smit E, Leeflang P et al (2001) Effect of genetically modified *Pseudomonas putida* WCS358r on the fungal rhizosphere micro-flora of field-grown wheat. Appl Environ Microbiol 67:3371–3378

Glick BR, Penrose DM, Li JP (1998) A model for the lowering of plant ethylene concentrations by plant growth-promoting bacteria. J Theor Biol 190:63–68

Govan JRW, Balendreau J, Vandamme P (2000) *Burkholderia cepacia* – friend and foe. ASM News 66:124–125

Graner G, Persson P, Meijer J, Alstrom S (2003) A study on microbial diversity in different cultivars of Brassica napus in relation to its wilt pathogen, *Verticillium longisporum*. FEMS Microbiol Lett 29:269–276

Grosch R, Faltin F, Lottmann J, Kofoet A, Berg G (2005) Effectiveness of 3 antagonistic bacterial isolates to control *Rhizoctonia solani* Kühn on lettuce and potato. Can J Microbiol 51:345–353

Hacker J, Carniel E (2001) Ecological fitness, genomic islands and bacterial pathogenicity. EMBO reports 2:376–381

Hacker J, Hentschel U, Dobrindt U (2003) Prokaryotic chromosomes and diseases. Science 301:790–793

Hagemann M, Ribbeck-Busch K, Klähn S, Hasse D, Steinbruch R, Berg G (2008) The plant-associated bacterium *Stenotrophomonas rhizophila* expresses a new enzyme for the synthesis of the compatible solute glucosylglycerol. J Bac 190:5898–5906

Hebbar KP, Martel MH, Heulin T (1998) Suppression of pre- and postemergence damping-off in corn by *Burkholderia cepacia*. Eur J Plant Pathol 104:29–36

Hornschuh M, Grotha R, Kutschera U (2002) Epiphytic bacteria associated with the bryophyte *Funaria hygrometrica*: effect of *Methylobacterium* strains on protonema development. Plant Biol 4:682–682

Jäderlund J (2008) Fates and impact of the genetically modified plant growth promoting bacterium *Pseudomonas fluorescens* SBW25. PhD thesis Swedish University of Agricultural Sciences Uppsala

Jones BD, Nichols WA, Gibson BW, Sunshine MG, Apicella MA (1997) Study of the role of the htrB gene in *Salmonella typhimurium* virulence. Infect Immun 65:4778–4783

Kiewitz C, Tümmler B (2000) Sequence diversity of *Pseudomonas aeruginosa*: impact on population structure and genome evolution. J Bacteriol 182:3125–3135

Köthe M, Antl M, Huber B, Stoecker K, Ebrecht D, Steinmetz I, Eberl L (2003) Killing of *Caenorhabditis elegans* by *Burkholderia cepacia* is controlled by the cep quorum-sensing system. Cell Microbiol 5:343–351

Lottmann J, Heuer H, de Vries J, Mahn A, Düring K, Wackernagel W, Smalla K, Berg G (2000) Establishment of introduced antagonistic bacteria in the rhizosphere of transgenic potatoes and their effect on the bacterial community. FEMS Microbiol Ecol 33:41–49

Lugtenberg BJJ, Chin-A-Woeng TFC, Bloemberg GV (2002) Microbe-plant interactions: principles and mechanisms. Antonie van Leewenhoek 81:373–383

Lugtenberg BJJ, Dekkers LC (1999) What makes *Pseudomonas* bacteria rhizosphere competent?. Environ Microbiol 1:9–13

Marschner P, Yang CH, Lieberei R, Crowley DE (2001) Soil and plant specific effects on bacterial community composition in the rhizosphere. Soil Biol Biochem 33:1437–1445

Miller KJ, Wood JM (1996) Osmoadaption by rhizosphere bacteria. Ann Rev Microbiol 50:101–136

Minkwitz A, Berg G (2001) Comparison of antifungal activities and 16S ribosomal DNA sequences of clinical and environmental isolates of *Stenotrophomonas maltophilia*. J Clin Microbiol 39:139–145

Morales G, Wiehlmann L, Gudowius P, van Deldenb C, Tümmler B, Martinez JL, Rojo F (2004) Structure of *Pseudomonas aeruginosa* populations analyzed by single nucleotide polymorphisms and pulsed-field gel electrophoresis. J Bacteriol 186:4228–4237

Nakayama T, Homma Y, Hashidoko Y, Mitzutani J, Tahara S (1999) Possible role of xanthobaccins produced by *Stenotrophomonas* sp. strain SB-K88 in suppression of sugar beet damping-off disease. Appl Environ Microbiol 65:4334–4339

Opelt K, Berg C, Berg G (2007a) The bryophyte genus *Sphagnum* is a reservoir for powerful and extraordinary antagonists and potentially facultative human pathogens. FEMS Microb Ecol 61:38–53

Opelt K, Chobot V, Hadacek F, Schönemann S, Eberl L, Berg G (2007b) Investigations of the structure and function of bacterial communities associated with *Sphagnum* mosses. Environ Microbiol 9:2795–2809

Parke JL, Gurian-Sherman D (2001) Diversity of the *Burkholderia cepacia* complex and implications for risk assessment of biological control strains. Ann Rev Phytopathol 39: 225–258

Pieterse CMJ, van Pelt JA, Verhagen BWM, Ton J, van Wees SCM, Lon-Klosterziel KM, van Loon LC (2003) Induced systemic resistance by plant growth promoting rhizobacteria. Symbiosis 35:39–54

Preston GM, Bertrand N, Rainey PB (2001) Type III secretion in plant growth promoting *Pseudomonas fluorescens* SBW25. Mol Microbiol 41:999–1014

Raaijmakers JM, Paulitz CT, Steinberg C, Alabouvette C, Moenne-Loccoz Y (2008) The rhizosphere: a playground and battlefield for soilborne pathogens and beneficial microorganisms. Plant Soil. doi:10.1007/s11104-008-9568-6

Rahme LG, Stevens EJ, Wolfort SF, Shoa J, Tompkins RG, Ausubel FM (1995) Common virulence factors for bacterial pathogenicity in plants and animals. Science 268:1899–1902

Rasche F, Hodl V, Poll C, Kandeler E, Gerzabek MH, van Elsas JD, Sessitsch A (2006) Rhizosphere bacteria affected by transgenic potatoes with antibacterial activities compared with the effects of soil, wild-type potatoes, vegetation stage and pathogen exposure. FEMS Microbiol Ecol 56:219–235

Ribbeck-Busch K, Roder A, Hasse D, de Boer W, Martínez JL, Hagemann M, Berg G (2005) A molecular biological protocol to distinguish potentially human pathogenic *Stenotrophomonas maltophilia* from plant–associated *Stenotrophomonas rhizophila*. Environ Microbiol 7:1853–1858

Scherwinski K, Grosch R, Berg G (2007) Root application of bacterial antagonists to field-grown lettuce: effects on non-target micro-organisms and disease suppression. IOBC/WPRS Bull 30:255–257

Scherwinski K, Grosch R, Berg G (2008) Effect of bacterial antagonists on lettuce: active biocontrol of *Rhizoctonia solani* and negligible, short-term effect on non-target microbes. FEMS Microb Ecol 64:106–116

Scherwinski K, Wolf A, Berg G (2006) Assessing the risk of biological control agents on the indigenous microbial communities: *Serratia plymuthica* HRO-C48 and *Streptomyces* sp. HRO-71 as model bacteria. BioControl 52:87–112

Schulenburg H, Ewbank JJ (2004) Diversity and specificity in the interaction between *Caenorhabditis elegans* and the pathogen *Serratia marcescens*. BMC Evol Biol 4:49–56

Schwieger F, Tebbe CC (2000) Effect of field inoculation with *Sinorhizobium meliloti* L33 on the composition of bacterial communities in rhizospheres of a target plant (*Medicago sativa*) and a non-target plant (*Chenopodium album*)-linking of 16S rRNA gene-based single-strand conformation polymorphism community profiles to the diversity of cultivated bacteria. Appl Environ Microbiol 66:3556–3565

Siciliano DS, Forin N, Mihoc A et al (2001) Selection of specific endophytic bacterial genotypes by plant in response to soil contamination. Appl Environ Microbiol 67:2469–2475

Smalla K (2004) Culture-independent microbiology. In: Bull AT (ed) Microbial diversity and bioprospecting. ASM, Washington DC, pp 88–99

Smalla K, Wieland G, Buchner A, Zock A, Parzy J, Roskot N, Heuer H, Berg G (2001) Bulk and rhizosphere soil bacterial communities studied by Denaturing Gradient Gel Electrophoresis: plant dependent enrichment and seasonal shifts. Appl Environ Microbiol 67: 4742–4751

Sokol PA, Malott RJ, Riedel K, Eberl L (2007) Communication systems in the genus *Burkholderia*: global regulators and targets for novel antipathogenic drugs. Future Microbiol 2:555–563

Steinkamp G, Wiedemann B, Rietschel E, Krahl A, Giehlen J, Barmeier H, Ratjen F (2005) Prospective evaluation of emerging bacteria in cystis fibrosis. J Cyst Fibros 4:41–48

Sørensen J (1997) The rhizosphere as a habitat for soil microorganisms. In: Van Elsas JD, Trevors JT, Wellington EMH (eds) Modern Soil Microbiology. Marcel Dekker, Hongkong, pp 21–45

Tan MW, Ausubel FM (2000) *Caenorhabditis elegans* a model genetic host to study *Pseudomonas aeruginosa* pathogenesis. Curr Opin Microbiol 3:29–34

Tan MW, Rahme LG, Sternberg JA, Tompkins RG, Ausubel FM (1999) *Pseudomonas aeruginosa* killing of *Caenorhabditis elegans* used to identify *P. aeruginosa* virulence factors. Proc Natl Acad Sci USA 96:2408–2413

Viebahn M, Glandorf DCM, Ouwens TWM, Smit E, Leeflang P, Wernars K, Thomashow LS, Van Loon LC, Bakker PAHM (2003) Repeated introduction of genetically modified *Pseudomonas putida* WCS358r without intensified effects on the indigenous microflora of field-grown wheat. Appl Environ Microbiol 69:3110–3118

Vincent JL, Bihari DJ, Suter PM, Bruining HA, White J, Nicolas-Chanoin MH, Wolff M, Spencer RC, Hemmer M (1995) The prevalence of nosocomial infection in intensive care units in Europe. Results of the European Prevalence of Infection in Intensive Care (EPIC) Study. EPIC International Advisory Committee. JAMA 274:639–644

Walsh UF, Moënne-Loccoz Y, Tichy H-V, Gardner A, Corkery DM, Lorkhe S, OGara F (2003) Residual impact of the biocontrol inoculant *Pseudomonas fluorescens* F113 on the resident population of rhizobia nodulating a red clover rotation crop. Microbiol Ecol 45:145–155

Whipps J (2001) Microbial interactions and biocontrol in the rhizosphere. J Exp Bot 52:487–511

Wilson RK (1999) How the worm was won The *C. elegans* genome sequencing project. Trends Genet 15:51–58

Winding A, Binnerup SJ, Pritchard H (2004) Non-target effects of bacterial biological control agents suppressing root pathogenic fungi. FEMS Microbiol Ecol 47:129–141

Wolfgang MC, Kulasekara BR, Liang X, Boyd D, Wu K, Yang Q, Miyada CG, Lory S (2003) Conservation of genome content and virulence determinants among clinical and environmental isolated of *Pseudomonas aeruginosa*. PNAS 100:88484–8489

Zabetakis I (1997) Enhancement of flavour biosynthesis from strawberry (*Fragaria x ananassa*) callus cultures by *Methylobacterium* species. Plant Cell Tissue Organ Culture 50:179–183

Zachow C, Pirker H, Westendorf C, Tilcher R, Berg G (2009) *Caenorhabditis elegans* provides a valuable tool to evaluate the human pathogenic potential of bacterial biocontrol agents. Eur J Plant Pathol 125:367–376

Zachow C, Tilcher R, Berg G (2008) Sugar beet-associated bacterial and fungal communities show a high indigenous antagonistic potential against plant pathogens. Microb Ecol 55:119–129

# Chapter 9
# Metabolite Toxicology of Fungal Biocontrol Agents

**Hermann Strasser, Stefan Hutwimmer, and Wolfgang Burgstaller**

**Abstract** Up to 300,000 potentially unique mycotoxins were reported by the Council of Agricultural Science Technology in 2003. Nevertheless, little information is available on fungal biological control agents (FBCAs) that have been developed or are being developed. The knowledge on fungal secondary metabolites and their toxicological significance depends on what is already known in the published scientific literature or in few cases becomes apparent by chance from high throughput screening programmes during product development. The purpose of this chapter is to describe what data basis is most frequently used for a routine evaluation of fungal metabolites and their residues in FBCAs. A decision scheme is discussed, which should be used to assess metabolite toxicity in the context of a worst-case scenario testing. This risk assessment procedure will help to identify the low risk, if any, of old and newly developed microbial pest control agents, and give support to applicants to market their FBCAs.

## Contents



## 9.1 Introduction

Fungi secrete an array of natural metabolites (extrolites), mostly products of secondary metabolism, which serve different functions depending on the ecological habitat of the fungus (Butt 2002; Calvo et al. 2002; Vey et al. 2001). Secondary

H. Strasser (✉)
Faculty of Biology, Institute of Microbiology, University of Innsbruck, Innsbruck A 6020, Austria
e-mail: hermann.strasser@uibk.ac.at

R.-U. Ehlers (ed.), *Regulation of Biological Control Agents*,
DOI 10.1007/978-90-481-3664-3_9, © Springer Science+Business Media B.V. 2011

metabolism is commonly associated with fungal development (i.e. sporulation, cell differentiation). Calvo et al. (2002) classified the function of these products into three groups: (i) metabolites that activate sporulation; (ii) pigments required for spore structures; and (iii) toxic metabolites secreted by growing fungi (mycotoxins). The latter are of great interest to scientists and legislators, not only because of their powerful and varied biological effects, which can be used for medical or industrial purpose (Baker et al. 2007; Hoffmeister and Keller 2007; Nielson and Smedsgaard, 2003), but also to overcome obstacles in the registration and subsequent commercialization of fungal biological control agents (FBCAs).

The authorisation of BCAs is strictly regulated but there is still much debate in defining the criteria for registration of FBCAs, especially with respect to the potential of the micro-organism to produce metabolites, including toxins (points IIM 2.4 and IIM 3.5.2. OECD 2004; Strasser et al. 2000; Goettel et al. 2001; Strasser et al. 2008).

A global harmonization of registration procedures is still not realised, just because of the fact, that all "old" active BCAs in the EU will be put on the "green track (1097/2007/EC; EU 2007a)" and therefore will be listed in Annex I after examination by the rapporteur member state and the commission (Directive 91/414/EC, EU 1991; see 2008/113/EC; EU 2008a). The peer review will be organised by EFSA and will allow member states and EFSA to comment further on the scientific conclusions of the Draft assessment report. This "fast-track" authorisation has been given because unacceptable negative effects to humans and the environment were excluded (see also Table 9.1: Remarks and specific provisions). There is still no official "lesson learned document" available, which was expected to be published based on the experience of the judgement of 4$^{th}$ list substances (2229/2004/EC, EU 2004). Nevertheless, for the first time the Commission gave a clear indication that all already notified old active biocontrol agents do not have any harmful effects on human or animal health, or on groundwater or any unacceptable influence on the environment.

The focus of this chapter is to provide an overview about information available on fungal secondary metabolites (i.e. toxicants) and their effects, and to suggest how to deal with the registration of fungal BCAs for which no information about metabolites is available. A decision scheme for the assessment of potential relevant metabolites of fungal BCAs is presented, which will help to identify and to characterise low risk FBCAs.

## 9.2 Mycotoxins of Fungal Biocontrol Agents

In recent years, significant progress has been made in the development of fungal biocontrol agents (BCAs) for the suppression of pests (insects, nematodes), weeds and diseases of a wide range of forest, horticultural and agricultural crops (Butt et al. 2001). Nevertheless, relatively few of these products have reached the market: for example, at the time of writing this article only 24 FBCAs have been registered or

**Table 9.1** Secondary metabolites from fungal biopesticide active products mentioned in the "Handbook of secondary fungal metabolites" (Cole and Schweikert 2003a, 2003b; Cole et al. 2003). All fungal biocontrol agents from the online forum of US EPA (2007a) and/or in Annex I (91/414/EC; EU 1991) or currently suggested for inclusion in Annex I (EU 2008a) were considered

| Fungal biocontrol agent (intended use) | Permit | Secondary metabolites (biological activity) | Remarks/specific provisions | Product name(s) |
|---|---|---|---|---|
| *Alternaria destruens* 059 (Herbicide) | US | Dehydroaltenusin (Anti-tumor effects) | *A. destruens* is minimally toxic and nonpathogenic to mammals. Consequently, no cumulative effects from the residues of this product with other related microbial pesticides are anticipated (US EPA 2005a) | Smolder G, Smolder WP |
| *Ampelomyces quisqualis AQ 10* (Fungicide) | EU/US | No relevant metabolites | Standing Committee concluded . . . [no toxins are involved (EU 2005a)] | AQ 10 |

**Table 9.1** (continued)

| Fungal biocontrol agent (intended use) | Permit | Secondary metabolites (biological activity) | Remarks/specific provisions | Product name(s) |
|---|---|---|---|---|
| *Aspergillus flavus* AF36 (Cottonseed protector) | US | Ditryptophenaline<br>Aflavinine; (antiinsectan activity)<br>Seven indol derivates, e.g. 20-Hydroxyaflavinine, Aflazvazole; (antifeeding activity)<br>Paspalinine and two more related compounds, e.g. Alfatrem; (tremorgenic in rodents)<br>Aflatoxin B1, B2, $B_{2a}$, G1, $G_{2a}$, M1, M2<br>(hepatocarciogen)<br>Parasiticol (cause biliary hyperplasia in organs)<br>O-Methylsterigmatocystin<br>Dihydro- O-methylsterigmatocystin (negligible effects in cell culture)<br>Aspertoxin (in developing chicken embryo – 100% killed after injection of 2.0 μg/egg)<br>Ergosta-4,6,8(14),22-tetraen-3-one<br>24-Ethylcholesta-4,6,8(14),22-tetraen-3-one<br>Aspervalvin<br>Cyclopiazonic acid (potent inhibitor of $Ca^{2+}$ activated ATPase) | This strain of *A. flavus* does not produce aflatoxin (atoxigenic strain; US EPA 2003a) | *Aspergillus flavus* AF36 |
| *A. flavus* NRRL 21882 (Peanut protector) | US | See above | This strain of *A. flavus* does not produce aflatoxin (US EPA2004) | Aflaguard |

**Table 9.1** (continued)

| Fungal biocontrol agent (intended use) | Permit | Secondary metabolites (biological activity) | Remarks/specific provisions | Product name(s) |
|---|---|---|---|---|
| *Beauveria bassiana* ATCC 74040 (Insecticide) | EU*/US | Isoleucylisoleucyl anhydride<br>cyclo-(L-Isoleucyl-L-valine)<br>cyclo-(L-Alanyl-L-proline) | No risks expected. Residues of the fungus are not expected to remain on treated food or feed (US EPA 1999a). Based on the Commission review report – the BCA will be listed in Annex I (SANCO/1546/2008; EU 2008a) | Troy Boverin, Fermone Naturalis L-225 |
| *B. bassiana* GHA (Insecticide) | EU*/US | See above | No risks expected (US EPA 2000a). Based on the Commission review report – the BCA will be listed in Annex I (SANCO/1547/2008; EU 2008a) | Mycotrol ES, Botanigard 22WP, Organigard ES |
| *B. bassiana* 447 (Insecticide) | US | See above | No risks expected. EPA will require additional studies if the bait stations are proposed for outdoor or other additional uses (US EPA 2002a) | "Baits motel stay a while-rest forever" |
| *B. bassiana* HF23 (Insecticide) | US | See above | Residues do not remain on treated food (US EPA 2006a)<br>Confirmatory analyses of five batches were requested to demonstrate that beauvericin and other unintentional ingredients, are within levels required for quality assurance and quality control (US EPA 2006b) | *Beauveria bassiana* HF23 |

**Table 9.1** (continued)

| Fungal biocontrol agent (intended use) | Permit | Secondary metabolites (biological activity) | Remarks/specific provisions | Product name(s) |
|---|---|---|---|---|
| *Candida oleophila* isolate I-182 (Fungicide) | US | Ergosterol<br>5,6-Dihydroergosterol<br>4-alpha-Methylzymosterol<br>Sitosterol | No risks expected. Fact Sheet (US EPA 2000b) | Aspire |
| *Candida oleophila strain O* (Fungicide) | EU* | See above | 2007/380/EC: no open questions published (EU 2007b) | Bionext |
| *Chondrostereum purpureum* HQ1 (Herbicide) | US | No relevant metabolites | No information available (US EPA 2005c) | Myco-tech paste |
| *Chondrostereum purpureum* PFC2139 (Herbicide) | US | No relevant metabolites | No known reports that fungus produce any toxins harmful to animals, and which has caused adverse effects in people (CAN 2007) | Chontrol paste |
| *Colletotrichum gloeosporioides* f. sp. *Aeschynomene* (Herbicide) | US | No relevant metabolites | No harmful endpoints when tested in laboratory rodents for toxicity and pathogenicity by the oral, dermal, and pulmonary routes (US EPA 2006c) | Collego |
| *Coniothyrium minitans* CON/M/91-08 (Fungicide) | US | Palmarumycin $C_1$ to $C_6$<br>Palmarumycin $C_8$ to $C_{16}$ (biological activity against bacteria, fungi and garden cress) | The review did not reveal any open questions or concerns (2003/79/EC; EU 2003)<br>*C. minitans* has not been reported to produce toxins or antibiotics (US EPA 2001c) | Contans WG |
| *Gliocladium virens* GL- 21 (Fungicide) | US | Viridin (strong antifungal activity)<br>Viridol | No risks expected (US EPA 2000c) | WRC-AP-1 |

**Table 9.1** (continued)

| Fungal biocontrol agent (intended use) | Permit | Secondary metabolites (biological activity) | Remarks/specific provisions | Product name(s) |
|---|---|---|---|---|
| *G. catenulatum* J1446 (Fungicide) | EU/US | No relevant metabolites | The review did not reveal any open questions or concerns (EU 2005a). According to a literature search on April 2000 there is no indication of toxin production by *G. catenulatum*. No cell toxicity was detected in an FL cell tests or in the *Escherichia coli* WP2/CM871 DNA-repair test. No gliotoxin was detected by HPLC analysis (EU 2005b). There are no reports indicating that *G. catenulatum* strain J1446 produces any toxins or antibiotics (US EPA 2002b) | Primastop, Prestop |
| *Lecanicillium muscarium* (formerly *Verticillium lecanii)* (Insecticide) | EU* | 23,24,25,26,27-Pentanorlanost-8-ene-3beta,22-diol | Based on the Commission review report – the BCA will be listed in Annex I (SANCO/1861/2008; EU 2008a) | Mycotal |
| *Metarhizium anisopliae* ESF1 (Insecticide) | US | Cytochalasin C (strong inhibitor of mammalian cellular functions)<br>Zygosporin A (Teratogenicity effects in chicken, see also cytochalasin C) | No information available (US EPA 2001a) | Bioblast |

**Table 9.1** (continued)

| Fungal biocontrol agent (intended use) | Permit | Secondary metabolites (biological activity) | Remarks/specific provisions | Product name(s) |
|---|---|---|---|---|
| *M. anisopliae* F52 (Insecticide, Acaricide) | EU*/US | See above | Although destruxin A, B and E secretion, often in significant quantities, are reported, .... [no harm is expected to humans or the environment when BCAs containing *M. anisopliae* strain F52 are used according to label instructions (US EPA 2003b)]. BCA will be listed in Annex I (SANCO/1862/2008; EU 2008a) | Taenure, Tick-Ex (G, EC); |
| *M. anisopliae* BIPESCO5 (Insecticide, Acaricide) | EU*/US | See above | Based on the Commission review report the BCA will be listed in Annex I (SANCO/1862/2008; EU 2008a) | GRANMET-P (GR), WP |
| *Muscodor albus* QST20799 (Bactericide, Fungicide, Nematicide) | US | A number of volatiles, mainly alcohols, acids, and esters (inhibit and kill plant pathogenic and other organisms) | Re-evaluation in March 2008 (US EPA 2005d) | Arabesque, Andante, Glissade |
| *Paecilomyces fumosoroseus* Apopka strain 97 or PFR 97 or CG 170, ATCC20874 (Insecticide) | EU/US | Paecilospirone | Each fermentation broth should be checked by HPLC to ensure that no secondary metabolites are present (2001/47/EC; EU 2001b).<br>There are no expected health risks from use of this fungus as a pesticide (US EPA 1999b) | PFR-97 (Preferal) |

**Table 9.1** (continued)

| Fungal biocontrol agent (intended use) | Permit | Secondary metabolites (biological activity) | Remarks/specific provisions | Product name(s) |
|---|---|---|---|---|
| *P. fumosoroseus* Fe 9901 (Insecticide) | EU | No relevant metabolites | Completeness check of dossier passed (2008/565/EC; EU 2008b) | FUTURECO – NOFLY |
| *P. lilacinus* strain 251 (Nematicide) | EU/US | No relevant metabolites | Although many species of *Paecilomyces* release potent toxins, no such toxins (mycotoxins or paecilotoxins) have been detected in this strain (. . ..as shown by chromatographic analyses and lack of toxicity to mammals and other organisms; US EPA 2005e). No relevant toxic metabolite/residue mentioned (2008/44/EC; EU 2008c) | Melocon WG |
| *Phlebiopsis gigantea* (Fungicide) | EU* | No relevant metabolites | Based on the Commission review report – the BCA will be listed in Annex I (SANCO/1863/2008; EU 2008a) | ROTSTOP, ROTEX |

**Table 9.1** (continued)

| Fungal biocontrol agent (intended use) | Permit | Secondary metabolites (biological activity) | Remarks/specific provisions | Product name(s) |
|---|---|---|---|---|
| *Pseudozyma flocculosa* PF-A22 (Fungicide) | EU*/US | Three fungitoxic unsaturated C-17 fatty acids: (9-heptadecenoic acid, 6-methyl-9-heptadecenoic acid and 4-methyl-7,11-heptadecadienoic acid) fungitoxins acyclic norterpene (2, 6, 10, 14, 18-pentamethyl-2, 6, 8, 10, 12, 14, 17-nonadecaheptene-1,19-diol) | No harm is expected if humans eat food containing residues of *Pseudozyma flocculosa* strain PF-A22 UL (US EPA 2002c) | Sporodex L |
| *Puccinia thlaspeos* strain woad (dyer's woad rust) (Herbicide) | US | No relevant metabolites | There is a reasonable certainty that no harm will result from human exposure to *P. thlaspeos* "strain woad". All toxicity study requirements concerning human risk assessment were waived (US EPA 2002d) | Woad Warrior |
| *Pythium oligandrum* DV74 (Fungicide) | EU*/US | No relevant metabolites | No harmful health effects to humans are expected: no evidence that the fungus is toxic to humans or other mammals, .... and no adverse environmental effects are expected (US EPA 2007b)<br>Based on the Commission review report – the BCA will be listed in Annex I (SANCO/1864/2008; EU 2008a) | DV 74 |

**Table 9.1** (continued)

| Fungal biocontrol agent (intended use) | Permit | Secondary metabolites (biological activity) | Remarks/specific provisions | Product name(s) |
|---|---|---|---|---|
| *Trichoderma asperellum (*formerly *T. harzianum* ICC012 (Fungicide) | EU* | No relevant metabolites | Based on the Commission review report – the BCA will be listed in Annex I (SANCO/1842/2008; EU 2008a) | Remedier |
| *T. asperellum (*formerly *T. viride)* T11, TV1, T25 (Fungicide) | EU* | No relevant metabolites | Based on the Commission review report – the BCA will be listed in Annex I (SANCO/1868/2008; EU 2008a ) | TV1 and T11 |
| *T. atroviride* (formerly *T. harzianum*) T-11 (Fungicide) | EU* | No relevant metabolites | Based on the Commission review report – the BCA will be listed in Annex I (SANCO/1841/2008; EU 2008a) | Binab T pellets component, (see also *T. polysporum*) |
| *T. atroviride* I-1237 (Fungicide) | EU* | No relevant metabolites | Completeness check of dossier passed (2008/565/EC; EU 2008b) | I-1237 (Agrauxine) |
| *T. atroviride* (formerly *T. harzianum*) IMI 206040; (= *T. viride* ATCC 20476) (Fungicide) | EU*/US | No relevant metabolites | Based on the Commission review report – the BCA will be listed in Annex I (SANCO/1866/2008; EU 2008a); see also US EPA (2000e) | Binab component, (see also *T. polysporum*) |

**Table 9.1** (continued)

| Fungal biocontrol agent (intended use) | Permit | Secondary metabolites (biological activity) | Remarks/specific provisions | Product name(s) |
|---|---|---|---|---|
| *T. polysporum* ATCC 20475 + (*T. atroviride* ATCC 20476) (Fungicide) | EU*/US | Viridin, (strong antifungal activity)<br>Viridol | Based on the Commission review report – the BCA will be listed in Annex I (SANCO/1867/2008; EU 2008a); see also US EPA (2000e) | Binab T WP |
| *T. harzianum RIFAI ITEM 908* (Fungicide) | EU* | Koninginin A, (inhibits growth of etiolated wheat coleoptiles at $10^{-3}$ M)<br>Koninginin C, (inhibits growth of etiolated wheat coleoptiles at $10^{-3}$ M) | Based on the Commission review report – the BCA will be listed in Annex I (SANCO/1840/2008; EU 2008a) | TRICOVER |
| *T. harzianum RIFAI* T-39 (Fungicide) | US | See above | Based on its low toxicity potential, an additional FQPA safety factor is not required for residues of *T. harzianum* strain T-39 (US EPA 2000d) | TRICHODEX |
| *T. harzianum RIFAI* T-22 (KRL-AG2) (Fungicide) | EU*/US | See above | No risks ecpected (US EPA 2001b)<br>Based on the Commission review report – the BCA will be listed in Annex I (SANCO/1839/2008; EU 2008a) | TRIANUM P, T-22 G, T-22 WP, Fafard Growing mix |
| *T. gamsii (formerly T. viride)* ICC080 (Fungicide) | EU* | Trichodermin, (strong inhibitor of protein synthesis in mammalian cells, antifungal antibiotic)<br>Viridin (strong antifungal activity)<br>Viridol | Based on the Commission review report – the BCA will be listed in Annex I (SANCO/1868/2008; EU 2008a) | Remedier |

**Table 9.1** (continued)

| Fungal biocontrol agent (intended use) | Permit | Secondary metabolites (biological activity) | Remarks/specific provisions | Product name(s) |
|---|---|---|---|---|
| *Verticillium albo-atrum (formerly V. dalia) WCS850* (Fungicide) | EU*/US | No relevant metabolites | No mechanism of pathogenicity or toxicity in mammals has been identified for this organism, no cumulative effects from the residues of this product with other related microbial pesticides are anticipated (US EPA 2005b) | Dutch Trig |

EU Active ingredients and products listed in Annex I Directive 91/414 (Europe)
EU* Active ingredients and products currently suggested for inclusion in Annex I (2008/113/EC 2008a)
US Active ingredients and products containing them US EPA (2007a)

will be registered in the near future in the European Union under the harmonized registration procedure of Council Directive 91/414/EEC (EU 1991). Likewise, only 33 FBCAs have been approved under the Pest Control Products Act in Canada and U.S. Pesticide Data Requirements, respectively (Kabaluk and Gazdik 2005). Today, only 39 FBCAs (i.e. insecticides, fungicides and herbicides), comprising 21 fungal genera are registered in the EU, USA and Canada (Table 9.1).

The existence of fungal secondary metabolites and their toxicological significance, apart from accessible information in the literature, often becomes known only by chance, when acute toxicity studies with products based on FBCAs had led to negative effects (Rochon and Belliveau 2006) and applicants in the registration process were committed to isolate and characterise the toxicants.

Based on two examples, the difficulties related to the registration of potential mycotoxin-producing FBCAs are presented. The type of data that need to be produced and will later be available to the target audience will be described. The circumstances under which authorities will ask for data on relevant metabolites (mycotoxins) will be detailed.

In 2003 Cole and co-authors published the comprehensive standard work "Handbook of fungal secondary metabolites" (Cole and Schweikert 2003a, 2003b; Cole et al. 2003). The authors' aim was to offer data on all major groups of secondary fungal metabolites assigned to fungal species. Although economically important groups (e.g. the aflatoxins, trichothecenes, fumonisins) have been included in the series, most of the relevant secondary metabolites of commercialised FBCAs (e.g. beauvericine, beauveriolide, destruxins, gliovirin, glioprennins, heptelidic acids, oosporein and viridian) are missing (Table 9.1). This comprehensive handbook is an example for all other relevant publications and demonstrates how difficult it is to get a precise overview on secondary metabolites produced by specific, potential fungal biocontrol agents.

A literature search in the database ISI Web of Knowledge (Thomson Reuters 2008) for all reported secondary fungal metabolites and for all FBACs listed in the online forum of US EPA (2007a) and/or in Annex I (91/414/EC, EU 1991) or currently suggested for inclusion in Annex I (2008/113/EC, EU 2008a) resulted in one new identified secondary metabolite per FBCA and per year on average (Table 9.2). The timeframe for this research was defined for the last 5 years, from January 2003 until August 2008.

Since fungi are the subject of numerous high-throughput screening programmes (Nielsen and Smedsgaard 2003), and metabolic profiling techniques are used to build up and complete comprehensive metabolite data bases (i.e. using NMR spectroscopy and chemometric tools; Seger and Sturm 2007; Holmes et al. 2006), new fungal metabolites are found permanently. One example is the recently re-opened discussion for the safety of the genus *Metarhizium.* Krasnoff et al. (2006, 2008) presented data on two compounds (i.e. fusarin C analogues) which exhibited mutagenic activity in the Ames assay. The authors concluded that "the impact of these findings on the use of *M. anisopliae* as biocontrol agents is currently unknown and need further investigation". The subjects of this study were knockout mutant strains of *M. anisopliae,* which were characterised as metabolite overproducing isolates. From

**Table 9.2** New reported secondary metabolites from fungal biopesticide active products published in peer reviewed Journals. All fungal biocontrol agents from the online forum of US EPA 12/2007 and/or in Annex I (91/414/EC; EU 1991) or currently suggested for inclusion in Annex I (EU 2008a) were considered. The following search terms were always checked in the database "ISI web of knowledge" (Thomson Reuters 2008) for the last 5 years: Species name combined with the general search terms (toxin* OR mycotoxin* OR metabolite* OR bioactive; the asterisk stands for the plural, e.g. toxins)

| Fungal biocontrol agent | ISI hits/hits with new metabolites | Novel secondary metabolites | References |
|---|---|---|---|
| *Alternaria destruens* | No records found | | |
| *Ampelomyces quisqualis* | 6/2 | Two new sulfur-containing phenolic compounds: (7-hydroxy-5-hydroxymethyl-2H-benzo[1,4]thiazin-3-one and 2,5-dihydroxy-3-methanesulfinylbenzyl alcohol) | Zhang et al. (2008) |
| | | Ampelopyrone, Desmethyldiaportinol, Desmethyldichlorodiaportim, Macrosporin-7-O-sulfate, 3-O-Methylalaternin-7-O-sulfate, Ampelanol | Aly et al. (2008) |
| *Aspergillus flavus* | 336/0 | | |
| *Beauveria bassiana* | 65/1 | Novel beauvericin derivate (beauvericins G1–3, beauvericins H1–3) cytotoxicity (haptotaxis). | Xu et al. (2007) |
| *Candida oleophila* | 2/0 | | |
| *Chondrostereum purpureum* | 1/0 | | |
| *Colletotrichum gloeosporioides* f. sp. *Aeschynomene* | 23/2 | Tetraol (-)-1<br>Cis-4-hydroxy-6-deoxyscytalone and (4R)-4,8-dihydroxy-a-tetralone | Femenia-Rios et al. (2006), Inacio et al. (2006) |

**Table 9.2** (continued)

| Fungal biocontrol agent | ISI hits/hits with new metabolites | Novel secondary metabolites | References |
|---|---|---|---|
| *Coniothyrium minitans* | 5/1 | Macrosphelide A<br>Palmarumycin $C_1$ (MW 350.03; Palmarumycins) | McQuilken et al. (2003) |
| *Gliocladium virens* | 6/1 | Gliotoxin | Anitha and Murugesan (2005) |
| *G. catenulatum* | No records found | | |
| *Lecanicillium muscarium (formerly Verticillium lecanii)* | 10/0 | | |
| *Metarhizium anisopliae* | 64/4 | Helvolic acid and 1,2 dihdrohelvolic acid (antibacterial compound)<br>Two analogues of fusarin C (i.e. 7-desmethyl and (8Z)-fusarin C)<br>Serinocyclins A and B (cyclic heptapeptides) | Lee et al. (2008)<br>Krasnoff et al. (2006, 2008, Moon et al. (2008)<br>Krasnoff et al. (2007) |
| *Muscodor albus* | 8/1 | Bioactive volatile organic compounds (VOCs) | Strobel (2006) |
| *Paecilomyces fumosoroseus* | 8/0 | | |
| *P. lilacinus* strain | 4/0 | | |
| *Phlebiopsis gigantea* | 1/0 | | |
| *Pseudozyma flocculosa* | 1/1 | Flocculosin (novel glycolipid) | Cheng et al. (2003) |
| *Puccinia thlaspeos* | No records found | | |
| *Pythium oligandrum* | 3/0 | | |
| *Trichoderma asperellum* | 3/0 | | |
| *T. atroviride* | See above | | |

**Table 9.2** (continued)

| Fungal biocontrol agent | ISI hits/hits with new metabolites | Novel secondary metabolites | References |
|---|---|---|---|
| *T. polysporum* | 25/4 | Oxiranyldecene; viridenepoxydiol<br>Viridepyronone<br>Trichovirin I<br>Trichodermamides A and B | Chovanec et al. (2005), Evidente et al. (2003), Brückner and Koza (2003), Reino et al. (2008) |
| *T. harzianum* | 75/4 | T39 butenolide, T22 azaphilone<br>Anthraquinones and xanthone derivatives<br>Nonanoic acid<br>Antibiotics | Vinale et al. (2006), Reino et al. (2008)<br>Aneja et al. (2005), El-Hasan et al. (2008) |
| *T. gamsii* | No records found | | |
| *Verticillium albo-atrum (formerly V. dalia)* | 6/0 | | |

the scientific point of view the authors raised the right questions, and based on this case study it is understandable why strain-specific data on relevant metabolites are required for a risk assessment of FBCAs. Simply because *Metarhizium anisopliae* var. *anisopliae* relevant data could be presented to regulation authority in spring 2008 (i.e. PMRA Canada), the ongoing safety evaluation for this potential BCA in Canada was not suspended. Otherwise, new labour- and cost-intensive studies would have to be presented by the applicant to demonstrate, that their specific production strain is safe and does not produce these fusarin-like mycotoxins.

It is understandable that regulatory authorities want to have a stringent procedure for the assessment of potentially toxic metabolic byproducts by candidate fungi (Rochon and Belliveau 2006). In order not to hinder the marketing of potentially low risk products (see Laengle and Strasser 2010), REBECA experts proposed a tiered scheme to be able to assess the risks of potential metabolites of FBCAs.

## 9.3 Standard Procedure for Toxic Metabolite Assessment

REBECA proposed a decision scheme (see Chapter 13), which has been tested initially in case studies on fungal toxic metabolites [i.e. alamethicin, antiamoebins, destruxins (type A, B, E), elsionchrome A, gliotoxin, paracelsin, oosporein (Boss et al. 2007; Favilla et al. 2006; Ganassi et al. 2007; Skrobek and Butt 2005; Skrobek et al. 2006) and selected crude extracts from *Beauveria brongniartii, Metarhizium anisopliae*, *Paecilomyces lilacinus*, *Stagonospora convolvuli*, *Trichoderma harzianum* and *Verticillum lecanii* (Boss et al. 2007; Butt et al. 2004; Skrobek and Butt 2005; Skrobek et al. 2006)].

Based on RAFBCA results (EU funded project QLK1-CT2001-01391; Strasser et al. 2007; 2008) and the outcome of REBECA workshops and conferences held in Innsbruck (April 2006), Salzau (September 2006), Alès (June 2007) and Brussels (September 2007), REBECA experts emphasized as a final recommendation that potential effects of relevant metabolites produced by micro-organisms should be handled according to the pre-submission data request (see Chapter 15) and the tiered scheme presented in Chapter 15.

## 9.4 Conclusion

While microbials are often reported to pose low risks to the environment (OECD 2007), it is critical for the credibility of microbial pest control products to underline such generic statements with solid data. The "decision tree", presented in Chapter 15, permits the unbiased generation of a risk assessment of metabolites produced by FBCAs, which have a high toxicological relevance on the basis of scientific data. Unfortunately, only few standardised bioassays are available (such as effect-based ones) to evaluate metabolites and their specific toxicity. To provide a sound basis for a verification of the proposed decision scheme for future risk assessment of FBCAs, crude extracts from a number of well-known mycoparasitic, entomopathogenic or phytopathogenic fungi as well as from new isolates (bacteria included) should be tested (Strasser et al. 2008). Crude extracts, produced from fractions of polar and/or non-polar solvents from different production batches, and selected metabolites displaying different structural nature and mode of actions, should be made available from isolates where analyses showed activity of suspected toxic metabolites. Cultivation methods and extraction protocols have to be developed to meet the conditions for high-level production of toxins. Furthermore, bioassays have to be established and/or standardised to be able to define tolerance levels of metabolite toxicity (i.e. genotoxicity, cytotoxicity and ecotoxicity), because crude extracts are not expected to show zero toxicity.

The benefits and pitfalls in the use of crude extract analysis were extensively discussed by RAFBCA and REBECA experts (Strasser et al. 2008, Chapter 15). Rationales to increase the budget for more publicly funded projects (possibly with matching funds from the industry) for metabolite analysis and their risk assessment have been proposed for many years because consumers, animals and the environment require more protection. These new long-term projects will result in a generic safety registration of each particular agent and will also help to define the basis for new alternative regulation concepts.

**Acknowledgements** This work was supported by the European Commission, Specific Support Action SSPE-022709. The authors are indebted to the RAFBCA team (QLK1-CT2001-01391) to provide the actual reference list on metabolites. We also wish to thank Claudio Altomare (ISPA Bari), Rüdiger Hausschild (GAB Consulting, Lamstedt), and Olaf Strauch (University Kiel) for their helpful discussion and kindly reviewing the manuscript

## References

Aly AH, Edrada-Ebel R, Wray V, Muller WEG, Kozytska S, Hentschel U, Proksch P, Ebel R (2008) Bioactive metabolites from the endophytic fungus *Ampelomyces* sp. isolated from the medicinal plant urospermum picroides. Phytochem 69:1716–1725

Aneja M, Gianfagna TJ, Hebbar PK (2005) *Trichoderma harzianum* produces nonanoic acid, an inhibitor of spore germination and mycelial growth of two cacao pathogens. Physiol Molecular Plant Pathol 67:304–307

Anitha R, Murugesan K (2005) Production of gliotoxin on natural substrates by *Trichoderma virens*. J Basic Microbiol 45:12–19

Baker DD, Chu M, Oza U, Rajgarhia V (2007) The value of natural products to future pharmaceutical discovery. Nat Prod Rep 24:1225–1244

Boss D, Maurhofer M, Schläpfer E, Défago G (2007) Elsinochrome A production by the bindweed biocontrol fungus *Stagonospora convolvuli* LA39 does not pose a risk to the environment or the consumer of treated crops. FEMS Microbiol Ecol 59:194–205

Bruckner H, Koza A (2003) Solution phase synthesis of the 14-residue peptaibol antibiotic trichovirin I. Amino Acids 24:311–323

Butt TM (2002) Use of entomogenous fungi for the control of insect pests. In: Kempken F (ed) The Mycota XI agricultural applications. Springer, Berlin, pp 111–134

Butt TM, Jackson C, Magan N (2001) Fungi as biocontrol agents: progress, problems and potential. CABI, Wallingford, CT

Butt TM, Skrobek A, Wang C, Shah FA, El Hadj NB (2004) RAFBCA Report P1 Final. EU RTD project (QLK1-2001-01391) Risk assessment of fungal biological control agents. University Swansea, Swansea

CAN (2007) Registration decision. Health Canada Pest Management Regulatory Agency. RD2007-06. http://www.pmra-arla.gc.ca/english/pdf/rdd/rd2007-06-e.pdf. Accessed 4 Jun 2007

Calvo AM, Wilson RA, Bok JW, Keller NP (2002) Relationship between secondary metabolism and fungal development. Microbiol Mol Biol Rev 66:447–459

Cheng YL, Mcnally DJ, Labbe C, Voyer N, Belzile F, Belanger RR (2003) Insertional mutagenesis of a fungal biocontrol agent led to discovery of a rare cellobiose lipid with antifungal activity. Appl Environ Microbiol 69:2595–2602

Chovanec P, Kalinak M, Liptaj T, Pronayova N, Jakubik T, Hudecova D, Varecka L (2005) Study of *Trichoderma viride* metabolism under conditions of the restriction of oxidative processes. Can J Microbiol 51:853–862

Cole RJ, Jarvis BB, Schweikert MA (2003) Handbook of secondary fungal metabolites, vol III. Academic, Amsterdam

Cole RJ, Schweikert MA (2003a) Handbook of secondary fungal metabolites, vol I. Academic, Amsterdam

Cole RJ, Schweikert MA (2003b) Handbook of secondary fungal metabolites, vol II. Academic, Amsterdam

EU (1991) EC Directive 91/414/EC. Off J Eur Union 1999 L 0414

EU (2001a) EC Directive 2001/36/EC. Off J Eur Union L 164/1

EU (2001b) EC Directive 2001/47/EC. Off J Eur Union L 175/21

EU (2003) EC Directive 2003/79/EC. Off J Eur Union 14.6.2003

EU (2004) EC Regulation 2229/2004. Off J Eur Union L 379/13

EU (2005a) EC Directive 2005/2/EC. Off J Eur Union L 20/15

EU (2005b) SANCO/10383/2004 – rev. 4. Off J Eur Union OJ No L 020

EU (2007a) EC Regulation 1097/2007. Off J Eur Union L 246/31

EU (2007b) EC Directive 2007/380/EC. Off J Eur Union 2.6.2007

EU (2008a) EC Directive 2008/113/EC. Off J Eur Union L 330/6

EU (2008b) EC Directive 2008/565/EC. Off J Eur Union L 181/49

EU (2008c) EC Directive 2008/44/EC. Off J Eur Union L 94/13

El-Hasan A, Walker F, Buchenauer H (2008) *Trichoderma harzianum* and its metabolite 6-pentyl-alpha-pyrone suppress fusaric acid produced by *Fusarium moniliforme*. J Phytopathol 156: 79–87

Evidente A, Cabras A, Maddau L, Serra S, Andolfi A, Motta A (2003) Viridepyronone, a new antifungal 6-substituted 2 h-pyran-2-one produced by *Trichoderma viride*. J Agri Food Chem 51(24):6957–6960

Favilla M, Macchia L, Gallo A, Altomare C (2006) Toxicity assessment of metabolites of fungal biocontrol agents using two different (*Artemia salina* and *Daphnia magna*) invertebrate bioassays. Food Chem Toxicol 44:1922–1931

Femenia-Rios M, Garcia-Pajon CM, Hernandez-Galan R, Macias-Sanchez AJ, Collado IG (2006) Synthesis and free radical scavenging activity of a novel metabolite from the fungus *Colletotrichum gloeosporioides*. Bioorg Med Chem Let 16:5836–5839

Ganassi S, De Cristofaro A, Grazioso P, Altomare C, Logrieco A, Sabatini MA (2007) Detection of fungal metabolites of various *Trichoderma* species by the aphid *Schizaphis graminum*. Entomol Exper Applic 122(1):77–86

Goettel MS, Hajek AE, Siegel JP, Evans HC (2001) Safety of fungal biocontrol agents. In: Butt TM, Jackson C, Magan N (eds) Fungi as biocontrol agents: progress, problems and potential. CABI, Wallingford, CT, pp 347–375

Hoffmeister D, Keller NP (2007) Natural products of filamentous fungi: enzymes, genes, and their regulation. Nat Prod Rep 24:393–416

Holmes E, Tang HR, Wang YL, Seger C (2006) The Assessment of plant metabolite profiles by NMR-based methodologies. Planta Med 72:771–785

Inacio ML, Silva GH, Teles HL, Trevisan HC, Cavalheiro AJ, Bolzani VDS, Young MCM, Pfenning LH, Araujo AR (2006) Antifungal metabolites from *Colletotrichum gloeosporioides*, an endophytic fungus in *Cryptocarya mandioccana* Nees (Lauraceae). Biochem Syst Ecol 34:822–824

Kabaluk T, Gazdik K (2005) Directory of microbial pesticides for agricultural crops in OECD countries. http://www4.agr.gc.ca/resources/prod/doc/pmc/pdf/micro_e.pdf. Accessed 14 Feb 2008

Krasnoff SB, Keresztes I, Gillilan RE, Szebenyi DME, Donzelli BGG, Churchill ACL, Gibson DM (2007) Serinocyclins A and B, cyclic heptapeptides from *Metarhizium anisopliae*. J Nat Prod 70:1919–1924

Krasnoff SB, Sommers CH, Moon YS, Donzelli BGG, Vandenberg JD, Churchill ACL, Gibson DM (2006) Production of mutagenic metabolites by *Metarhizium anisopliae*. J Agri Food Chem 54:7083–7088

Krasnoff SB, Sommers CH, Moon YS, Donzelli BGG, Vandenberg JD, Churchill ACL, Gibson DM (2008) Production of mutagenic metabolites by *Metarhizium anisopliae* (Vol 54, Pg 7083, 2006). J Agri Food Chem 56:1158

Laengle T, Strasser H (2010) Developing a risk index to comparatively assess environmental risks posed by microbial and conventional pest control agents. Biocont Sci Technol 20(7):659–681

Lee SY, Kinoshita H, Ihara F, Igarashi Y, Nihira T (2008) Identification of novel derivative of helvolic acid from *Metarhizium anisopliae* grown in medium with insect component. J Biosci Bioeng 105:476–480

McQuilken MP, Gemmell J, Hill RA, Whipps JM (2003) Production of macrosphelide a by the mycoparasite *Coniothyrium minitans*. FEMS Microbiol Let 219:27–31

Moon YS, Donzelli BGG, Krasnoff SB, Mclane H, Griggs MH, Cooke P, Vandenberg JD, Gibson DM, Churchill ACL (2008) *Agrobacterium*-mediated disruption of a nonribosomal peptide synthetase gene in the invertebrate pathogen *Metarhizium anisopliae* reveals a peptide spore factor. Appl Environ Microbiol 74(14):4366–4380

Nielsen KF, Smedsgaard J (2003) Fungal metabolite screening: database of 474 mycotoxins and fungal metabolites for dereplication by standardised liquid chromatography-UV-mass spectrometry methodology. J Chrom A 1002:111–136

OECD (2004) Appendix 6b part 4. Dossier Guidance for Microbials, Series on Pesticides No. 23. http://www.oecd.org/dataoecd/26/57/27766483.pdf. Accessed 16 Sept 2008

OECD (2007) Biological pesticide registration. Environment Directorate. http://www.oecd.org/document/8/0,2340,en_2649_34383_31962760_1_1_1_1,00.html. Accessed 20 Jun 2007

Reino JL, Guerro RF, Hernández-Galán R, Collado IG (2008) Secondary metabolites from species of the biocontrol agent *Trichoderma*. Phytochem Rev 7:89–123

Rochon D, Belliveau B (2006) OECD Issue Paper – Discussion on microbial metabolite residues in treated food crops. http://www.rebeca-net.de/downloads/DRAFT%20OECD%20ISSUE%20PAPER%20CONTAMINATIONS.pdf. Accessed 7 Aug 2007

Seger C, Sturm S (2007) Analytical aspects of plant metabolite profiling platforms: current standings and future aims. J Proteome Res 6:480–497

Skrobek A, Boss D, Defago G, Butt TM, Maurhofer M (2006) Evaluation of different biological test systems to assess the toxicity of metabolites from fungal biocontrol agents. Toxicol Lett 161:43–52

Skrobek A, Butt TM (2005) Toxicity testing of destruxins and crude extracts from the insect-pathogenic fungus *Metarhizium anisopliae*. FEMS Microbiol Lett 251:23–28

Strasser H, Altomare C, Typas M, Butt TM (2008) REBECA proposal on the assessment of microbial metabolites. In: Ehlers RU, Enkerli J, Glazer I, Lopez-Ferber M, Thaczuk C (eds) Insect pathogens and Insect Nematodes. IOBC Bull 31:21–26

Strasser H, Typas M, Altomare C, Butt TM (2007) Annex 7: Position paper on microbial metabolite assessment. In: Strauch O, Strasser H, Ehlers RU, Hausschild R (eds) Deliverable 10: Proposals for improved regulatory procedures for microbial BCAs. http://www.rebeca-net.de/downloads/report/deliverable%2010.pdf. Accessed 28 Feb 2008

Strasser H, Vey A, Butt TM (2000) Are there any risks in using entomopathogenic fungi for pest control, with particular reference to the bioactive metabolites of *Metarhizium*, *Tolypocladium* and *Beauveria* species?. Biocont Sci Technol 10:717–735

Strobel G (2006) *Muscodor albus* and its biological promise. J Ind Microbiol Biotechnol 33: 514–522

Thomson Reuters (2008) ISI Web of knowledge. http://isiwebofknowledge.com/. Accessed 16 Sept 2008

US EPA (1999a) *Beauveria bassiana* ATCC 74040 (128818) Fact sheet. http://www.epa.gov/oppbppd1/biopesticides/ingredients/factsheets/factsheet_128818.htm. Accessed 17 Jan 2011

US EPA (1999b) *Paecilomyces fumosoroseus* Apopka Strain 97 (115002) Fact sheet. http://www.epa.gov/pesticides/biopesticides/ingredients/factsheets/factsheet_115002.htm. Accessed 17 Jan 2011

US EPA (2000a) *Beauveria bassiana* strain GHA (128924) Fact sheet. http://www.epa.gov/oppbppd1/biopesticides/ingredients/tech_docs/tech_128924.htm. Accessed 17 Jan 2011

US EPA (2000b) *Candida oleophila* isolate I-182 (021008) Fact sheet. http://www.epa.gov/oppbppd1/biopesticides/ingredients/factsheets/factsheet_021008.htm. Accessed 17 Jan 2011

US EPA (2000c) *Gliocladium virens* GL-21 (129000) Fact sheet. http://www.epa.gov/oppbppd1/biopesticides/ingredients/factsheets/factsheet_129000.htm. Accessed 17 Jan 2011

US EPA (2000d) *Trichoderma harzianum* Rifai Strain T-39 (119200) Fact sheet. http://www.epa.gov/opp00001/biopesticides/ingredients/factsheets/factsheet_119200.htm. Accessed 17 Jan 2011

US EPA (2000e) *Trichoderma polysporum* ATCC 20475 (128902) *Trichoderma harzianum* ATCC 20476 (128903) Fact sheet. http://www.epa.gov/oppbppd1/biopesticides/ingredients/factsheets/factsheet_128902.htm. Accessed 17 Jan 2011

US EPA (2001a) *Metarhizium anisopliae* strain ESF1 (129056) Fact sheet. http://www.epa.gov/oppbppd1/biopesticides/ingredients/factsheets/factsheet_129056.htm. Accessed 17 Jan 2011

US EPA (2001b) *Trichoderma harzianum* Rifai Strain T-22 (ATCC # 20847) (119202) Fact sheet. http://www.epa.gov/pesticides/biopesticides/ingredients/factsheets/factsheet_119202.htm. Accessed 17 Jan 2011

US EPA (2001c). *Coniothyrium minitans* CON/M/91-08 (028836) Fact sheet. http://www.epa.gov/oppbppd1/biopesticides/ingredients/factsheets/factsheet_028836.htm. Accessed 17 Jan 2011

US EPA (2002a) *Beauveria bassiana* strain 447 (128815) Fact sheet. http://www.epa.gov/oppbppd1/biopesticides/ingredients/factsheets/factsheet_128815.htm. Accessed 17 Jan 2011

US EPA (2002b) *Gliocladium catenulatum* strain J1446 (021009) Fact sheet. http://www.epa.gov/oppbppd1/biopesticides/ingredients/factsheets/factsheet_021009.htm. Accessed 17 Jan 2011

US EPA (2002c) *Pseudozyma flocculosa* strain PF-A22 UL (119196) Fact http://www.epa.gov/oppbppd1/biopesticides/ingredients/factsheets/factsheet_119196.htm. Accessed 17 Jan 2011

US EPA (2002d) *Puccinia thlaspeos* strain woad (dyer's woad rust) (006489) Fact sheet. http://www.epa.gov/pesticides/biopesticides/ingredients/factsheets/factsheet_006489.htm. Accessed 17 Jan 2011

US EPA (2003a) *Aspergillus flavus* strain AF36 (006456) Fact sheet. http://epa.gov/oppbppd1/biopesticides/ingredients/factsheets/factsheet_006456.htm. Accessed 17 Jan 2011

US EPA (2003b) *Metarhizium anisopliae* strain F52 (029056) Biopesticide fact sheet. http://www.epa.gov/pesticides/biopesticides/ingredients/factsheets/factsheet_029056.htm. Accessed 17 Jan 2011

US EPA (2004) *Aspergillus flavus* NRRL 21882 (006500) Fact sheet. http://www.epa.gov/oppbppd1/biopesticides/ingredients/factsheets/factsheet_006500.htm. Accessed 17 Jan 2011

US EPA (2005a) *Alternaria destruens* Strain 059 (028301) Fact sheet. http://www.epa.gov/oppbppd1/biopesticides/ingredients/factsheets/factsheet_028301.htm. Accessed 17 Jan 2011

US EPA (2005b) Biopesticide Registration Action Document *Verticillium* Isolate WCS850 (PC Code). http://epa.gov/pesticides/biopesticides/ingredients/tech_docs/brad_081305.pdf. Accessed 17 Jan 2011

US EPA (2005c) *Chondrostereum purpureum* strain HQ1 (081309) Fact sheet. http://www.epa.gov/oppbppd1/biopesticides/ingredients/factsheets/factsheet_081309.htm. Accessed 17 Jan 2011

US EPA (2005d) *Muscodor albus* QST 20799 (006503) Fact sheet. http://www.epa.gov/oppbppd1/biopesticides/ingredients/factsheets/factsheet_006503.htm Accessed 17 Jan 2011

US EPA (2005e) *Paecilomyces lilacinus* strain 251 (028826) Fact sheet. http://www.epa.gov/oppbppd1/biopesticides/ingredients/factsheets/factsheet_028826.htm. Accessed 17 Jan 2011

US EPA (2006a) *Beauveria bassiana* HF23 (090305) Fact sheet. http://www.epa.gov/oppbppd1/biopesticides/ingredients/factsheets/factsheet_090305.htm. Accessed 17 Jan 2011

US EPA (2006b) Biopesticide registration Document *Beauveria bassiana* HF23 (PC Code 090305). http://www.epa.gov/oppbppd1/biopesticides/ingredients/tech_docs/brad_090305.pdf. Acessed 17 Jan 2011

US EPA (2006c) *Colletotrichum gleoesporiodes* f.sp. aeschynomene (226300) Biopesticides registration action document. http://www.epa.gov/oppbppd1/biopesticides/ingredients/tech_docs/brad_226300.htm. Accessed 17 Jan 2011

US EPA (2007a) New Biopesticide Active Ingredients – 2007. http://www.epa.gov/oppbppd1/biopesticides/product_lists/new_ai_2007.htm. Accessed 24 Jan 2008

US EPA (2007b) *Pythium oligandrum* DV 74 (028816) Fact sheet. http://epa.gov/oppbppd1/biopesticides/ingredients/factsheets/factsheet_028816.htm. Accessed 17 Jan 2011

Vey A, Hoagland RE, Butt TM (2001) Toxic metabolites of fungal biocontrol agents. In: Butt TM, Jackson C, Magan N (eds) Fungi as biocontrol agents: progress, problems and potential. CABI, Wallingford, CT, pp 311–346

Vinale F, Marra R, Scala F, Ghisalberti EL, Lorito M, Sivasithamparam K (2006) Major secondary metabolites produced by two commercial *Trichoderma* strains active against different phytopathogens. Lett Appl Microbiol 43:143–148

Xu YQ, Zhan JX, Wijeratne EMK, Burns AM, Gunatilaka AAL, Molnar L (2007) Cytotoxic and antihaptotactic beauvericin analogues from precursor-directed biosynthesis with the insect pathogen *Beauveria bassiana* ATCC 7159. J Nat Prod 70:1467–1471

Zhang HY, Xie HH, Qiu SX, Xue JH, Wei XY (2008) Heteroatom-containing antibacterial phenolic metabolites from a terrestrial *Ampelomyces* fungus. Biosci Biotechnol Biochem 72:1746–1749

# Chapter 10
# Risks of Biocontrol Agents Containing Compounds of Botanical Origin or Semiochemicals

**Catherine Regnault-Roger**

**Abstract** Semiochemicals and botanicals have the potential to control plant pests or diseases but before they can be used as plant protection products they have to be registered. In the registration process, the risk assessments associated with their properties and their uses have to be evaluated. These risks are linked to the toxicity on the organisms and populations, as well as the exposure. Potential hazards for humans (operators, bystanders, consumers), wildlife and the environment (fate in air, soil and water, non target organisms including the routes to which they are exposed) must be identified and evaluated depending on the uses of the end-products. Semiochemicals and botanicals are currently involved in several approaches for pest biocontrol: insect detection and monitoring, mating disruption and mass trapping for pheromones, insecticide formulations and enhancement of plant resistance for plant allelochemicals and botanicals. These approaches are discussed according to the risk assessment.

## Contents



C. Regnault-Roger (✉)
Université de Pau et des Pays de l'Adour-France, Pau Cedex BP 1155-64012, France
e-mail: catherine.regnault-roger@univ-pau.fr

R.-U. Ehlers (ed.), *Regulation of Biological Control Agents*,
DOI 10.1007/978-90-481-3664-3_10, © Springer Science+Business Media B.V. 2011

## 10.1 Introduction

Today, the problems of crop protection are still frequent and varied as they have been throughout the history of agriculture for more than 10 000 years. Many examples illustrate that the survival of mankind depends on its capacity to protect the plants that constitute its food, in the fields or after harvest (Philogène et al. 2008). The protection of crops has always been: (i) the result of a competition between consumers, mainly man, insect and rodent, according to Ferron (1989), and (ii) the resistance of the plant to the pathogens.

The development of the chemical pesticides, whose production was easy and the costs relatively low, constituted a technological revolution for crop protection during the second half of the 20th century; but successes in control of harmful species in the field and successes in human (malaria) and animal (cattle disease) health as well, led to their intensive and often inappropriate use. The consequences are well known: ecological disorders at numerous levels. The current ideal pesticide was defined with the following qualities: selectivity and absence of toxicity toward non-target species, biodegradability and lack of resistance (Regnault-Roger 2005c). The need to implement an agricultural system taking into account sustainable development has fostered many initiatives to develop alternative methods in order to reduce the use of chemical synthetic pesticides. Among these alternatives, the use of semiochemicals and plant compounds, resulting essentially from secondary metabolism of the organisms, has aroused increasing interest because of their ecological advantages. Improvement of scientific knowledge about biological and toxicological data resulted in re-registration procedures for pesticides in several developed countries. However, it remains to be evaluated whether these natural compounds do not present any risk for health or environment according to the regulatory rules that are currently applied to pesticides. Consequently here are some questions to be discussed: (i) Would the use of semiochemicals and botanicals be without any potential danger? (ii) What is the probability that exposure to these alternative products generates harmful effects for the organisms or the environment? (iii) What is the ratio benefit/risk of these alternative substances compared with conventional pesticides? Answers to these questions will evaluate the opportunity for these substances to promote sustainable agriculture and development. We will examine these main points in this chapter. After defining semiochemicals and botanicals and giving key points for risk assessment, we will detail the characteristics, the successes and the limits of semiochemicals and botanicals linked with the risk assessment of their uses.

## 10.2 Botanicals and Semiochemicals: Definitions for a Concept in Evolution

Defining semiochemicals and botanicals clearly is not as easy as it would appear.

### *10.2.1 Historical Background*

The first definition for this kind of compounds was given by Whittaker and Feeny (1971) at the time when Sondheimer and Simeone (1970) and then Harborne (1972), set the foundations of a new discipline, Chemical Ecology. In a pioneering observation, Whittaker (1970) pointed out that there were chemical mediators biosynthesised by organisms and which affected the behaviour or the physiology of other organisms "for reasons other than food as such". The definition indicated that these chemicals, also called semiochemicals, were involved in interrelationships between organisms belonging to the same species (intraspecific) or to different species (interspecific). They were subdivided into pheromones acting for intraspecific relationships and allelochemicals for interspecific relationships.

The concept of semiochemicals was the result of several observations. Karlson and Lüsher (1959), observing the chemical control of caste development in termites, proposed the term pheromone to describe a chemical that an animal secretes or excretes, and that "releases a specific reaction, for example, a definite behaviour or development process". Wilson and Bossert (1963) later divided pheromones into releasers, which induced an immediate change, and primers, which initiate changes in development, sexual maturation or physiological state. According to Howse (1998), the concept of pheromone "draws attention to a means of communication not previously suspected in animals".

Whittaker (1970), who studied the biochemical ecology of higher plants, suggested the term allelochemics underlined that "ecologists may think of community metabolism in term of three major group substances – inorganic nutrient, foods and allelochemics – by which the species are in a community linked with one another and their environment". These allelochemicals were further divided into allomones, which were advantageous to the emitter (e.g. defensive secretions), and kairomones, which were advantageous to a receiver (e.g. secretions that can be detected by a predator or parasite) (Blum 1977). The study of semiochemicals and the interactions they mediate contributes to an understanding of the behaviour and the evolution of organisms.

Through the modification of behaviour and physiology of other organisms that they cause, semiochemicals presented interesting features not only for intra- and interspecific communication, but also potentially for controlling pest populations. The early identification of lepidopterous sex pheromones and the aggregation pheromones of Coleoptera gave several approaches for using these compounds in pest control. Thus, many commercial systems for monitoring or slow release formulations were developed. Some plant volatiles that played an important role in host location, not only for pollinators but also for predators, and also involved in plant defence, were considered to be semiochemicals (Agelopoulos et al. 1999). Through these examples, because the use of semiochemicals was a component of a non toxic pest management system, it had been understood that semiochemicals constituted an alternative to the use of broad-spectrum toxicant pesticides. In this mindset, semiochemicals (including allelochemicals) were defined as non toxic agents to control pests, although the formal definition of Whittaker (1970) did not indicate this aspect. This pioneering definition included not only volatile compounds but also all chemicals biosynthesised by organisms and involved in the communication between organisms.

### *10.2.2 Current Definitions*

When the NAFTA Technical Working Group on Pesticides updated procedures for semiochemicals, semiochemicals were defined as “a message-bearing substance produced by a plant or animal, or a synthetic analogue of that substance that evokes behavioural response in individuals of the same or other species. Some examples of semiochemicals are allomones, kairomones, pheromones, and synomones” (PMRA-ARLA 2002). This definition is supplemented by information provided by several documents on protocols and data requirements for the registration of these products. The U.S. EPA 40CFR 158.690 Biochemical Pesticides Data Requirements is one of these references. The US EPA (Environment Protection Agency) states that biopesticides are derived from “natural materials as animals, plants, bacteria, and certain minerals” and fall into three classes of compounds: (i) microbial pesticides; (ii) Plant-Incorporated-Protectants (PIPs) that are “pesticidal substances that plants produce from genetic material that has been added to the plant”; (iii) Biochemical pesticides that are “naturally occurring substances that control pests by non-toxic mechanisms”. This definition mentions that “conventional pesticides, by contrast, are generally synthetic materials that directly kill or inactivate the pest. Biochemical pesticides include substances, such as insect sex pheromones, that interfere with mating, as well as various scented plant extracts that attract insect pests to traps” (EPA 2008b).

According to these definitions, semiochemicals clearly belong to this third category in that they are considered to be non toxic to pests. However, as an example, what should happen to the monoterpene linalool identified within many essential oils from Lamiaceae? Pascual-Villalobos and Balesta-Acosta (2003) emphasised that it was known to be a repellent in 1981 (Chapman et al. 1981), then demonstrated to be an effective reproduction inhibitor of the bruchid *Acanthoscelides obtectus* Say (i.e. ovicide and larvicide) in 1995 (Regnault-Roger and Hamraoui 1995). Now it is identified as neurotoxic as it suppresses voltage-gated currents in sensory neurons (Narusuye et al. 2005) and reduces the post-hyperpolarization phase (Huignard et al. 2008). Should linalool, which is today known to act not only by repellency but also by toxic mechanisms, still be considered as a semiochemical and a biochemical pesticide according the EPA definition? Another example is the protective effect of vegetable oils. Some of their volatile compounds develop toxicity by inhalation but non volatile compounds act by forming a film over the insect cuticle that suffocates it (i.e. a physical mechanism). Thus, what is the classification of linalool and such vegetable oils? Aware of the confusion this definition provides, EPA has established a special committee to determine "whether a substance meets the criteria for classification as a biochemical pesticide".

Regarding the EPA biopesticide definition, it has to be emphasised that the category of PIPs is not recognised by the European Union (EU) as biopesticides because PIPs are transgenic compounds. In Europe, Genetically Modified Organisms (GMOs) fall under EC Directive 2001/18/EC, which requires risk assessment, labelling, and public information on GMOs, although all pesticides of any kind are under Directive 91/414/EEC and micro-organisms under Directive 2001/36/EC.

The necessity to have an harmonised approach for data requirements for pheromones and other semiochemicals used for arthropod pest control, and discussions between Canada, United States, Japan, Australia and several member states of the European Union, through the European Crop Protection Association (ECPA), were carried out. A Working Group on Pesticides (WGP) was constituted. According to WGP, "harmonisation was considered as a means to encourage the development of new environmentally friendly pest control products for sustainable agriculture" (OECD 2002b). After the last workshop in Ottawa in 1999, the following definitions were adopted: "Semiochemicals (SCs) are chemicals emitted by plants, animals, and other organisms – and synthetic analogues of such substances – that evoke a behavioural or physiological response in individuals of the same or other species. They include pheromones and allelochemicals.

- Allelochemicals are semiochemicals produced by individuals of one species that modify the behaviour of individuals of a different species (i.e. an interspecific effect). They include allomones (emitting species benefits), kairomones (receptor species benefits) and synomones (both species benefit).
- Pheromones are semiochemicals produced by individuals of a species that modify the behaviour of other individuals of the same species (i.e. an intraspecific effect)."

This definition did not emphasise the necessity of having a non toxic mechanism in order to be classified as a semiochemical. According to this definition, linalool, which is an allomone emitted by plants for defence against pests, could clearly be classified as a semiochemical i.e. allelochemical whatever the effect (repellent or neurotoxic) it produces.

Another argument has to be taken into consideration for the definition of allelochemicals. During the last decade, some research showed interesting properties of plant cell chemical mediators, which participated in the response of the plant against bio-aggressors, or which stimulated as elicitors the natural defence of plants (Benhamou 1996; Walters et al. 2007). Some of these compounds are also called plant strengtheners. As their chemical structures belong to secondary metabolism of plant (e.g. phytoalexines) and because they are chemicals biosynthesised by the plant to defend itself (see Section 4.3.2), they could be considered to be allomones. However, these allelochemicals, despite their participation to the plant defence against pathogens and disease, are not volatile. Their biological activities indicate a strong potential for plant protection and they demonstrate interesting properties to be considered as Biocontrol Agents (BCAs). In this chapter, according to the pioneering definition, we will take into consideration both volatile and non volatile plant allelochemicals.

The definition of botanicals did not elicit such controversy and questioning because it seems evident that botanicals come from plants, as a part of it (e.g. dry leaves or roots) or as a fraction extracted by various processes and more or less purified. However, a question needs to be answered to clarify the definition of botanicals. Should a purified or technical plant extract containing mainly one compound (e.g. linalool extracted from essential oils) still be considered to be a botanical?

To finish this overview of definitions, an interesting point is mentioned by OECD definitions and should be discussed. Semiochemicals could be "synthetic analogues" of chemicals emitted by plants, animals and other organisms. This statement seems relevant to pheromones because it is not possible to extract from insects a sufficient quantity of pheromonal substances for agricultural use on a large scale. Consequently, the products based on pheromones contain synthetic analogues of sexual or aggregative pheromones. But what happens to allelochemicals and by extension to botanicals? Could a synthetic compound be considered to be a BCA if it takes the place of a natural molecule? Should it be considered to be biological despite its synthetic origin? The Natural Products Working Group of the commission "Alternative Methods" of the French Association for Plant Protection (AFPP, Association Française pour la Protection des Plantes) indicated that a chemically synthetic product, if it is very similar to the natural product, had to be considered as a natural product as well (Descoins et al. 2003). The key point is the notion of "to be very similar". Should we distinguish two compounds that are identical? Would they react differently because of their natural or synthetic origin if they are the same molecule? Is a natural product different from a synthetic analogue? To answer these questions, some chemists suggest that it is very difficult to obtain a degree of purity of a synthetic analogue so that no distinction could be made between the natural and synthetic products. They considerer that a "chemical signature" (solvent residues) would give the difference between natural products extracted from plants and synthetic compounds (Hubertus Kleeberg, personal communication). But if no distinction could be made, extracted compound and its synthetic copy must be considered in the same way, whatever their origin. This precision is important for further development of botanicals and semiochemicals as BCAs. When supplying resources are insufficient, the synthetic copy of an identified natural molecule becomes the last resort. This is the case for pheromones, but also for p-menthane-3,8-diol. This compound, which occurs naturally in the lemon eucalyptus plant (*Eucalyptus citriodora* Hook), is chemically synthesised for commercial use (Mosiguard Natural®) (Copping 2004). However, Isman et al. (2008) observed that the toxicity of natural rosemary essential oil for controlling agricultural pests was better than artificial one prepared by mixing the nine major constituents in proportions reflecting the average proportion of commercial oil. This debate is a key point that needs to be clarified in the near future for relevant regulatory propositions.

## 10.3 Risk Assessment

The risk assessment is based on the basic paradigm [Risk = Hazard X Exposure] by integrating hazard (toxicity) and exposure (distribution) data. According to Zubkoff (2008), the hazard components (toxicity) are usually based on experimental observations. The questions asked address the safety of populations and of the environment in which they are exposed to the crop protection materials. The populations embrace humans (including the most vulnerable – women and children) and wildlife (including endangered species). The safety of such populations depends on the environment in which they dwell (air, soil, and water) and on the routes by which they are exposed. The risk assessment has to take into account the answers regarding the populations, the habitats and the environment. The exposure components focus on the distribution, both in terms of concentration and of time, of the crop protection compounds in the environment. They are generally developed from experimental observations of concentrations in the environment and also from measurements of exposure with tested organisms. Modern and rigorous risk-assessment methods have to be applied to observe the nature of the effects and to predict further impact of the evaluated compound. This evaluation could be measurable by classical parameters:

- toxicity on mammalians : acute oral ($LC_{50}$ or $LD_{50}$), acute inhalation and acute dermal, primary dermal irritation, primary eye irritation, allergenicity and similarity to known allergens, mutagenicity, teratogenesis and reproductive inhibition.
- toxicity on the environment is examined, in relation to EPA, with a maximum hazard exposure test using a single dose of a Technical Grade of the Active Ingredient (TGAI) at a concentration of the standard environmental exposure multiplied by 10–20 (Zubkoff 2008). Standard tests to be used are those for evaluating fate in soils, fish environmental toxic effects, freshwater invertebrate tolerance, insect resistance, and honeybee monitoring.
- additional data relating to the modes of action of the compounds would be pertinent.

According to several national agencies, the risk assessment must take into account the active ingredient under the approved conditions of use and it must verify if the end-use products do not present an unacceptable risk to human health or to the environment. The Pest Control Products Act (Department of Justice Canada 2002) stated that a risk for health or for the environment is considered acceptable if "there is reasonable certainty that no harm to human health, future generations or the environment will result from use or exposure to the product under its conditions of registration".

Several approaches were investigated to define pesticide risk indicators (Devillers et al. 2005; Peterson 2006) and several OECD programs were focused on them (OECD 2002a; 2004; OECD 2007). Among the most achieved indicators, risk indicators for health and environment applied to the province of Québec, Canada, the Quebec Pesticide Risk Indicators (QPRI) were defined and based on the principles of indicators developed by Norway (Samuel et al. 2007). These risk indicators should be used "as a supplement of risk assessment" and as a tool for a simplified representation of reality and facilitation of decision making. Consequently, a Toxicological Risk Index (TRI) is calculated by taking into account the total of (i) acute risks (oral, dermal and inhalation $LD_{50}$, plus dermal and ocular irritations, plus sensitisation) evaluated by a score (0–8 points), added to (ii) the total of chronic risks (carcinogenicity, genotoxicity, endocrine disruption, reproductive effects and development inhibition) evaluated by a score from 0 to 16 points and modified by a scored factor (1–3 points) linked to environmental persistence and bioaccumulation potential in humans. This TRI is then adjusted by a weighting factor depending on end-use product characteristics (formulation, amount of active ingredient in end-use products) to calculate the HRI (Health Risk Index). Following the same principles, an Environmental Risk Index (ERI) is calculated. It takes into account the environmental parameters like (i) physicochemical and environmental fate properties e.g. half-life, organic carbon adsorption, water solubility, octanol-water partition coefficient; (ii) ecotoxicological indicators ($LC_{50}$, $LD_{50}$ and $EC_{50}$ for earthworms, honeybees, birds (North American sentinel species mallard duck and bobwhite quail), fish and aquatic invertebrates, algae and sentinel vascular plant (duckweed); (iii) end-use product parameters linked to Standardized Area Dose (SAD), quantity of active ingredient applied/or sold on a provincial scale, and types of crops. Thus defined, HRI and ERI take into consideration the toxicological characteristics of active ingredients and properties linked to an end-use product. These indicators enable comparison of pesticides from any origin in order to make informed choices to facilitate the selection of pesticides that would be the less harmful for health and environment.

Semiochemicals and botanicals are naturally occurring compounds, but some natural substances are commonly present in the environment without any noticeable adverse effect, whereas others are known to develop toxicity in particular conditions. Consequently, like conventional pesticides, it should be necessary to evaluate the impact of semiochemicals and botanicals on all the parameters taken into account by the HRI and ERI, particularly the biological parameters, biodegradability and biodisponibility, exposure to target, non-target and mammalian species, and potential development of resistance by target species. However, the specificity of these substances, in relation to known physiological responses of the target pest or to the absence of adverse effects for populations and environment, should be considered if this information is available from reliable sources. Thus, the advantages of semiochemicals and botanicals and the way they are used for biocontrol have also to be taken into account to assess risks.These data would be essential to clarify the procedure for registration of these compounds. We will examine the risks linked to pheromones and botanicals used as BCAs in this context.

## 10.4 Pheromones

### *10.4.1 Characteristics*

Pheromones are volatile chemicals which are emitted by individual of a species to give indication to others about territory and movement, aggregation, mating, oviposition and nest-building, sexual maturation, alarm etc. (Howse 1998). Their molecular weight is light, so the volatility of the pheromones gives to this chemical signal an advantage for intraspecific communication for social insects. Volatile chemicals can travel long distances in the wind, distributing the signal without energy expenditure for communication, except for the energy needed for biosynthesis of the pheromone by the emitter. They are information-specific and can be used in the dark. The signal travels around obstacle without any reflection and the emitter can remain hidden from the receiver. The chemical structure of the molecules contained within pheromone limits the durability of the pheromone signal to a short half-life. However, it is longer than a visual signal and its modulation is not as easy to change in amplitude or qualitatively compared with an acoustic signal.

### *10.4.2 Use of Pheromones in IPM: Successes and Limits*

The specificity of the pheromone and their benign environmental properties soon stimulated investigations on their potential for Integrated Pest Management (IPM). For over 20 years, several reviews and books focused on this topic (Jutsum and Gordon 1989; Cardé and Minks 1997; Howse et al. 1998; Renou and Guerrero 2000; Wyatt 2003; Picimbon 2005; Picimbon and Regnault-Roger 2008). The pheromones are used to lure insects and trap them following three main approaches: (i) detection and monitoring; (ii) mating disruption; (iii) attract and kill (or lure and kill) mass-trapping capture.

The principle of the use of insect pheromones for detection and monitoring is to attract insects to the trap in order to determine their occurrence in the field. Most often, the trap bait contains a female sex pheromone to attract males into the traps. Consistent trapping protocols are essential to have relevant information for identification of the insects, the evaluation of insect populations and year to year comparisons. This monitoring gives very useful information for decision making on insecticide treatments in the fields, to survey and sample low density populations.

The mating disruption approach involves confusing males by placing several point sources of female sex pheromones in the field. The male follows false trails and expands mating energy in pursuit of artificial pheromone sources. Consequently, the reproduction of the targeted population is reduced.

The attract-and-kill mass trapping is based on formulations containing a combination of pheromone which attracts the insect, and an insecticide that kills it. According to Flint and Doane (1996), damage to the target species was very limited, but success was reported against the Chinese tortrix *Cydia trasias* (Meyrick) to protect Chinese scholar-trees *Sophora japonica* L.; damage to the trees was reduced by about 70% following control of three generations (Zhang et al. 2002).

The efficiency of pheromones as BCAs is not the same for the three strategies. The detection and monitoring approach is certainly the most efficient because trapping insects is a tool for further insecticidal treatments for organic farming and classical agriculture as well. It is currently also used on a large scale for experimental or conventional cropping. Another application is the monitoring of insecticide resistance and distribution in a population, because of the difficulties of sampling by traditional methods (Suckling and Karg 1998). According to Royer and Delisle (2005), the use of pheromone traps to survey the density of an arrhenotoc species is inappropriate but is relevant to follow the change of geographical distribution.

The success of mating disruption strategy for control of insect pests depends on the quality of dispensers to deliver a homogenous emission of pheromone to achieve a sufficiently saturated area for male confusion and capture (Delisle and Royer 2005). Mating disruption strategies has been developed with success in the forests of North America as well as in arboriculture and viticulture in Europe (Frérot 2005).

The efficacy of attract and kill mass trapping strategy for control of insect pests largely depends on the targeted species. The knowledge of the biology of the species (monoginy, polyginy, protandry) as well as the density of population, the surface to be protected and the position of the traps are essential to the success of this method (Royer and Delisle 2005).

Biocontrol by pheromones is not as well developed as it could be, and their sales represent no more than 20% of European biopesticide market (Regnault-Roger et al. 2005b). There are several reasons for this situation: the quality of pheromone formulations, the motivation of the agricultural producers and the cost of treatments.

Most insect sex pheromones are multicomponent with precise ratios of components which may be expensive to manufacture. The current commercial formulations of pheromones do not always sufficiently mimic the natural chemical blends pheromones from females. One difficulty is that the chemical signal changes according to the geographical distribution of insect species (El Sayed et al. 2003) and to season generation renewal (Delisle 1992). Consequently, a comparison between a virgin female in a trap versus a commercial pheromone showed the superiority of the insect (Delisle and Royer 2005).

Another point to temper this approach is the high level of constraints for the farmer. Pheromone pest management needs the installation of many traps in which it is essential that the diffusers allow a regular and sufficient release of the pheromone. This also requires constant monitoring of the plots where the traps are distributed and these plots must be isolated from external contaminations using reinforcement of the treatments at the edges of the treated area. This requires increased vigilance by the farmer to monitor the phases of development of the various parasites to avoid the phenomena of resurgence. Only farmers who have a high degree of motivation put this plant protection approach into practice.

The cost of the products (insect sex pheromone formulations and traps) is another factor that restricts this approach for insects of economic importance. The use of pheromone becomes relevant in particular situations when conventional pesticides are not operating or when the environmental conditions (forests with high trees, arboriculture) do not facilitate the use of conventional pesticides.

### *10.4.3 Risk Assessment for Pheromones*

The risk for health, non target species and environment is considered to be weak because of the specificity of pheromones, and because they are mostly used in the manufacture of impregnated lures that are installed within traps for IPM.

Regarding health, two key factors have to be considered: the toxicity of the products and the levels to which people might be exposed. People are naturally exposed to insect pheromones in houses, gardens and fields without any particular adverse effect. But should this natural level be increased when pheromones are used as BCAs? Pheromones are released from dispensers at very low quantities in the environment. The lures are inside dispensers and the exposure to the active ingredient is unlikely. The amounts for end-use products are not significant and consequently the probability of water or soil pollution is negligible. Take the example of the lure with the German Cockroach Extract to follow the risk assessment evaluation (Health Canada 2008). The post application exposure risks to human health are negligible because the amount of active ingredient formulated into each trap is low (0.125 mg per trap). As the lures with active ingredient are attached to the inside of the traps during manufacturing, they are not available for direct exposure to applicators or bystanders. The way which the active ingredient is formulated ensures that the end-use products will not cause significant amounts of human exposure to the German Cockroach Extract. The dietary risks from food and water are not of concern because German Cockroach Extract is not used on food or feed crops.

Regarding the non target species, an unexpected sensitivity might occur in some cases. The EPA (1994) has reviewed and evaluated ecotoxicity data for a number of pheromones, and some of them demonstrated a high toxicity to aquatic invertebrates, a moderate toxicity to fish, but practically no toxicity to birds that were tested. These results demonstrated the potential toxicity of arthropod pheromones to aquatic organisms. However, this risk is not relevant if the pheromones are used in appropriate baits following the labelling indicated by registration. EPA recently registered California Red Scale Pheromone because it recognises the low toxicity, negligible expected exposure, and lack of expected adverse effects on humans and non-target organisms of arthropod pheromones when used in polymeric dispensers. Consequently, following the instructions given on the labelling, no available risk was expected (EPA 2008a). In the fields, some risks as a consequence of mating disruption technology were evaluated on the dynamic of associated fauna or natural enemies of the target species, but with contradictory results (Arakaki et al. 1997; Delisle and Royer 2005). These examples show that very low risks are expected for the use of pheromones as BCAs for human health and environment when registration recommendations and labelling are followed.

## 10.5 Botanicals and Plant Allelochemicals

### *10.5.1 Characteristics*

The botanical extracts come from fractionation of the plant by various processes and their composition varies depending on the botanical sample, the experimental conditions and the physicochemical properties of the compounds. Thus, the extracts from the same plant are not only complex, but also the molecular composition is very variable from one extraction to another.

Plants are rich in allelochemicals, which were formerly called "Secondary Compounds of the Plants (CSP)" because there was limited knowledge about them. The large majority of plant allelochemicals are from plant secondary metabolism though several pathways, shikimate, mevalonate, acetate or amino-acids. The complexity of this metabolism results in a large number of molecules, but they fit into a small number of chemical families. An estimation gave more than 500 000 plant allelochemicals (Mendelsohn and Balick 1995) from three main chemical families: (i) phenolics and phenylpropanoïds; (ii) terpenoïds and steroids; (iii) alkaloids and nitrogen compound (Harborne 1989). Some allelochemicals play a role in plant defence (allomones). They are repellent, antifeedent, antinutritional, or neurotoxic. More generally, they affect the biotic potential of parasites and pests. Plant allelochemicals act on a broad diversity of species: insects, which were the most studied because of a better visibility, and also nematodes, phytopathogene micro-organisms (fungi and bacteria), as well as other species plants (allelopathy). For decades, the use of plant allelochemicals and botanicals was more focused on the control of insects than other plants organisms (Regnault-Roger et al. 2008). In recent years, the improvement of knowledge of plant resistance mechanisms against bio-aggressors underlined that allelochemicals play an essential role in plant defence. Phytoalexines are low molecular-weight compounds of a non proteinaceous nature, mainly belonging to polyphenols, terpenoïds and polyacetylens (Lepoivre 2003). They are synthesised de novo in response to biotic or abiotic stresses and participate in plant induced resistance. Others, for example diferulates, are involved in the mechanical and biochemical barrier that constitutes the wall of maize grain (Bergvinson et al. 1994; Bily et al. 2003). Consequently, plants contain a true arsenal of allomones resulting from co-evolution of the species to defend themselves.

These allelochemicals have such high potential against bio-aggressors that they must be taken into consideration for plant protection biocontrol. It should be noted at this point that botanical and plant allelochemical BCAs have to be distinguished from other plant products that are not the result of the species co-evolution, i.e. according to the terminology of EPA, the Plant Incorporated Product (PIPs). Because BCAs are supposed to be environmentally friendly, and because of many gaps in the knowledge of environmental consequences of PIPs, they must not at this moment be considered to be BCAs until rigorous field bioassays clearly demonstrate that no adverse effect is noticeable.

In any case, the potential of plant allelochemicals and botanicals for plant protection could be used in two alternative strategies. The first one could be qualified as "external" approach because it aims at reinforcing the protection of the plant using traditional soaps with formulation including plant allelochemicals or plant extracts as active ingredients. It is the oldest use which has been made of plant extracts and allelochemicals. The second one is an "internal" approach. It is more recent and probably less risky than the first one. It aims at reinforcing the plant defence by developing its own mechanisms through allelochemicals.

## *10.5.2 Botanicals and Plant Allelochemicals Using in Insecticidal Formulation*

The commercialised pesticide soaps and specialities including plant allelochemicals and botanicals can be used in both organic and conventional agriculture depending on the formulation. If all the active ingredients and additives are of plant origin, thus satisfied the requirements for organic agriculture, they can be used that way. If they are associated with synthesised pesticides, they can be applied in conventional agriculture.

Plant allelochemicals and botanicals are still not used in plant protection to their full potential. Within the 835 actives substances that have been reviewed by the European Union under the Directive 91/414/EC, 680 were synthetic pesticides (82%), 69 pheromones (8%), 61 plant extracts (7%) and 25 microbial (3%) (Redbond 2003). To understand this situation, we have to examine which are the factors that hamper the development of these BCAs: Unfavourable background? Economical obstacles? Unacceptable risk assessments?

### 10.5.2.1 Historical Background

It is thus difficult to assess exactly where and when plants or plant extracts were systematically used in plant protection or, more generally, in agriculture. In the 18th century, some publications dealt with plant-based formulations to control insect pests (Balachowsky 1951). At the end of the 19th century, empirical methods including the use of toxic plants or minerals, oils, tars, sulfocalcic sprays and boiling water were commonly put into practice (Pesson 1990). Integration of empirical and scientific observations led to the development of plant extracts. The first botanicals and allelochemicals to be used as pesticides came from easily available products. Pest insects were targeted more than pathogens because they could be easily identified. Several recent books and chapters have reviewed biopesticides of plant origin (Prakash and Rao 1997; Weinzierl 1998; Koul and Dhaliwal 2001; Philogène et al. 2005; Regnault-Roger et al. 2005a; Regnault-Roger 2005b; Regnault-Roger and Philogène 2008).

Before the World War II, four main groups of compounds were commonly used: nicotine and alkaloïds, rotenone and rotenoïds, pyrethrum and pyrethrins, and vegetable oils. Some of them had several inconvenient properties because of their toxicity on non target species (nicotine) or the instability of the molecules (pyrethrum). As a consequence, the use of these substances decreased with the commercialisation of chemically synthesised insecticides developed during the World War II (organochlorides, organophosphates, and carbamates), which moreover were easier to produce and handle and were less expensive. This situation continued until the late 1960s, although some research on biopesticides of plant origin was pursued throughout the second half of the 20th century in order to improve their stability or to discover new molecules and new sources of molecules.

As a result of the many demonstrations of the ecological hazards of synthesised insecticides, there was renewed interest in the 1970s for botanicals, which was illustrated by the development of pyrethrinoïds derived from pyrethrum. Although these derivatives are products of chemistry synthesised from a natural source so as to be more stable in sunlight, today they have only slight resemblance to chemical structures of natural pyrethrins and to their mode of action. As they are, in some way, derived from a natural compound and despite their chemical background, pyrethrinoïds nevertheless benefit from the label of "green" products. Also, because they are efficient and less hazardous than the other synthesised pesticides, they are nowadays key products for IPM. They have, without any doubt, contributed to modify the perception that people had of botanical insecticides. The research on neem (Meliaceae) also illustrated this renewed interest (see Section 4.2.3). However, during the whole 20th century, only a limited number of botanicals or plant allelochemicals were used for crop protection.

#### 10.5.2.2 Botanicals and Plant Allelochemicals Today: Factors that Hamper Their Development

All the pros for using botanicals and allelochemicals in IPM can easily be negated by the environmental hazards generated by synthetic pesticides. As products of metabolism resulting from species co-evolution, they exhibit many environmental advantages (Regnault-Roger 2005c): (i) they possess a selectivity and a specificity in their effects on the target species; (ii) biosynthesised, they are enzymatically biodegradable with short half-lives in general; (iii) the association of several compounds can be synergistic, thus decreasing the effective amounts of active ingredients; and (iv) they belong to several different chemical families. By increasing the choice of the molecules available, they contribute to the diversification of the biochemical and molecular targets towards insects and hence limit or delay the resistance phenomenon.

However, Isman (2005b) indicated that only a few botanicals and plant extracts are currently commercialised. Four substances are mainly used, pyrethrum, rotenone, Neem, and essential oils, followed by nicotine, ryania and sabadilla for minor uses. Even the global biopesticide market is expected to go up by 4% (Philogène et al. 2008), the present situation is little different from the past. The question is to know what are the factors that impede development.

Several factors hamper the industrial development of insecticide formulations containing plant compounds (Isman 1997; 2005a; Regnault-Roger and Philogène 2008). Beside economical and commercial considerations such as availability of the raw material and its accessibility, or standardisation and refinement of plant commercial products, the toxicity of plant extract compounds on non targeted species is not negligible. Although they are natural, all products are not necessarily safe for people and for the environment. The current claims that plant protecting products or BCAs should not pose unreasonable risks to people or the environment, means that the evaluation of these compounds meet today's most stringent standards of scientific knowledge.

In this context, the risk assessment for botanicals and plant allelochemicals have to be evaluated by taking into account their toxicological nature and their ecological advantages, as well as the exposure scenario linked to the current use of the formulated end products. To illustrate these arguments, some current commercialised botanicals will be detailed.

#### 10.5.2.3 Main Commercialised Botanicals and Plant Allelochemicals: Uses and Risk Assessment

Botanical and allelochemical compounds are particularly well developed in the United States because of the historical background and conditions of regulatory approvals. Their uses and the risk assessment were re-evaluated recently according the Reregistration procedure under code 40 CFR (Code of Federal Regulations) of application of FIFRA (Federal Insecticide, Fungicide and Rodenticide Act). The Reregistration Eligibility Decision (RED) of EPA detailed clearly the risk assessments linked to uses and exposures. These considerations were based on experiments and reliable data. However, many plant oils (e.g. cinnamon, mint, geranium etc.) and allelochemicals (e.g. monoterpenes geraniol and eugenol) are considered to be minimum risk pesticides (40CFR 152.25f) and they are exempted from the requirement of FIFRA. In Europe, in the context of a harmoniation procedure for all European Union Member States, the Directive 91/414/EEC came into force in 1991 for re-evaluation of all Plant Protection Products except micro-organisms and GMOs (see Section 1.2.). All existing substances previously registered at national level were subjected to re-evaluation at EU level. For practical reasons, the review process was divided into 4 lists and natural substances including semiochemicals and botanicals were included on list 4 which was the last one to be examined. During the re-evaluation procedure of the active ingredients, formulations that were already registered in an EU Member State when Directive 91/414/EEC came into force were allowed to stay on the market but were subjected to re-evaluation after the active substance's inclusion in Annex I. This procedure now is over. Several decisions of the Commission of the European Communities were published in 2008 and 2009 to include or not several plant extracts and botanicals. Here follow some examples.

*Pyrethrum* is a powder obtained by crushing dried flowers of daisies belonging to the family of Asteraceae: *Chrysanthemum* spp., *Pyrethrum* spp., *Tanacetum* spp.

*Chrysanthemum cinerariaefolium* Benth & Hook was first used in Europe in the 1800s against lice and flies (Regnault-Roger and Philogène 2008). Other species of *Chrysanthemum, C. roseum, C. tamrutense* and *C. carneum* also contain significant amounts of pyrethrum. Pyrethrum or pyrethrins is a mixture of six esters pyrethrins I and II (the most abundant), cinerin I and II, and jasmoline I and II. Pyrethrin I is the most toxic ester. It alters nerve transmission by slowing the shutting of $Na^+$ channels during the recovery phase of neuronal action potentials. The insect consequently presents hyperactivity followed by convulsions. Pyrethrins are very toxic and act very quickly on insects. On the other hand, they have low to moderate toxicity towards mammals (oral $LD_{50} = 1400$ mg kg$^{-1}$ for rats, dermal $LD_{50} > 2000$ mg kg$^{-1}$ for rabbit, inhalation $LC_{50} = 3.4$ mg L$^{-1}$ for rats), moderate eye irritant, mild dermal irritant and no skin sensitisation. A massive intoxication by pyrethrum cause a tremor followed by convulsions and nervous system lesions were observed in rat and mouse following an acute exposure. A thyroid effect was observed following chronic exposure in rat and dog, and liver effects in rat, dog and mouse following a short or long exposure. Pyrethrum is quickly hydrolysed in the digestive tract, although it is more toxic by inhalation or if administered intravenously. Pyrethrins are classified as "suggestive evidence of carcinogenicity" because of occurrence of benign tumors in female rat. However, there is insufficient evidence to assess their human carcinogenis potential. Toxicity is mentioned for non-targeted species, especially fish, invertebrate and bees. However, its great instability in light, air and moisture considerably reduces risks related to its use. Despite its high production cost, it is a natural insecticide that is currently widely used (1000 tonnes of pyrethrum are sold every year with about 90% being used in non-agricultural sites in USA) (EPA 2006). It is recommended for the control of flying and crawling insects and arthropods and mites on fruits, field crops, ornamentals, greenhouse crops and house plants as well as stored products, domestic and farm animals. It is normally applied in combination with piperonyl butoxide, a synergist that inhibits detoxification (Copping 2004). Because pyrethrum is registered for use in agricultural, residential, commercial, industrial and public health sites in USA, several scenarios of uses to evaluate risks for human health (dietary risk, residential) and non target organisms (ecological risk on aquatic and terrestrial organisms) were tested and EPA (2006) concluded that the "currently registered uses of pyrethrins are eligible for reregistration provided mitigation measures... implemented through label amendments". These mitigations focus on the restriction for using the end-products in specific places (e.g. nursing homes, hospitals, schools etc.), and on the method of application of the end-products and the protection equipment required for applicators, and the number of application for agricultural use in relation to the season and the pest pressure. This example demonstrates that the most popular botanicals must be used cautiously. In EU, Pyrethrins were included for uses as insecticide only in Annex I of Council Directive 91/414/EEC in December 2008. It entered in force in September 2009 until August 2019 (OJEU 2008a).

*Rotenone* is widespread in Fabaceae (ex Papilionaceae) growing in Asia (*Derris* spp.) and in America (*Lonchocarpus* spp.). Rotenone is one of oldest insecticides used all over the world. The use of crushed roots of Fabaceae to catch freshwater fish by native populations of South America was mentioned as early as 1665, whilst it was reported that these extracts were added to insecticidal soaps in 1848 (McEwen and Stephenson 1979). The active ingredient belonging to flavonoïds, was isolated by Geoffroy in 1895 from *Lonchocarpus nicou*, an American Fabaceae and was called "nicouline". Similar work was carried out in 1902, with the roots of *Derris elliptica* (roten in Japanese) by Nagai who called the compound rotenone, whose name prevailed (Dajoz 1969).

Rotenone inhibits cellular respiration and energy metabolism at the level of the mitochondrial respiratory chain. Harmless for warm-blooded animals, it is very active against cold-blooded animals such as amphibians, fish and reptiles. Although some accidents were reported with enzymatic inhibition, rotenone was regarded for a long time as being moderately toxic for mammals. Although its acute dermal $LD_{50}$ (rabbit) was over 5000 mg $kg^{-1}$ and no skin sensitisation was noticed, its acute oral $LD_{50}$ (rat) was 39.5 mg $kg^{-1}$ for female and 102 mg $kg^{-1}$ for male. The acute inhalation $LC_{50}$ (rat) was 21.2 μg $kg^{-1}$. These latter figures resulted in rotenone being classified in the highest category of toxicity (EPA 2007). Cases of chronic toxicity leading to kidney and liver damage were noted, and it was also found to be carcinogenic for rodents (Weinzeirl 1998). More recently a link between rotenone and Parkinson's disease was hypothesised (Betarbet et al. 2000). Rotenone persists 3–5 days on the foliage after its application and is easily biodegradable. Its half-lives in warm and cold water are 1.5 and 20 days, respectively. Rotenone used alone is not toxic for bees but is lethal in combination with pyrethrum (Copping 2004).

This compound was used until the 40 s but, as with many other insecticides extracted from plant, its use declined at the end of the World War II (Philogène et al. 2005). Until recently, rotenone was used in organic agriculture alone or associated with other ingredients such as pyrethrinoïds, synergist (piperonyl butoxide), sulphur or copper to control a wide range of arthropod pests including aphids, thrips, moths, beetles and spider mites. However, following the regulatory update 46/2007 within the frame of Directive 91/414/EEC and EC Decision (2008/317/EC) published on 10 April 2008, because of a lack of required information, the rotenone substances should not be included in the Annex I to Directive 91/414/EEC and consequently was withdrawn from the European Union plant protection products market at the date of October 10th 2009 (OJEU 2008b). Nevertheless, rotenone has been granted essential use in the UK, Italy and France until 2011 on fruit trees, ornamentals and potatoes only. This derogation is limited to professional users with appropriate protective equipment. The uses of rotenone were also restricted in the USA for livestock, residential and home owner use, domestic pet uses, and all other uses except for piscicide uses. Consequently, rotenone is now registered to be applied directly to water to manage fish populations in lakes, ponds, reservoirs, rivers, streams, and in aquaculture, to eliminate completely or partially undesired fish species in the treatment area (EPA 2007). The risk assessment was evaluated according to human health, the occupational risks (workers), non target aquatic (fresh water fish other than target species, invertebrates) and terrestrial (piscivorous birds, wild mammals, plants, bees) species. It is classified by EPA (2007) as Restricted Use Pesticides (due to acute inhalation, acute oral and aquatic toxicity).

*Neem* is extracted from *Azadirachta indica* A.Juss which is native to arid regions of India. The ability of the oil to repel pests has been known for thousands of years. The oil has also been used on skin and medicinally. Neem is a part of the traditional practices in India. It is a mixture of more than one hundred limonoid compounds including azadirachtin, salannin, and nimbin and their analogues. All these compounds act differently and numerous effects of Neem on insects have been reported. Salannin causes repellence and feeding deterrence, while azadirachtins are the only compounds that have a significant activity as inhibitors of insect growth (Schmutterer 1990). This results from an inhibition of the synthesis of ecdysteroids with, as a consequence, a disruption of moults and of the reproductive cycle of the insect. Neem oil was classified by EPA (1995) in class IV because of its acute oral $LD_{50}$ on rat above $5000^{-1}$ (no mammal toxicity) and all route of exposure were classified the same. It had a mild (minimal) effect on skin sensitisation and eye irritation but was not cytotoxic and mutagenic according the test of Ames. However, Kleter et al. (2006) reported that, according to Boeke et al. (2004), some unknown hazards with new extraction methods would produce toxic effects of the Neem extract in mice and guinea-pigs with gastro-intestinal spasm, hypothermia and death with $200–400^{-1}$ of leaf extract. Neem and azadirachtin were recently suspected to be endocrine disruptors but with contradictory results (Falconer et al. 2006; Pfau et al. 2009; PAN 2010)

In relation to its environmental impact, Neem is sensitive to light and degrades in water (Isman 1997). Consequently, it has limited persistence in the environment. The half life of azadirachtin A after spraying on leaves of tomato or potato was 1 day (Kleeberg 2006). A study on six aquatic organisms (crayfish, shrimps, mosquitoes larvae, water fleas) concluded that the risk values of azadirachtin and neem-based insecticides (Neemix™ and Bioneem™) did not exceed the criteria. Consequently no ecological hazard was likely to result from their use (Goktepe et al. 2004) or from the forest pest management application on aquatic macroinvertebrates (Kreutzweiser 1997). Azadirachtin acts on a wide range of insects: balsam fir sawfly *Neodiprion abietis* (Harris), thrips, leaf miners, aphids, caterpillars, pine false webworms. It deters certain insects, such as locusts, from feeding, and it interferes with the normal life cycle of insects, including feeding, moulting, mating and egg laying.

Tested on over 300 species, it has been effective on 90% of susceptible species with a large variability of $DL_{50}$ (Philogène et al. 2008). Recommended by the National Research Council of "Tree for solving global problems" (NRC 1992), Neem is considered by many experts to be the superior BCA (Brahmachari 2004; Kleeberg 2008). Despite such qualities, the development of this insecticide is hampered by: (i) cultivating the plant on a commercial scale; (ii) extraction of the active ingredients; (iii) development of persistent formulations and shelf life (Philogène et al. 2005). According to Kleeberg and Ruch (2006), the standardisation of Neem seeds extracts, which show a large variation of azadirachtin content, is one of the key factors to enhance the commercialisation of Neem products. Neem and azadirachtin are currently registered in several countries. In USA, EPA (2001) considered that the Clarified Hydrophobic Extract of Neem Oil is a naturally occurring compound that displays a non toxic mode of action on the target and consequently classified this active ingredient as a biochemical pesticide. This Clarified Hydrophobic Extract of Neem Oil is obtained when the natural neem oil is removed from the seeds and treated with alcohol. The remaining oil does not contain azadirachtin because all of the azadirachtin and related substances are separated from the oil itself. The remaining oil – without the azadirachtin – is called Clarified Hydrophobic Extract of Neem Oil. It was first registered by EPA in 1995 and azadirachtin in 1985. The Commission of the European Communities decided to not include azadirachtin in Annex I of the Directive 91/414/EEC because the notifiers voluntarily withdrew their support for an inclusion. But as this non-inclusion was not "based on the presence of clear indication of harmful effect", the decision did not prejudice the submission of a new application (OJEU 2008c).

*Nicotine* was one of the first molecules used as an insecticide since the use of aqueous extracts of tobacco against the sucking-piercing insects of cereals was mentioned in 1690. But the active molecule of this plant, nicotine, was isolated only in 1828 and it was in 1904 that it was synthesised (Matsumara 1985; Ware 2000). This very stable alkaloid in its levogyre form is neurotoxic for insects, mammals and birds. It is an acetylcholine mimic, binding to postsynaptic receptors and interfering with the transmission of signals in nerves. This causes stimulation followed by depression of the vegetative nervous system, muscles and the central nervous system. Nicotine is acutely toxic (Category I) by all routes of exposure (oral, dermal and inhalation). The $LD_{50}$ of nicotine is 50 mg $kg^{-1}$ for rats and 3 mg $kg^{-1}$ for mice. A dose of 40–60 mg can be a lethal dosage for adult human beings through paralysis of respiratory muscles and doses as low as 1–4 mg can be associated with toxic effects in some individuals. Nicotine is neither an initiator nor a promoter of tumours in rodents but it is also toxic for birds. Some countries like China or Bolivia use nicotine to protect rice cultivation (by immersing the stems of tobacco in the plantations) and potato fields (spraying) (Thacker 2002). In the USA, since May 21st 2008, because of risks (i) for applicators both during and after application, (ii) for consumers of plants from treated greenhouses and (iii) for people who might be exposed to nicotine residues in treated greenhouses, the sole remaining nicotine registration is a restricted pesticide use only in greenhouse for ornamentals against adult whiteflies, aphids, and thrips (EPA 2008c). The Commission of the European Communities decided to not include nicotine in Annex I because it was not demonstrated "a safe use with respect to operators, workers, bystanders and consumers" (OJEU 2009).

It is apparent from these examples of the main botanicals that have been available on the market over the last decades that the situation is complex and that only a few compounds used in insecticide formulations really appear to have a future as BCAs.

### *10.5.3 Botanicals and Plant Allelochemicals Used in the Reinforcement of Plant Resistance*

The use of molecular biology as a tool improves knowledge of the mechanisms of resistance of plants to the attacks of bio-aggressors and of the role of allelochemicals. It opens prospects to use them in two approaches: (i) improving the plant resistance by increasing the level of constitutive allelochemicals in varieties; (ii) stimulating plant induced resistance.

These two approaches, which aim at reducing the use of synthetic pesticides in agriculture, involve not only plant allelochemicals, but also other microbial, fungal mineral molecules. We will focus this review on examples relating to the plant allelochemicals and extracts.

#### 10.5.3.1 Plant Allelochemicals Improving Resistant Varieties

One of the basis of modern agronomy is the selection of the characters of interest to improve quality of crop plants. Selection of varieties, which was carried out by traditional methods, now uses biotechnology tools. It was observed that there exists, in certain cases, a direct relation between the degree of tolerance of a plant to a disease and allelochemicals. Positive correlations between the content of allelochemicals and resistance to pathogens were highlighted. As an example, Bily et al. (2003) noted that a variety of maize that was resistant (CO387) to the ear rot disease caused by *Fusarium graminearum* had higher concentrations of diferulate than a susceptible variety (CG62). The subsequent research aims to establish if it is possible to build a hierarchy from diferulate levels between the various varieties of maize according to ear rot disease resistance. The aim is contribute to the development of corn varieties resistant to *F. graminearum*. It is necessary to determine if diferulates are implicated genetically in this resistance in order to achieve variety selection and breeding. This can be done by traditional techniques (crossings using the Mendelian laws) or by biotechnology tools with Marker Assisted Selection (MAS). MAS is a powerful tool to help the breeders to identify genes of resistance to the diseases by identification of the genomic regions contributing to pathogen resistance with characterisation of QTL (Quantitative Trait Loci). QTLs describe the roles of specific loci in genetically complex disease resistance traits and identify the genomic regions contributing to resistance function (Pandey et al. 2006). The QTLs mapping associated with overlapping regions, which are implicated in resistance and phenotypic variation, is a key for the identification of molecular markers linked to the genes for resistance. Considering the previous example with Gibberella ear rot resistance, a molecular linkage map contained 162 markers distributed over 10 linkage groups. Composite interval mapping identified 11 quantitative trait loci (QTLs) for Gibberella ear rot resistance following silk inoculation and 18 QTLs following kernel inoculation. The majority of the favourable alleles were derived from the resistant parent (CO387) (Ali et al. 2005). The germplasm and markers for QTLs with significant phenotypic effects may be useful for marker-assisted selection to incorporate *Gibberella* ear rot resistance into commercial corn cultivars by classical crossing or genetic engineering.

#### 10.5.3.2 Plant Extracts and Allelochemicals Enhancing Induced Resistance

This technology has been developed for over 15 years. This innovating approach enhances plant resistance to pathogen infection by treatment with a variety of biotic and abiotic inducers, also called elicitors. These agents could be virulent or avirulent pathogens, non pathogens, cell wall fragments, plant extracts, or synthetic chemicals. They can lead to the induction of resistance to subsequent pathogen attack, both locally and systemically. The activation of defence responses includes an oxidative burst, which can lead to cell death trapping the pathogen in dead cells, or changes in cell wall composition that can inhibit the pathogen (Walters et al. 2007). Plant allelochemicals, especially polyphenols, are strongly implicated in these mechanisms (El Modafar et al. 2008). The elicitors currently identified are mainly of microbial origin. However, in the list drawn up by the Scottish Crop Research Institute in 2004, some plant extracts and allelochemicals were included. Polyphenols (gallic, m-hydroxybenzoic and p-hydroxybenzoic acids, phloroglucinol) showed eliciting activities against several fruit and vegetable pathogens: *Colletotrichum lagenarium, Phytophthora infestans, Sclerotinia sclerotiorum, Pyricularia oryzae,* or *Xanthomonas oryzae* pv.*oryzae* (Regnault-Roger and Philogène 2008). Extracts from *Hedera helix L., Salix alba L., Viscum album L., Alchemilla vulgaris L., Reynoutria sacchalinensis* (F.Schmidt) were identified as inducers of resistance against Fire Blight of apple and of *Cotoneaster watererii* (Zeller 2006). *Reynoutria sacchalinensis* induced phenolic phytoalexines. Marketed under the name of Milsana® (KHH Bioscience), it is used particularly in North America for the protection of ornamental plants like roses and begonias, and also against various *Oïdium* of vegetables and fruit (Konstantinidou-Doltsinis et al. 2006). *Macleaya cordata* extract registered under the name of fungicide Qwel® (Camas Technologies Inc), induces increased amounts of polyphenolic phytoalexines and also SAR (Systemic Acquired Resistance) (Copping 2004). The extract laminarin, a polymer of glucane β-1.3-1.6 purified from the brown algae *Laminaria* sp., was included in Annex I of Directive 91/414/EEC. Because of the chemical properties and its mechanism of stimulating the plants' natural defence, only a reduced set of data was required during evaluation of the active ingredient (Hauschild et al. 2008). Another plant extract, *Trigonella foenum graecum* L., marketed under the name of Stifénia®, was recently approved in France against the vine oïdium (Pajot and Regnault-Roger 2008). Plant inducers act on a very broad spectrum of plant species and fungal and viral pathogens as well, whilst the expression of their efficacy can be cultivar dependent. In the same context, studies highlight that the physiological stage of the treated plants plays a significant role in the expression of the stimulation of plant defence; for example Stifénia® whose use is recommended before flowering. Elicitors to be efficient must be used at a receptive physiological stage of the plant.

The limit of this technology is probably the incomplete control of disease (20–85%) or non significant results under field conditions because the expression of induced resistance is influenced by environmental conditions, genotype and crop nutrition. An important challenge would be to convince farmers and growers that stimulation of natural plant defence will provide a useful and practical approach to be used in association with fungicides, by decreasing the frequency and amount of chemical treatment to enhance sustainable development (Walters et al. 2005). Another point is the cost of induced resistance. Should this technology limit the allocation costs for growth and metabolism? According to studies on the stimulation of plant defence by the chemical inducer ASM (S-methylbenzo[1,2,3]thiadiazole-7-carbothiate/acibenzolar-S-methyl), the response differed between two experiments.

The first experiment showed that treated beans and wheat gave a reduced biomass and reduced ears and grains (Heil et al. 2000), although in the second experiment an insignificant reduction in bean seed yield was noticed (Iriti and Faoro 2003). By contrast, some cases of yield increase were associated with induced resistance to powdery mildew infection in barley when compared with plants receiving no inducing treatment (Walters and Boyle 2005). At present, data on fitness costs are contradictory, depending on plant, elicitor, micrometeorological and environmental conditions, plant nutritional and growth status (Iriti and Faoro 2006). Further work on the influence of climate and soil, as well as agronomic factors will provide a better understanding on induced resistance applications as BCAs.

## 10.6 Conclusion

All IPM approaches for plant protection are needed and are essential for reducing the amounts of chemical pesticides in agriculture and for using them in appropriate situations when the pest pressure is so great that no alternative is possible. These IPM strategies include the renewal of agronomical and prophylactic practices, physical control methods, genetic engineering and BCAs in which semiochemicals and botanicals are involved (Regnault-Roger 2005a).

During the 20th century, semiochemicals and botanicals clearly developed and their successes in diversifying approaches for plant protection were noticeable. However, they faced two main factors that hampered progress: (i) more stringent standards for risk assessment according to the claims that plant protection products should not pose unreasonable risks to people or the environment; (ii) an economic challenge because alternative approaches do not control pests as perfectly as chemical pesticides and are also more expensive.

The technologies involving (i) pheromones for detection and monitoring insects and (ii) allelochemicals to enhance plant resistance to bio-aggressors by selection of improved varieties and elicitation are probably the most promising, because of the lack of associated significant risks for human health and environment. The use of pheromones in mating disruption and mass trapping are niche markets with undoubted constraints but they cannot be ignored when chemical pesticide control fails. The use of botanicals in insecticide formulations is more questionable. Their lack of persistence because of their biodegradability linked to more specific modes of action of some of them that improve selectivity to target species (e.g. insect growth regulators) is probably the key for their future in IPM. Because they diversify the active ingredients for plant protection, they decrease the risk of insect resistance as far as they are used in rational agricultural practices. The renewal of agriculture for the sake of sustainable development requires that pollutant practices are reduced. Among alternative approaches for plant protection, semiochemicals and botanicals certainly have a role to play.

## References

Agelopoulos N, Birkett MA, Hick AJ, Hooper AM, Pickett JA, Pow EM, Smart LE, Smiley DWM, Wadhams LJ, Woodcock CM (1999) Exploiting semiochemicals in insect control. Pesticide Sci 55:225–235

Ali ML, Taylor JH, Jie L, Sun G, William M, Kasha KJ, Reid LM, Pauls KP (2005) Molecular mapping of QTLs for resistance to Gibberella ear rot, in corn, caused by *Fusarium graminearum*. Genome 48:521–533

Arakaki N, Wakamura S, Yasuda T, Yamagishi K (1997) Two regional strains of a phoretic egg parasitoid, *Telenomus euproctidis* (Hymenoptera: Scelionidae), that use different sex pheromones of two allopatric tussock moth species as kairomones. J Chem Ecol 23:153–161

Balachowsky AS (1951) La lutte contre les insectes. Payot, Paris, 380 pages

Benhamou N (1996) Elicitor-induced plant defence pathways. Trends Plant Sci 1:233–240

Bergvinson DJ, Arnason JT, Hamilton RI, Tachibana S, Towers GHN (1994) Putative role of photodimerized phenolic acids in maize resistance to *Ostrinia nubilalis* (Lepidoptera: Pyralidae). Environ Entomol 23:1516–1523

Betarbet R, Sherer TB, MacKenzie G, Garcia-Osuna M, Panov AV, Greenamyre JT (2000) Chronic systematic pesticide exposure reproduces features of Parkinson's disease. Nat Neurosci 3:1301–1306

Bily A, Reid LM, Taylor JH, Johnston D, Malouin C, Burt AJ, Bakan B, Regnault-Roger C, Pauls KP, Arnason JT, Philogène BJR (2003) Dehydrodimers of ferulic acid in maize grain pericarp: resistance factors to *Fusarium graminearum*. Phytopathology 93:712–719

Blum MS (1977) Behavioural responses of Hymenoptera to pheromones, allomones and kairomones. In: Soney HH, Mc Kelvey TJ (eds) Chemical control of insect behaviour, theory and applications. Wiley, New York, NY, pp 149–168

Boeke SJ, Boersma MG, Alink GM, van Loon JJ, van Huis A, Dicke M, Rietjens IM (2004) Safety evaluation of neem (*Azadirachta indica*) derived pesticides. J Ethnopharmacol 94:25–41

Brahmachari G (2004) Neem–an omnipotent plant: a retrospection. ChemBioChem 5: 408–421

Cardé RT, Minks AK (1997) Insect pheromone research: new directions. Chapman & Hall, New York, NY, 681 pages

Chapman RF, Bernays EA, Simpson SJ (1981) Attraction and repulsion of the aphid *Cavariella aegopodii* by plant odors. J Chem Ecol 7:881

Copping LG (2004) The manual of biocontrol agents. BCPC, Alton, 528 pages

Dajoz R (1969) Les insecticides. Presses universitaires de France, Paris, 128 pages

Delisle J (1992) Monitoring the seasonal male flight activity of *Choristoneura rosaceana* (Lepidoptera: Tortricidae) in eastern Canada using virgin females and several pheromone blends. Environ Entomol 21:1007–1012

Delisle J, Royer L (2005) Y a-t-il un avenir pour la lutte par confusion sexuelle contre les ravageurs forestiers?. In: Regnault-Roger C (ed) Enjeux phytosanitaires pour l'agriculture et l'environnement. Lavoisier, Paris, pp 679–696

Department of Justice Canada (2002) Pest control Products Act 2002 c 28 (last updated: 2008-08-11) http://laws.justice.gc.ca/en/ShowFullDoc/cs/P-9.01///en. Accessed 11 Aug 2008

Descoins C, Francois P, Rousel G, Trouvé C, Joubert JM, Mercier T (2003) Les produits naturels : une chance à saisir. Phytoma 564:9–12

Devillers J, Farret R, Girardin P, Rivière JL, Soulas G (2005) Indicateurs pour évaluer les risques liés à l'utilisation des pesticides. Lavoisier, Paris, 288 pages

El Modafar C, El Boustani ES, El Aabidine AZ (2008) Rôle des polyphénols dans la défense naturelle des plantes contre les phytopathogènes. Résistance du palmier dattier à la fusariose. In: Regnault-Roger C, Philogène BJR, Vincent C (eds) Biopesticides d'origine végétale, 2nd edn. Lavoisier, Paris, pp 259–276

El-Sayed AM, Delisle J, De Lury N, Gut LJ, Judd GJR, Legrand S, Reissig WH, Roelofs WL, Unelus CR, Trimble RM (2003) Geographic variation in pheromone chemistry, antennal electrophysiology, and pheromone-mediated trap catch of North American populations of the oblique-banded leaf-roller. Environ Entomol 32:470–476

EPA (1994) Federal Register 59 FR 3681; January 26, 1994

EPA (1995) Neem oil, tolerance exemptions. FR Doc 95-29991. http://www.epa.gov/EPA-PEST/1995/December/Day-13/pr-494.html. Accessed 1 Aug 2008

EPA (2001) Azadirachtin(121701) Clarified Hydrophobic Extract of Neem Oil (025007) Fact Sheet. http://www.epa.gov/pesticides/biopesticides/ingredients/factsheets/factsheet_025007.htm. Accessed 1 Aug 2008

EPA (2006) Reregistration Eligibility Decision for Pyrethrins RED List B Case No. 2580, EPA 738-R-06-004, June 2006

EPA (2007) Reregistration Eligibility Decision for Rotenone EPA 738-R-07-005, March 2007

EPA (2008a) California Red Scale Pheromone BRAD. www.epa.gov/opp00001/biopesticides/ingredients /tech_docs/brad_017703.pdf - 27/07/08. Accessed 6th Aug 2008

EPA (2008b) Pesticides: regulating pesticides ? What are biopesticides? http://www.epa.gov/opp00001/biopesticides/whatarebiopesticides.htm (Last updated July 21st, 2008). Accessed 11 Aug 2008

EPA (2008c) Reregistration Eligibility Decision for nicotine list B case no. 2460 OPP-2005-0231 March 2008

Falconer IR, Chapman HF, Moore MR, Ranmuthugala G (2006) Endocrine-disrupting compounds: a review of their challenge to sustainable and safe water supply and water reuse. Environ Toxicol 21:181–191

Ferron P (1989) Préface. In: Riba G, et Silvy C (eds) Combattre les ravageurs des cultures. INRA, Paris, PX111–XV

Flint HM, Doane DC (1996) Understanding semiochemicals with emphasis on insect sex pheromones in integrated pest management programs. In: Radcliffe EB, Hutchinson WD (eds) Radcliffe's IPM world textbook. http://ipmworld.umn.edu, University of Minnesota, St. Paul, MN. Assessed on 11th Aug 2008

Frérot B (2005) Les phéromones sexuelles des Lépidoptères : nouvelles approches et perspectives en protection intégrée. In: Regnault-Roger C (ed) Enjeux phytosanitaires pour l'agriculture et l'environnement. Lavoisier, Paris, pp 697–710

Goktepe I, Portier R, Ahmedna M (2004) Ecological risk assessment of neem-based pesticides. J Environ Sci Health B 39:311–320

Harborne JB (1972) Phytochemical ecology. Academic, London, 272 pages

Harborne JB (1989) Introduction to ecological biochemistry, 3rd edn. Academic, London, 356 pages

Hauschild R, Speiser B, Tamm L, Ehlers R-U (2008) Réglementation et homologation des produits phytopharmaceutiques à base d'extraits végétaux dans la législation de l'Union Européenne (UE): présent et perspectives d'évolution. In: Regnault-Roger C, Philogène BJR, Vincent C (eds) Biopesticides d'origine végétale, 2nd edn. Lavoisier, Paris, pp 477–496

Health Canada (2008) Registration Decision RD2008-02 German Cockroach Extract. Catalogue number: H113-25/2008-2E (H113-25/2008-2E-PDF). Minister of Health Canada, Ottawa

Heil M, Hilpert A, Kaiser W, Linsenmair E (2000) Reduced growth and seed set following chemical induction of pathogen defence: does systemic acquired resistance (SAR) incur allocation costs?. J Ecol 88:645–654

Howse PE (1998) Pheromones and behaviour. In: Howse PE, Stevens IDR, Jones OT (eds) Insect pheromones and their uses in pest management. Chapman & Hall, London, pp 1–37

Howse PE, Stevens IDR, Jones OT (1998) Insect pheromones and their use in pest management. Chapman & Hall, New York, NY, 369 pages

Huignard J, Lapied B, Dugravot S, Magin-Robert M, Koffivi Kétoh G (2008) Modes d'actions neurotoxiques des dérivés soufrés et de certaines huiles essentielles, et risques liés à leur utilisation.

In: Regnault-Roger C, Philogène BJR, Vincent C (eds) Biopesticides d'origine végétale, 2nd edn. Lavoisier, Paris, pp 187–206

Iriti M, Faoro F (2003) Does benzothiadiazole-induced resistance increase fitness cost in bean?. J Plant Pathol 85:265–270

Iriti M, Faoro F (2006) Fitness costs of chemically-induced resistance: double edged sword or (unstable) equilibrium? J Plant Pathol 88:5–6

Isman MB (1997) Neem insecticides. Pesticide Outlook 8:32–38

Isman MB (2005a) Problems and opportunities for the commercialization of botanical insecticides. In: Regnault-Roger C, Philogène BJR, Vincent C (eds) Biopesticides fo plant origin. Lavoisier Paris & Intercept, Andover (U.K.), pp 283–292

Isman MB (2005b) Botanical insecticides, deterrents and repellents in modern agriculture and an increasingly regulated world. Annu Rev Entomol 50:45–66

Isman MB, Wilson JA, Bradbury R (2008) Insecticidal activities of commercial rosmary oils, (*Rosmarinus officinalis*) against larvae of *Pseudaletia unipuncta* and *Trichoplusia ni* in relation to their commercial compositions. Pharm Biol 46(1–2):82–87

Jutsum AR, Gordon RFS (1989) Insect pheromones in plant protection. Wiley, New York, NY, 369 pages

Karlson P, Lüsher M (1959) Pheromone, ein Nomenklatur-vorschlag fur ein Wirkstoffklasse. Naturwiss 46:63–64

Kleeberg H (2006): Demands for plants protection products- Risk assessment botanicals and semiochemicals REBECA workshop, Brussels, 13–14 June 2006; http://www.rebeca.net.de/downloads/Risk%20assessment%20botanicals%20semiochemicals%20Minutes.pdf. Accessed 26th Jul 2007

Kleeberg H (2008) Le Neem, un agent de contrôle biologique à base d'extraits végétaux. In: Regnault-Roger C, Philogène BJR, Vincent C (eds) Biopesticides d'origine végétale, 2nd edn. Lavoisier, Paris, p 14

Kleeberg H, Ruch B (2006) Standardization of Neem-extracts. In: Proceedings of International Neem Conference, Kunming, China, 11–15 November 2006

Kleter GA, Poelman M, Groot MJ, Marvin HJP (2006) Inventory of possible emerging hazards to food safety and an analysis of critical factors. Report 2006.010 RIKILT. htpp://www.rikilt.wur.nl. Accessed 1st Aug 2008

Konstantinidou-Doltsinis S, Markellou E, Kasselaki AM, Fanouraki MN, Koumaki CM, Schmitt A, Liopa-Tsakalidis A, Malathrakis NE (2006) Efficacy of Milsana (R), a formulated plant extract from *Reynoutria sachalinensis*, against powdery mildew of tomato (*Leveillula taurica).* BioControl 51:375–392

Koul O, Dhaliwal GS (2001) Phytochemical biopesticides. Harwood Academic, Amsterdam, 223 pages

Kreutzweiser DP (1997) Nontarget effects of neem-based insecticides on aquatic invertebrates. Ecotox Environ Safe 36:109–117

Lepoivre P (2003) Phytopathologie. Editions De Boeck Université, Bruxelles, 427 pages

Matsumura F (1985) Toxicology of insecticides, 2nd edn. Plenum Press, New York, NY, 503 pages

McEwen FL, Stephenson GR (1979) The use and significance of pesticides in the environment. Wiley, New York, NY, 538 pages

Mendelsohn R, Balick M (1995) The value of undiscovered pharmaceuticals in tropical forest. Econom Bot 49:223–228

NRC (1992) National Research Council. Neem tree for solving global problems. National Academic Press, Washington, DC, 139 pages

Narusuye K, Kawai F, Matsuzaki K, Mivachi E (2005) Linalool suppresses voltage-gated currents in sensory neurons and cerebral purkinje cells. J Neural Transm 112: 193–203

OECD (2002a) Evaluating progress in pesticide risk reduction: summary report of the OECD project on pesticide aquatic risk indicators. http://www.oecd.org/dataoecd/6/3/2753049.pdf. Accessed 11 Aug 2008

OECD (2002b) OECD Series on pesticides Number 12 Guidance for Registration Requirements for Pheromones and Other Semiochemicals Used for Arthropod Pest Control JT00121481 ENV/JM/MONO(2001)12

OECD (2004) Summary report of the OCDE project on pesticide terrestrial risk indicators (TERI) http://www.oecd.org/env/pesticides. Accessed 11 Aug 2008

OECD (2007) OECD Series on pesticides Number 40 Draft report on the joint OECD/EC Seminar on Harmonised Environmental Indicators for pesticide risk (HAIR) ENV/JM/MONO (2007)27

OJEU (2008a) Official Journal of European Union. Commission Decision of 18 December 2008 amending Council Directive 91/414/EEC to include several active substances. L 344/89. 20.12.2008

OJEU (2008b) Official Journal of European Union. Commission Decision of 10 April 2008 concerning the non-inclusion of rotenone, extract from equisetum and chinin-hydrochlorid in Annex I to Council Directive 91/414/EEC and the withdrawal of authorisations for plant protection products containing these substances. L 108/30. 18.4.2008

OJEU (2008c) Official Journal of European Union. Commission Decision of 8 December 2008 concerning the non-inclusion of certain active substances in Annex I to Council Directive 91/414/EEC and the withdrawal of authorisations for plant protection products containing these substances. L 335/91.13.12.2008

OJEU (2009) Official Journal of European Union. Commission Decision of 8 December 2008 concerning the non-inclusion of nicotine in Annex I to Council Directive 91/414/EEC and the withdrawal of authorisations for plant protection products containing that substance. L5/7. 9.1.2009

Pajot E, Regnault-Roger C (2008) Stimulation des défenses naturelles des plantes et résistance induite : une nouvelle stratégie phytosanitaire ?. In: Regnault-Roger C, Philogène BJR, Vincent C (eds) Biopesticides d'origine végétale, 2nd edn. Lavoisier, Paris, pp 231–258

Pandey SK, Sankar D, Sharma S (2006) Quantitative Trait Loci (QTLs) for plant disease response. Proc Ind Natl Sci Acad 72:179–200

PAN (2010) Pesticide Action Network Pesticide Database Chemicals.Azadirachtin- Identification, toxicity, use, water pollution potential, ecological toxicity and regulatory information http://www.pesticideinfo.org/Detail_Chemical.jsp?Rec_Id=PC35467. Accessed 4 Apr 2010

Pascual-Villalobos MJ, Ballesta-Acosta MC (2003) Chemical variation in an *Ocimum basilicum* germplasm collection and activity of the essential oils on *Callosobruchus maculates*. Biochem Syst Ecol 31:673–679

Pesson P (1990) Protection des cultures de la Révolution à nos jours: de l'empirisme à l'organisation de la lutte intégrée. In: Académie de l'Agriculture (ed) Deux siècles de progrès pour l'agriculture et l'alimentation (1789–1989). Lavoisier, Paris, pp 169–190

Peterson RKD (2006) Comparing ecological risks of pesticides: the utility of a Risk Quotient ranking approach across refinements of exposure. Pest Manag Sci 62:46–56

Pfau W, Bieler C, Neurath G (2009) Safety of azadirachtin. Toxicol Let 189(Sup1):S214

Philogène BJR, Regnault-Roger C, Vincent C (2005) Botanicals : yesteday's and today's promises. In: Regnault-Roger C, Philogène BJR, Vincent C (eds) Biopesticides of plant origin. Intercept - Lavoisier, Andover and Paris, pp 1–15

Philogène BJR, Regnault-Roger C, Vincent C (2008) Biopesticides d'origine végétale :bilan et perspectives. In: Regnault-Roger C, Philogène BJR, Vincent C (eds) Biopesticides d'origine végétale, 2nd edn. Lavoisier, Paris, 546 pages

Picimbon JF (2005) Synthesis of odorant reception-suppressing agents, Odorants-Binding Proteins (OBPs) and Chemosensory Proteins (CSPs): molecular targets for pest management. In: Regnault-Roger C, Philogène BJR, Vincent C (eds) Biopesticides of plant origin. Intercept-Lavoisier, Andover and Paris, pp 245–266

Picimbon JF, Regnault-Roger C (2008) Composés sémiochimiques volatils, phytoprotection et olfaction : cibles moléculaires pour la lutte intégrée. In: Regnault-Roger C, Philogène BJR, Vincent C (eds) Biopesticides d'origine végétale, 2nd edn. Lavoisier, Paris, pp 383–416

PMRA-ARLA (2002) Updated Procedures for Joint Review of Microbials and Semiochemicals. Joint Review (NAFTA-JR Series). http://www.pmra-arla.gc.ca/english/pubs/jnt_rev-e.html. Accessed 11 Aug 2008

Prakash A, Rao J (1997) Botanical pesticides in agriculture. CRC Press, Boca Raton, FL, 461 pages

Redbond M (2003) Biocontrol 2003. Pesticide Outlook 4:168–170

Regnault-Roger C (2005a) Enjeux phytosanitaires pour l'agriculture et l'environnement. Lavoisier, Paris, 1013 pages

Regnault-Roger C (2005b) Molécules allélochimiques et extraits végétaux dans la protection des plantes: nature, rôle et bilan de leur utilisation au XX$^{e}$ siècle. In: Regnault-Roger C (ed) Enjeux phytosanitaires pour l'agriculture et l'environnement. Lavoisier, Paris, pp 625–650

Regnault-Roger C (2005c) New insecticides of plant origin for the third millennium?. In: Regnault-Roger C, Philogène BJR, Vincent C (eds) Biopesticides of plant origin. Intercept -Lavoisier, Andover and Paris, pp 17–36

Regnault-Roger C, Hamraoui A (1995) Fumigant toxic activity and reproductive inhibition induced by monoterpenes on *Acanthoscelides obtectus* (Say) (Coleoptera), a bruchid of kidney bean (*Phaseolus vulgaris* L.). J Stored Prod Res 31:291–299

Regnault-Roger C, Philogène BJR (2008) Past and current prospects for the use of botanicals and plant allelochemicals in integrated pest management. Pharm Biol 46:41–51

Regnault-Roger C, Philogène BJR, Vincent C (2005a) Biopesticides of plant origin. Intercept-Lavoisier, Andover and Paris, 313 pages

Regnault-Roger C, Philogène BJR, Vincent C (2008) Biopesticides d'origine végétale, 2nd edn. Lavoisier, Paris, 546 pages

Regnault-Roger C, Silvy C, Alabouvette C (2005b) Biopesticides: réalités et perspectives commerciales. In: Regnault-Roger C (ed) Enjeux phytosanitaires pour l'agriculture et l'environnement. Lavoisier, Paris, pp 849–880

Renou M, Guerrero A (2000) Insect parapheromones in olfaction research and semiochemical-based pest control strategies. Annu Rev Entomol 45:605–630

Royer L, Delisle J (2005) Utilisations des pièges à phéromones sexuelle dans la gestion des ravageurs. In: Regnault-Roger C (ed) Enjeux phytosanitaires pour l'agriculture et l'environnement. Lavoisier, Paris, pp 663–678

Samuel O, Dion S, St-Laurent L, April MH (2007) Québec (Pesticide Risk Indicator – QPRI – Health and Environment (on line) Québec Ministère de l'Agriculture, des Pêcheries et de l'Alimentation/Ministère du Développement durable, de l'Environnement et des Parcs/Institut national de santé publique du Québec, Québec, 44pp. http://mapaq.qc.ca, http://www.mddep.gouv.qc.ca and http://www.inspq.qc.ca

Schmutterer H (1990) Properties and potential of natural pesticides from the neem tree. Ann Rev Entomol 35:271–297

Sondheimer E, Simeone JB (1970) Chemical ecology. Academic, New York, NY, 336 pages

Suckling DM, Karg G (1998) Pheromones and other semiochemicals. In: Rechcigl JE, Rechcigl NA (eds) Biological and biotechnological control of insect pests. Lewis, Boca Raton, FL, pp 63–100

Thacker JMR (2002) An introduction to arthropod pest control. Cambridge University Press, Cambridge, 343 pages

Walters DR, Boyle C (2005) Induced resistance and allocation costs: what is the impact of pathogen challenge?. Physiol Mol Plant Pathol 66:40–44

Walters D, Lyon GD, Newton AC (2007) Induced resistance for plant defence: a sustainable approach to crop protection. Blackwell Science, Oxford, 272 pages

Walters DR, Walsh D, Newton A, Lyon G (2005) Induced resistance for plant disease control: maximizing the efficacy of resistance elicitors. Phytopathology 95:1368–1373

Ware GW (2000) The pesticide book, 5th edn. Thomson, Fresno, CA, 418 pages

Weinzeirl R (1998) Botanicals insecticides, soaps and oils. In: Rechcigl JE, Rechcigl NA (eds) Biological, biotechnological control of insect pest. Lewis Publishers, Boca Raton, FL, pp 101–121

Whittaker RH (1970) The biochemical ecology of higher plants. In: Sondheimer E, Simeone JB (eds) Chemical ecology. Academic, New York, NY, pp 43–70

Whittaker RH, Feeny P (1971) Allelochemicals: chemicals interactions between species. Science 171:757–770

Wilson EO, Bossert WH (1963) Chemical communication among animals. Recent Prog Hormone Res 19:673–716

Wyatt TD (2003) Pheromones and animal behaviour. Cambridge University Press, Cambridge, 391 pages

Zeller W (2006) Status of biocontrol methods againt fire blight. Phytopathol Pol 39:71–78

Zhang GF, Meng XZ, Yan H, Sheng CF (2002) Chinese tortrix Cydia trasias (Lepidoptera: Olethreutidae): Suppression on street-planting trees by Mass trapping with sex pheromone traps. Environ entomol 31:602–607

Zubkoff PL (2008) L'homologation des biopesticides botaniques et microbiologiques aux Etats-Unis. In: Regnault-Roger C, Philogène BJR, Vincent C (eds) Biopesticides d'origine végétale, 2nd edn. Lavoisier, Paris, pp 497–514

# Chapter 11
# Risks of Invertebrate Biological Control Agents – *Harmonia axyridis* as a Case Study

**Patrick De Clercq and Jeffrey S. Bale**

**Abstract** Despite the well documented economic and ecological benefits of biological control as a pest management strategy, there are increasing concerns about the risks associated with the use of non-native natural enemies. These concerns have prompted several countries in Europe and elsewhere to set up a regulatory system for non-native invertebrate biological control agents (IBCAs). To date, however, there is no coordinated system of regulation for IBCAs in Europe. Potential risks can be considered in three categories: risks to human and animal health, plant health, and most importantly, the environment. In this chapter we use the invasive harlequin ladybird *Harmonia axyridis* as a model to illustrate the potential negative impacts of non-native IBCAs. Its history of use in biological control is reviewed and the potential and realized adverse impacts in its adventive range are assessed. The case of the harlequin ladybird shows that there is an urgent need for a harmonized regulation of IBCAs in Europe, which should be based on appropriate risk analysis procedures. The development of such procedures should be a joint effort of biocontrol practitioners, scientists from different backgrounds and regulators alike.

## Contents



P. De Clercq (✉)
Department of Crop Protection, Ghent University, 9000 Ghent, Belgium
e-mail: patrick.declercq@ugent.be

R.-U. Ehlers (ed.), *Regulation of Biological Control Agents*,
DOI 10.1007/978-90-481-3664-3_11, © Springer Science+Business Media B.V. 2011

## 11.1 Benefits and Risks of Biological Control

Biological control with invertebrate species (IBCAs) has been described as the most successful, cost effective and environmentally-friendly method of pest management (van Lenteren 2008). Over the past 100 years there have been many successes in biological control with remarkably few environmental problems. In comparison with chemical pesticides, IBCAs are much cheaper to develop and have much higher cost-benefit ratios (Bale et al. 2008). The vast majority of pests and potential pests are suppressed below economic levels by 'natural control' – the activities of naturally occurring predators and parasitoids without human intervention. To put this into a quantitative perspective, van Lenteren (2008) reports that natural control occurs over all the world's terrestrial ecosystems (land with vegetation) on 8.95 billion ha of which 4.44 billion ha is used for some form of agricultural activity (including forestry and grassland); also, 95% of all potential arthropod pests (around 100,000 in total) are controlled 'naturally'. All other control methods are directed against the other 5,000 pest species. Over the past 120 years of modern biological control there have been around 5,000 introductions of 2,000 non-native (exotic) agents into 196 countries or islands (van Lenteren et al. 2006). Classical biocontrol operates over 350 million ha (8% of land under cultivation) and augmentative biocontrol over 16 million ha, equivalent to 0.4% of cultivated land growing crops on which this type of control potentially could be used, (van Lenteren 2008). The 'ecosystem services' provided by natural control have an estimated value of at least 400 billion US$ per year (Costanzo et al. 1997) and augmentative biocontrol has an annual value of 280 million US$.

At the present time there around 20 countries world-wide with some form of regulation in place for the import and release of non-native species, with the likelihood of more countries adopting the same or similar regulatory controls in the future. In Europe there is no coordinated system of regulation for IBCAs: some countries have regulation, others none. In those countries with regulation, the 'legal' basis may be related to legislation on pesticides, plant health or the environment. This raises an interesting question: if biological control is safe with so few reported negative effects, why do some countries (US, Canada, Australia, New Zealand) have such stringent regulatory systems, and other countries likely to adopt the same procedures?

Whereas the risks of native natural enemies used in augmentative biological control are often considered less important, they should not be simply ignored. Inundative releases of native agents may lead to transient non-target effects, and mass reared native species may carry contaminants (Goettel and Inglis 2006) or may interbreed with wild individuals of the same or closely related species, affecting their overall fitness (Hopper et al. 2006). Concerns for unwanted effects are, however, usually centred around exotic (non-native) organisms used in classical or augmentative biological control. Although the majority of releases for classical or augmentative biological control have not resulted in unwanted side effects, there have been reports of serious impacts caused by exotic invertebrate natural enemies of insects and weeds (see Follett and Duan 2000; Wajnberg et al. 2000; van Lenteren et al. 2006).

The debate around the safety of non-native biocontrol agents is often expressed in terms of 'risks', and whether these risks are real or perceived. A risk can be defined as the 'probability of an adverse event and the magnitude of the consequences'. However, there must be a potential hazard for risk to exist, and each risk has a number of component parts: the probability that the hazardous event will occur, the consequences (impact) of the event, the level of uncertainty in the system, and the ability for the risk to be managed. There is increasing public awareness and concerns about the risks posed to native biodiversity by alien invasive species, which may have contributed to a 'precautionary principle' approach in the regulatory systems for non-native biocontrol agents. Although the potential for undesired side effects of biological control agents has long been recognised, discussion has increased through the 1990s and several workers have since reviewed the risks associated with their use (Howarth 1991, 2001; Simberloff and Stiling 1996; Lockwood et al. 2001; Lynch et al. 2001; van Lenteren 2001; De Clercq 2002).

Discussions at REBECA meetings on the risks of IBCAs highlighted the need to separate 'true risks' from 'perceived risks', and then identify those risks that needed to be 'regulated'. Potential risks can be considered in three categories: Human and animal health, Plant and crop damage, and the Environment.

Personnel involved in the production of IBCAs are most likely to be exposed and protection measures can be introduced to minimise such risks. However, there have been very few reports of allergies in personnel working in the biocontrol industry, and any species that caused problems would be rapidly withdrawn. The probability of risks to humans is therefore remote and limited to allergic reactions and bites and stings. Likewise, there are only a few reports of crop damage by zoophytophagous species (e.g. *Macrolophus caliginosus* on cherry tomato), or related problems such as the contamination of crop products (e.g. *Harmonia axyridis* in grapes).

It is widely recognised that the most important potential risk with IBCAs is to the environment. A range of possible undesirable or deleterious outcomes can be identified: establishment in a new country, parasitism or predation of non-target species, competition or displacement of native species, perturbation of ecosystem functions (e.g. pollination), introduction of contaminating agents (pathogens, hyperparasites), and interbreeding with native species. In some countries such as the UK it is illegal to introduce and release a non-native species, irrespective of whether the organism is likely to cause any environmental damage; introduction and release is allowed, but only after the granting of a licence, and it would have to be shown that the intended release would not lead to permanent establishment (except where this was the aim) or exert any adverse effects on native species. Whilst predation, parasitism or competitive displacement of a native species would be regarded as undesirable, especially by environmental protection agencies, in some respects the greater damage is to the reputation of biocontrol as a technology, not least because it lends support to the 'precautionary principle', and in political terms, makes it more difficult to develop a balanced regulatory system, which safeguards the environment without placing too high a regulatory burden on the biocontrol industry.

Over the last decade, one species more than any other, the ladybird *Harmonia axyridis*, has highlighted the need for some form of regulation, underpinned by evidence-based methods of environmental risk assessment.

## 11.2 *Harmonia axyridis* as a Model High Risk Biological Control Agent

### *11.2.1 A Ladybird in the Spotlights*

One of the most high profile mediagenic cases that has raised increasing concerns in recent years in different parts of the planet, is that of the harlequin ladybird or multicoloured Asian lady beetle, *Harmonia axyridis*. Not only has its potential adverse impact on biodiversity alarmed ecologists and environmentalists, but it also has raised the interest and concern of the wider public due to its conspicuous nature (i.e., being a large brightly coloured ladybird) and its role as a nuisance pest in homes. Its downfall from a once promising beneficial insect to an obnoxious invasive alien species has been extensively documented in Roy & Wajnberg (2008). Once considered a potential biological control agent of aphid and coccid pests in a wide range of agricultural habitats, the species is now regarded as 'a model to prevent, or mitigate against, releases of high risk organisms' (Roy and Wajnberg 2008, in their foreword). In this chapter we use the *H. axyridis* model to illustrate potential negative impacts of non-native IBCAs. We will review its history of use in biological control, assess the potential and realised adverse impacts in its adventive range and evaluate the risks and benefits of its introduction.

### *11.2.2 Intentional and Accidental Introductions*

The harlequin ladybird, *H. axyridis*, is a coccinellid predator native to temperate and subtropical parts of east and central Asia (Iablokoff-Khnzorian 1982). Since the beginning of the twentieth century, *H. axyridis* was introduced repeatedly for the biological control of aphid and coccid pests into different parts of the world. In addition, there have been several reports of interceptions of the beetle in the international trade of agricultural produce (Poutsma et al. 2008). As a matter of fact, considering the failure of several of the early introductions into North America for classical biological pest control and the pattern of the coccinellid's spread after the first signs of establishment in the late 1980s, Day et al. (1994) assumed that establishment in North America may have been caused by an accidental introduction at the seaport of New Orleans, Louisiana. The assumption that the North American populations come from a single source is backed up by morphological and molecular studies indicating a large similarity among these populations (Krafsur et al. 1997; Koch 2003). However, recent molecular evidence suggests that there may have been two independent introductions into North America from its native area (Lombaert et al. 2010). The same study indicates that the European population of *H. axyridis*, which established in the late 1990s to early 2000s (Brown et al. 2008), is genetically a mixture of individuals most likely originating from the eastern United States and individuals introduced in Europe for biological control. The American material may either have reached Europe via international trade or passenger traffic, or may have been intentionally introduced. Further, the insect was released in the Mendoza province of Argentina in 1986–1987 and later in 1999, which may have led to further spread into the country and other parts of the South American continent. Interestingly, however, three out of four locations in South America where *H. axyridis* was reportedly recovered by 2007 were coastal areas with a nearby seaport (Poutsma et al. 2008). Feral populations of the coccinellid have also been found in South Africa, where it was first collected in 2002. However, it remains obscure how the insect gained entry into the country (Stals and Prinsloo 2007).

In summary, it remains uncertain whether the invasive populations of *H. axyridis* in different parts of the world stem from biological control introductions or accidental introductions, or from a combination of both.

### *11.2.3 Use in Classical Biological Control*

Classical biological control involves the introduction of a non-native natural enemy to control an exotic or sometimes also a native pest. The objective is to achieve establishment of the natural enemy and long term control of the pest in its new range with little further assistance (Van Driesche and Bellows 1996; van Lenteren 2008).

In North America, *H. axyridis* has been released repeatedly for classical biological control, for the first time in 1916, and later on also in the 1960s, 1970s and 1980s, in crops as diverse as pecans and red pines (Koch and Galvan 2008). In Europe, the ladybird was first introduced in Georgia in 1927, and later in the Ukraine from 1964 till 1971, and in Kazakhstan and Belarus from 1968 on. However, none of these introductions was successful (Poutsma et al. 2008 and references therein). In 1982, the Institut National de la Recherche Agronomique (INRA) in France introduced material from an unknown location in China (most probably the north east) but the insect was reportedly kept in quarantine until its first (experimental) release in 1990 (Coutanceau 2006). This stock was the source population of later introductions in Portugal (Algarve, Azores) in 1984–1985, in Greece from 1994 on and also in Argentina in 1986–1987 and 1999 (Brown et al. 2008; Poutsma et al. 2008). Interestingly, *H. axyridis* featured on a list of 'successfully introduced classical biological control agents' (Annex II of Standard PM 6/3(2)) of the European and Mediterranean Plant Protection Organisation (EPPO 2002), but the alleged successful establishment in the Azores, which was the basis for its inclusion on that list, was likely erroneous (Soares et al. 2008). The target pests for these biological control introductions were usually aphids.

### *11.2.4 Use in Augmentative Biological Control*

Augmentative biological control comprises the release of mass produced natural enemies where these are absent or too scarce to provide control. Commercial biological control is usually based on augmentation. This may be done by inoculative or inundative releases. In (seasonal) inoculative releases, small numbers are introduced in the crop with the expectation that they will reproduce and their offspring will continue to provide control of the target pest for an extended period of time. In inundative releases, the crop is swamped with large numbers of a natural enemy and pest control will be achieved primarily by the released individuals themselves (Van Driesche and Bellows 1996; van Lenteren 2008).

The first commercial releases of *H. axyridis* (mainly for aphid control in open air) were done in France in 1995. From the mid 1990s, the beetle was commercialised by a number of biocontrol suppliers in Western Europe (and in North America) for aphid control in greenhouse crops and urban ecosystems (Coutanceau 2006; Poutsma et al. 2008). All commercial populations in Europe presumably originated from the INRA stock established with Chinese material in 1982. The ladybird was never sold officially, however, in the UK, Switzerland and Germany (where it was recorded for the first time in the wild in Northwest Europe in 1999, Brown et al. 2008). Commercialisation was stopped in the Netherlands and Belgium in late 2003 to mid 2004, with the first reports of nuisance problems with the ladybird in homes and increasing concerns about the environmental effects of its use and establishment in the low lands (e.g., Adriaens et al. 2003). In France, the original flying strain of *H. axyridis* used since 1995 for commercial biocontrol was replaced in 2000 with a flightless strain developed by INRA, based on its greater effectiveness (Tourniaire et al. 2000; Coutanceau 2006). In 2010, this flightless strain of *H. axyridis* was still commercially available.

### *11.2.5 Beneficial Traits*

Its high prey searching ability, great voracity, polyphagous feeding, climatic adaptability, relative ease of rearing and the positive public image of ladybirds in general, are all properties that make *H. axyridis* attractive as a biological control agent. Koch (2003), Majerus et al. (2006) and Pervez & Omkar (2006) have reviewed the beneficial impacts of the harlequin ladybird as a classical and augmentative biocontrol agent of aphid and coccid pests.

Both in its native and introduced range, *H. axyridis* has provided effective control of aphid pests in pecans, apples, citrus, hops, strawberries, roses and several vegetable crops (Koch 2003; Pervez and Omkar 2006). Less than two decades after its presumed arrival, the predator was demonstrated to be a key factor in the natural control of the soybean aphid, *Aphis glycines*, an invasive pest of soybean in North America (Fox et al., 2004; Mignault et al. 2006). Records show that augmentative releases of the ladybird also effectively controlled scale pests in pine forests and bamboo in Asia (Pervez and Omkar 2006).

As dispersal of adults from release sites was considered to impinge on the effectiveness of augmentative biological control with *H. axyridis*, a flightless strain was developed by INRA in France (Tourniaire et al. 2000) and was subsequently commercialised by a French company since 2000 (Coutanceau 2006). According to Weissenberger et al. (1999), this strain can be effectively used to control aphids in hops.

Despite favourable reports on its efficacy as an augmentative biological control agent, the species was never a major player on the biocontrol market. At the peak of its commercialisation in Europe, it took perhaps 5% of the market share of aphidophagous natural enemies (J. Klapwijk, Koppert BV, personal communication).

## 11.2.6 Adverse Impacts

Adverse impacts of *H. axyridis* have been documented to some extent in North America, where the species established around 1988 and spread at an estimated rate of 400 km per year (Koch 2003; Koch and Galvan 2008). In Europe, where the ladybird established and rapidly expanded its range since the late 1990s to early 2000s (Brown et al. 2008), negative impacts are only beginning to be identified. For detailed discussions on the (potential) consequences of the use and establishment of *H. axyridis*, we refer to Koch (2003), Majerus et al. (2006) and several papers in Roy & Wajnberg (2008). Here we only provide a brief overview of the potential and realised adverse impacts of the harlequin ladybird in its adventive range.

### 11.2.6.1 Environmental Impact

A detailed environmental risk assessment was presented by van Lenteren et al. (2008). In that study, a stepwise procedure was used to assess the environmental risks of *H. axyridis* (see also Chapter 16). An environmental risk index was calculated, combining the likelihood and magnitude of the different risk components: establishment, dispersal, host range, and direct and indirect effects. The high risk index value attained indicated that *H. axyridis* is potentially risky for Northwest Europe. In short, the ladybird's eurytopic nature, polyphagous feeding habits (including the potential to use plant foods, see below), climatic adaptability, high degree of phenotypic plasticity, effective chemical and physical defence strategies and good dispersal abilities contribute to its high establishment potential (Majerus et al. 2006; van Lenteren et al. 2008; Berkvens et al. 2009). Due to its aggressive nature, great voracity, wide food range and high fecundity, *H. axyridis* may impact on other aphidophagous species by interspecific competition and intraguild predation. This may lead to declines in the diversity of the native aphidophagous guild (Pell et al. 2008). Furthermore, the ladybird may have adverse impacts on non-pest herbivores in its adventive range, including rare or valued species (Koch and Galvan 2008).

There is a great body of literature on the bio-ecology of *H. axyridis* and its potential interactions with other organisms, primarily based on laboratory experiments, but far fewer field studies have been done focusing on the realised direct and indirect environmental effects of *H. axyridis* in the areas it has invaded. Field surveys in both North America and Europe have shown that *H. axyridis* has become a prominent or even dominant member of the coccinellid community in many of the agricultural and (semi-) natural habitats it has invaded (Adriaens et al. 2008; Koch and Galvan 2008). The establishment of *H. axyridis* alone or in combination with other exotic coccinellids (like the seven-spot *Coccinella septempuncata* in North America), has been associated with a numerical and/or proportional decline of certain native coccinellids. One of the main species of concern is the two-spot ladybird *Adalia bipunctata,* the niche of which strongly overlaps with that of *H. axyridis* (Harmon et al. 2007; Adriaens et al. 2008). In the United States and Canada, analyses of long-term data were, however, not able to show a significant overall adverse effect of exotic coccinellids, including *H. axyridis*, on the populations of native coccinellids (Harmon et al. 2007; Koch and Galvan 2008). Also in Europe, there is currently little published information on the realised adverse impacts of *H. axyridis* on the native coccinellid fauna, but sampling campaigns have shown that the invasion by *H. axyridis* has been accompanied with declines in *A. bipunctata* numbers in the UK (Brown 2010) and Belgium (T. Adriaens, pers. comm.). Harmon et al. (2007), however, pointed out that sampling studies alone are not sufficient to demonstrate a causal relationship between the establishment of adventive species and the decline of native species.

Several laboratory studies have focused on intraguild predation involving *H. axyridis* and other aphidophagous organisms (for a review see Pell et al. 2008). In most of the interactions studied, *H. axyridis* generally dominated. Although there have been frequent records of intraguild predation in the field, the actual impact of this phenomenon on populations of guild members in managed and natural ecosystems remains poorly understood. The development of molecular tools (e.g., PCR-based gut analysis) may assist in understanding the ecological relevance of intraguild predation.

Further, there have been field reports of predation by the ladybird on non-pest insect prey, including aphids and a chrysomelid weed biocontrol agent (Koch and Galvan 2008). In North America, concerns were raised about the predation on eggs and caterpillars of the monarch butterfly *Danaus plexippus* based on observations in laboratory and field cage studies. A quantitative risk assessment indicated the potential for *H. axyridis* to have an impact on populations of the monarch in agricultural ecosystems (Koch et al. 2006).

#### 11.2.6.2 Impact on Plant Health

*Harmonia axyridis* may affect plant health both directly and indirectly. Via competition and intraguild predation the predator may in theory interfere with other biocontrol programmes which may result in reduced pest suppression (Pell et al. 2008), although there are as yet few indications of this in agricultural practice. Due to its facultatively phytophagous feeding habit, however, *H. axyridis* may also cause direct damage to plants. The ladybird not only has the ability to use pollen (Berkvens et al. 2008) and nectar (Spellman et al. 2006) as alternative foods, but also has been reported feeding on fruits such as grapes, apples, peaches, plums, pears, pumpkins and raspberries (Koch and Galvan 2008). Feeding on fruits by *H. axyridis* has been hypothesised to build up reserves in autumn for overwintering. It is worth noting here that other predaceous coccinellids, including *A. bipunctata* and *C. septempunctata*, do on occasion also feed on fruits (Hodek and Honěk 1996). In many cases, frugivory by *H. axyridis* does not appear to be primary damage, as the ladybird seems to have a preference for previously damaged fruits. The main plant health problem with *H. axyridis*, however, is its role as a contaminant in wine grapes. In particular the alkylmethoxypyrazines released by the insect during harvesting and processing of the grapes have been noted to taint wine in the eastern United States and the Great Lakes region (Galvan et al. 2007; Koch and Galvan 2008). As the taint cannot be completely removed from the wine, control measures against *H. axyridis* have been proposed for reducing its economic impact on the North American wine industry. Up to now, no problems have been reported with *H. axyridis* in wine producing areas in other parts of the world, including France where the ladybird is widely established (Coutanceau 2006).

#### 11.2.6.3 Impact on Human Health

In the insect's adventive and native range, there have been numerous reports of *H. axyridis* adults invading houses and other human made structures to overwinter, causing nuisance to their inhabitants. Aggregations at hibernation sites may consist of thousands of insects, but also much smaller aggregations of 10 insects or fewer may be found. There are fewer reports of very large aggregations (with over 1000 insects) in Northwest Europe than in Canada and the United States (e.g., Adriaens et al. 2008). Overall, nuisance problems with overwintering adults appear to be less serious in the more densely inhabited areas of Northwest Europe than in some rural areas in North America, which may be related to the availability of suitable overwintering sites.

Infestations inside homes cause problems when reflex bleeding adults stain furnishings and walls. Further, the insect can be a contaminant in the food industry and in health and research institutions (Koch and Galvan 2008, and references therein). Some cases of seasonal allergic reactions in humans to *H. axyridis* have been documented in the United States. Allergic reactions include rhinoconjunctivitis and less frequently asthma, urticaria and angiodema (Koch and Galvan 2008). In addition, *H. axyridis* adults have been reported in North America to bite humans when migrating to or aggregating in their overwintering sites. This was confirmed in a laboratory experiment by Kovach (2004).

Vacuuming appears to be the primary method of managing infestation problems inside homes (Huelsman and Kovach 2004), but the experimental use of repellents like N,N-diethyl-meta-toluamide (DEET), camphor and menthol has also yielded good results (Koch and Galvan 2008).

Also other coccinellids have been noted to cause nuisance to humans. For instance, *A. bipunctata* regularly enters houses to overwinter (Majerus and Kearns 1989) and has been observed to bite humans (Svihla 1952). However, at least in North America the magnitude of nuisance problems with the harlequin ladybird is greater than that reported for these other coccinellid species.

#### 11.2.6.4 Balancing Risks and Benefits

In an attempt to balance the 'good' and 'bad' functional traits of *H. axyridis,* Berkvens et al. (2009) concluded that the very same traits underlying the value of the species for pest suppression contribute to the risk of unwanted effects, including its undesired establishment in non-target habitats coupled with direct and indirect effects on non-target species and its status as an occasional plant pest.

There is ample evidence that *H. axyridis* can currently be categorised as an invasive species in North America and Northwest Europe and there are concerns that it will behave similarly in other parts of the world. The species spreads by human assistance both intentionally (through introductions for biocontrol purposes) and accidentally (via transport of agricultural produce and passenger traffic), but it is still not fully understood what the main pathway of introduction was for the invasive populations. Furthermore, the economic and ecological consequences of the invasion are as yet uncertain. Laboratory studies provide strong indications that this predatory coccinellid may negatively affect both guild members and non-pest herbivores, but such effects have only sparsely been demonstrated in the field. Whether the voracious nature of this predator and its interactions with the native predatory arthropod guild will weaken, or conversely strengthen, pest suppression in agricultural ecosystems is also highly uncertain, even in North America where the insect established some 20 years ago. Long-term field studies both in agricultural and natural habitats are essential to clarify the impact of this ladybird in its introduced range.

Risk analysis is the appropriate tool to weigh beneficial versus adverse impacts of exotic biological control agents like *H. axyridis*. Risk assessment is a first step in this process, which can be followed by an analysis of management options – for an overview of current and potential management strategies against *H. axyridis* see Kenis et al. (2008). A risk assessment by van Lenteren et al. (2008) concluded that *H. axyridis* never should have been released as a biological control agent in Europe, given that evidence indicating the potential risk was available at the time of its commercialisation in 1995. Although some of the potential adverse effects may not have been realised in the field at present, few workers will challenge the conclusions of this risk assessment (Berkvens et al. 2009).

## 11.3 Concluding Remarks

Biological control is a primary component of many cost-effective and environmentally friendly integrated pest management schemes. Ideally, biological control agents (microbial and macrobial organisms alike) should be effective in suppressing the target pest and have no risk to humans, crops and the environment. It is clear, however, that no biological control agent possesses all ideal states of desirable attributes (Mason et al. 2009) and that zero risk is not achievable. Mason et al. (2009) also pointed out that when assessing the efficacy and risks of a biological control programme, not only the functional attributes of the agents themselves, but also the ecological context in which the agents are used should be considered.

The case of the harlequin ladybird shows that there is an urgent need for a harmonised regulation of invertebrate biological control agents in Europe, which should be based on appropriate risk analysis procedures. Fortunately, through the REBECA project and similar initiatives, we have come to realize that the development of such procedures should be a joint effort of biocontrol practitioners, scientists from different backgrounds and regulators alike.

## References

Adriaens T, Branquart E, Maes D (2003) The multicoloured Asian ladybird *Harmonia axyridis* Pallas (Coleoptera: Coccinellidae), a threat for native aphid predators in Belgium?. Belgian J Zool 133:195–196

Adriaens T, San Martin y, Gomez G, Maes D (2008) Invasion history, habitat preferences and phenology of the invasive ladybird *Harmonia axyridis* in Belgium. BioControl 53:69–88

Bale JS, van Lenteren JC, Bigler F (2008) Biological control. In 'Sustainable Agriculture', special issue of Philosophical Transactions of the Royal Society B. Biol Sci 363:761–776

Berkvens N, Baverstock J, De Clercq P, Majerus MEN, Pell JK, Roy HE, Tirry L, Wells PM (2009) Good and bad traits of *Harmonia axyridis* – from biological control to invasion. In: Mason PG, Gillespie DR, Vincent C (eds) Proceedings of the 3rd international symposium on biological control of arthropods, 8–13 February 2009. Christchurch, New Zealand, pp 394–402

Berkvens N, Bonte J, Berkvens D, Deforce K, Tirry L, De Clercq P (2008) Pollen as an alternative food for *Harmonia axyridis*. BioControl 53:201–210

Brown PMJ (2010) The spread of the harlequin ladybird *Harmonia axyridis* (Coleoptera: Coccinellidae) in Europe and its effects on native ladybirds. PhD thesis, Anglia Ruskin University, Cambridge, UK

Brown PMJ, Adriaens T, Bathon H, Cuppen J, Goldarazena A, Hägg T, Kenis M, Klausnitzer BEM, Kovar I, Loomans AJM, Majerus MEN, Nedved O, Pedersen J, Rabitsch W, Roy HE, Ternois V, Zakharov IA, Roy DB (2008) *Harmonia axyridis* in Europe: spread and distribution of a non-native coccinellid. BioControl 53:5–21

Costanza R, dArge R, deGroot R, Farber S, Grasso M, Hannon B, Limburg K, Naeem S, O'Neill RV, Paruelo J, Raskin RG, Sutton P, vandebBelt M (1997) The value of the world's ecosystem services and natural capital. Nature 387:253–260

Coutanceau JP (2006) *Harmonia axyridis* (Pallas, 1773): une Coccinelle asiatique introduite, acclimatéallas, et en extension en France. B Soc Entomol Fr 111:395–401

Day WH, Prokrym DR, Ellis DR, Chianese RJ (1994) The known distribution of the predator *Propylea quatuordecimpunctata* (Coleoptera: Coccinellidae) in the United States, and thoughts on the origin of this species and five other exotic lady beetles in eastern North America. Entomol News 105:224–256

De Clercq P (2002) Dark clouds and their silver linings: exotic generalist predators in augmentative biological control. Neotrop Entomol 31:169–176

EPPO (2002) PM6/3(2) List of biological control agents widely used in the EPPO region. EPPO Bull 32:447–461

Follett PA, Duan JJ (2000) Nontarget effects of biological control. Kluwer, Dordrecht

Fox TB, Landis DA, Cardoso FF, Difonzo CD (2004) Predators suppress *Aphis glycines* Matsumura population growth in soybean. Environ Entomol 33:608–618

Galvan TL, Burkness EC, Hutchison WD (2007) Enumerative and binomial sequential sampling plans for the multicolored Asian lady beetle (Coleoptera: Coccinellidae) in wine grapes. J Econ Entomol 100:1000–1010

Goettel MS, Inglis GD (2006) Methods for assessment of contaminants of invertebrate biological control agents and associated risks. In: Bigler F, Babendreier D, Kuhlmann U (eds) Environmental impact of invertebrates for biological control of arthropods. CABI Publishing. Wallingford, CT, pp 145–165

Harmon JP, Stephens E, Losey J (2007) The decline of native ladybirds (Coleoptera: Coccinellidae) in the United States and Canada. J Insect Conserv 11:85–94

Hodek I, Honěk A (1996) Ecology of Coccinellidae. Kluwer, Dordrecht

Hopper KR, Britch SC, Wajnberg E (2006) Risks of interbreeding between species used in biological control and native species, and methods for evaluating their occurrence and impact. In: Bigler F, Babendreier D, Kuhlmann U (eds) Environmental impact of invertebrates for biological control of arthropods. CABI Publishing, Wallingford, CT, pp 78–97

Howarth FG (1991) Environmental impacts of classical biological control. Annu Rev Entomol 36:485–509

Howarth FG (2001) Environmental issues concerning the importation of nonindigenous biological control agents. In: Lockwood JA, Howarth FG, Purcell MF (eds) Balancing nature: assessing the impact of importing non-native biological control agents (an international perspective). Entomol Soc Am Proc. Thomas Say Publ. Entomol., Lanham, MD, pp 70–99

Huelsman M, Kovach J (2004) Behavior and treatment of the multicolored Asian lady beetle (*Harmonia axyridis*) in the urban environment. Am Entomol 50:163–164

Iablokoff-Khnzorian SM (1982) Les coccinelles Coléoptères – Coccinellidae: Tribu Coccinellini des régions Paléarctique et Orientale. Société Nouvelle des Editions Boubée, Paris

Kenis M, Roy HE, Zindel R, Majerus MEN (2008) Current and potential management strategies against *Harmonia axyridis*. BioControl 53:235–252

Koch RL (2003) The multicolored Asian lady beetle, *Harmonia axyridis*: a review of its biology, uses in biological control, and non-target impacts. J Insect Sci 3:1–16

Koch RL, Galvan TL (2008) Bad side of a good beetle: the North American experience with *Harmonia axyridis*. BioControl 53:23–35

Koch RL, Venette RC, Hutchison WD (2006) Predicted impact of an invasive generalist predator on monarch butterfly (Lepidoptera: Nymphalidae) populations: quantitative risk assessment. Biol Invasions 8:1179–1193

Kovach J (2004) Impact of multicolored Asian lady beetles as a pest of fruit and people. Am Entomol 50:159–161

Krafsur ES, Kring TJ, Miller JC, Nariboli P, Obrycki JJ, Ruberson JR, Schaefer PW (1997) Gene flow in the exotic colonizing lady beetle *Harmonia axyridis* in North America. Biol Control 8:207–214

Lockwood JA, Howarth FG, Purcell MF (2001) Summary: common ground, great divides, and building bridges. In: Lockwood JA, Howarth FG, Purcell MF (eds) Balancing nature: assessing the impact of importing non-native biological control agents (an international perspective). Entomol Soc Am Proc Thomas Say Publ Entomol, Lanham, MD, pp 15–30

Lombaert E, Guillemaud T, Cornuet J-M, Malausa T, Facon B, Estoup A (2010) Bridgehead effect in the worldwide invasion of the biocontrol Harlequin Ladybird. PLoS One 5(3):e9743. doi:10.1371/journal.pone.0009743

Lynch LD, Hokkanen HMT, Babendreier D, Bigler F, Burgio G, Gao ZH, Kuske S, Loomans A, Menzler-Hokkanen I, Thomas MB, Tommasini G, Waage JK, van Lenteren JC, Zeng QQ (2001) Insect biological control and non-target effects: a European perspective. In: Wajnberg E, Scott JK, Quimby PC (eds) Evaluating indirect ecological effects of biological control. CABI Publishing, Wallingford, CT, pp 99–125

Majerus M, Kearns P (1989) Ladybirds. Richmond Publishing, Slough

Majerus M, Strawson V, Roy H (2006) The potential impacts of the arrival of the harlequin ladybird, *Harmonia axyridis* (Pallas) (Coleoptera: Coccinellidae), in Britain. Ecol Entomol 31:207–215

Mason PG, De Clercq P, Heimpel GE, Kenis M (2009) Attributes of biological control agents against arthropods: what are we looking for? In: Mason PG, Gillespie DR, Vincent C (eds) Proceedings of the 3rd international symposium on biological control of arthropods, 8–13 February 2009. Christchurch, New Zealand, pp 385–392

Mignault M-P, Roy M, Brodeur J (2006) Soybean aphid predators in Quebec and the suitability of *Aphis glycines* as prey for three Coccinellidae. BioControl 51:89–106

Pell JK, Baverstock J, Roy HE, Ware RL, Majerus MEN (2008) Intraguild predation involving *Harmonia axyridis*: a review of current knowledge and future perspectives. BioControl 53: 147–168

Pervez A, Omkar A (2006) Ecology and biological control application of multicoloured Asian ladybird, *Harmonia axyridis*: a review. Biocontrol Sci Technol 16:111–128

Poutsma J, Loomans AJM, Aukema B, Heijerman T (2008) Predicting the potential geographical distribution of the harlequin ladybird, *Harmonia axyridis*, using the CLIMEX model. BioControl 53:103–125

Roy HE, Wajnberg E (2008) From biological control to invasion: the ladybird *Harmonia axyridis* as a model species. Springer, Dordrecht

Simberloff D, Stiling P (1996) How risky is biological control?. Ecology 77:1965–1974

Soares AO, Borges I, Borges PAV, Labrie G, Lucas E (2008) What will stop the invader?. BioControl 53:127–145

Spellman B, Brown MW, Mathews CR (2006) Effect of floral and extrafloral resources on predation of *Aphis spiraecola* by *Harmonia axyridis* on apple. BioControl 51:715–724

Stals R, Prinsloo G (2007) Discovery of an alien invasive, predatory insect in South Africa: the multicoloured Asian ladybird beetle, *Harmonia axyridis* (Pallas) (Coleoptera: Coccinellidae). S Afr J Sci 103:123–126

Svihla A (1952) Two-spotted lady beetles biting man. J Econ Entomol 45:134

Tourniaire R, Ferran A, Giuge L, Piotte C, Gambier J (2000) A natural flightless mutation in the ladybird, *Harmonia axyridis*. Entomol Exp Appl 96:33–38

Van Driesche RG, Bellows TS Jr (1996) Biological control. Chapman & Hall, New York, NY

van Lenteren JC (2001) Harvesting safely from biodiversity: natural enemies as sustainable and environmentally friendly solutions for pest control. In: Lockwood JA, Howarth FG, Purcell MF (eds) Balancing nature: assessing the impact of importing non-native biological control agents (an international perspective). Entomol Soc Am Proc Thomas Say Publ Entomol, Lanham, MD, pp 15–30

van Lenteren JC (2008) IOBC Internet book of biological control. http://www.iobc-global.org/downlaod/IOBC%20InternetBookBiCoVersion5January2008.pdf

van Lenteren JC, Bale J, Bigler F, Hokkanen HMT, Loomans AJM (2006) Assessing risks of releasing exotic biological control agents of arthropod pests. Annu Rev Entomol 51:609–634

van Lenteren JC, Loomans AJM, Babendreier D, Bigler F (2008) *Harmonia axyridis*: an environmental risk assessment for Northwest Europe. BioControl 53:37–54

Wajnberg E, Scott JK, Quimby PC (2000) Evaluating indirect ecological effects of biological control. CABI Publishing, Wallingford, CT

Weissenberger A, Brun J, Piotte C, Ferran A (1999) Comparison between the wild type and flightless type of the coccinellid *Harmonia axyridis* (Pallas) in control of the damson hop aphid. Fifth International Conference on Pests in Agriculture. Association Nationale pour le Protection des Plantes (ANPP). Paris & Montpellier, pp 727–734

# Part III
# Proposals for Balanced Regulation Procedures

# Chapter 12
# Facilitations in the Regulation of Plant Protection Products Containing Baculoviruses

**Rüdiger Hauschild**

**Abstract** Baculoviruses represent a family of double stranded DNA viruses that exclusively infect Arthropoda used in plant protection for control of insect pests in agriculture, horticulture and forestry. In the EU baculoviruses are regulated as microorganisms with data requirements laid down in the EU legislation. The OECD "Consensus document on information used in the assessment of environmental applications involving Baculoviruses" (2002) revised all publicly available information relevant for safety assessments of baculoviruses and concludes "the use of baculoviruses is safe". Potential risks from baculovirus products are minimal and can occur only indirectly through product components other than the baculovirus itself. Based on the safety assessment of different baculoviruses, REBECA experts proposed a simplified procedure for the inclusion into Annex I (91/414/EEC). Baculoviruses shall be evaluated at species level and new isolates shall be included with a reduced data set that is presented. This proposal resulted in the "Guidance Document on the assessment of new isolates of baculovirus species already included in Annex I of Council Directive 91/414/EEC" (SANCO/0253/2008 rev. 2 from January 22, 2008).

## Contents



R. Hauschild (✉)
GAB Consulting GmbH, Hinter den Höfen 24, 21769 Lamstedt, Germany
e-mail: ruediger.hauschild@gab-consult.de

R.-U. Ehlers (ed.), *Regulation of Biological Control Agents*,
DOI 10.1007/978-90-481-3664-3_12, © Springer Science+Business Media B.V. 2011

## 12.1 Introduction

This document was initiated by a working group at the REBECA conference in Kiel/Salzau in September 2006. The working group consisted of representatives of regulatory authorities involved in the evaluation of baculoviruses, producers of plant protection products based on baculoviruses, and independent research institutions.

The document is intended as a proposal to the Commission and member states in order to facilitate the registration procedure for plant protection products containing baculoviruses as the active ingredient. It aims at to facilitate the procedure for Annex I inclusion and to facilitate the national registrations.

Baculoviruses represent a family of double stranded DNA viruses that exclusively infect Arthropoda. The vast majority of the known species are confined to insects, predominantly *Lepidoptera*, with fewer species in *Diptera* and *Hymenoptera*. Some baculoviruses are used in plant protection products for the biological control of insect pests in agriculture, horticulture and forestry. Baculoviruses used as active substances in plant protection products in the EU are regulated as microorganisms according to the EU Council Directive 91/414/EEC. Data requirements for the registration of baculoviruses as active substances and of products based on baculoviruses are laid down in the Council Directive 91/414/EEC, amended by the Commission Directive 2001/36/EC. The Uniform Principles for evaluation and authorisation of plant protection products containing microorganisms are laid down in the Council Directive 2005/25/EC.

## 12.2 The OECD Consensus Document

In 2002, the OECD released the "Consensus Document on information used in the assessment of environmental applications involving Baculoviruses". This document revised all publicly available information relevant for safety assessments of baculoviruses. This includes the biology of baculoviruses, infection mechanisms in the host, host range determination, methods for molecular characterisation of isolates, and the history of use in plant protection products. Extensive information was gathered on effects of baculoviruses on human health including infectivity, replication in vertebrate cells, genotoxicity and carcinogenicity. Ecological information summarised in the OECD consensus document includes persistence and dissemination in the environment, host specificity and effects on non-target organisms.

The following characteristics of baculoviruses were outlined:

- Baculovirus species are extremely host-specific, with their host range limited to one or a few species of the same genus. Larger host ranges covering different genera or even different families are rare (e.g. *Autographa californica* NPV). Baculoviruses probably represent the most specific pesticidal agents, of all biologicals and chemicals.
- Baculoviruses occur only in arthropods, predominantly in the insect orders Lepidoptera, Diptera, and Hymenoptera.
- Baculoviruses are not infective for mammals and replication does not occur in mammalian cells.
- No pathogenic, genotoxic, mutagenic, or carcinogenic effect of baculoviruses was ever observed in mammals.
- Baculoviruses do not produce metabolites.
- Effects on non-target species can be excluded, especially for vertebrates, microorganisms and plants.

It should be noted that the document was developed under the OECD Working Group on Harmonization of Regulatory Oversight in Biotechnology and not all countries may have involved specialists for risk assessment concerning plant protection products during the development of the document. Nevertheless, this document was reviewed by a very large number of OECD member states. Taken together, the OECD consensus document concludes, "the use of baculoviruses is safe". Even if the document does not specify what uses are considered safe, human safety is reasonably specified in the document (page 45): "safety tests of more than 51 entomopathogenic viruses including more than 30 baculoviruses resulted in a long and complete safety record (Ref....). No adverse effect on human health has been observed in any of these investigations indicating that the use of baculoviruses is safe and does not cause any health hazard." However, the OECD conclusion should first be confirmed through the ongoing evaluations of baculoviruses under 91/414/EEC.

## 12.3 Genetic Composition of Baculovirus Isolates

Microorganisms are generally registered at strain level. Bacterial and fungal strains used in plant protection products derive from single colonies or spores and are consequently genetically homogenous. Different bacterial and fungal strains from the same species may have significant differences in their biology, especially in the production of secondary metabolites. Concerning their genetics, baculoviruses represent a unique case among microorganisms used in plant protection products in that they consist of a mixture of different, often very similar genotypes. These variations may influence some biological properties, such as the virulence to their specific target host, but they do not have consequences on the safety towards non-target organisms or on the environment. The composition of this mixture depends among other factors on the genotype of the host used to multiply the baculovirus. Isolation of a single genotype is extremely difficult if not impossible and even not desirable, since genetic variation is needed to account for variation in the target organisms. Therefore, the demand to evaluate microorganisms at strain level is not applicable for baculoviruses.

## 12.4 Potential Risks from Plant Protection Products Containing Baculoviruses

Due to the recorded safety of baculoviruses, no risks from the baculovirus itself for man or the environment are expected from plant protection products containing baculoviruses. Potential risks from baculovirus products are minimal and can occur only indirectly through product components other than the baculovirus itself.

All baculoviruses have to be produced in vivo in order to be infective to larvae. Host insect or media components might be allergenic, as with any other biological molecule. Hairs from some lepidopteran larvae (caterpillars) are known for their irritating and sensitising potential. Sensitisation through baculovirus-containing products was tested and no effects were found for products containing CpGV (produced in *Cydia pomonella* larvae, non-hairy), SpliNPV (*Spodoptera littoralis*, non-hairy larvae), and LdMNPV (*Lymantria dispar*, hairy larvae). To date, all larvae used to produce baculoviruses for use in plant protection products in the EU are not hairy. Also, microbial contaminants cannot be excluded in the products, but have to be controlled. A detailed proposal on contamination thresholds in baculovirus products can be found attached to this document. Antibiotics potentially included in the media to suppress bacteria and fungi will only end up in very small proportions in the final product.

## 12.5 Current Regulatory Situation in the EU

Two baculoviruses species are included in Annex I of Council Directive 91/414 EEC: *Cydia pomonella* Granulovirus (CpGV) Mexican Isolate is the only one classified as an "existing substance". *Spodoptera exigua* Nucleopolyhedrovirus (SeNPV) strain F1 was included as new active substance. Three further baculovirus species (all represented by at least one isolate) are currently being evaluated by authorities of EU member states for the inclusion in Annex I of Council Directive 91/414 EEC. *Adoxophyes orana* Granulovirus (AoGV, Swiss isolate, BV-0001), *Helicoverpa armigera* Nucleopolyhedrovirus (HearNPV, isolate BV-0003), and *Spodoptera littoralis* Nucleopolyhedrovirus (SpliNPV) are treated as new active substances.

It is expected that after evaluation of the isolates of CpGV, SeNPV, AoGV, and HearNPV by the member states and EFSA, these baculovirus isolates can be included in Annex I of Council Directive 91/414 EEC. Likewise, it is expected that the corresponding products can be used safely in relation to good agricultural practice. As detailed above, baculoviruses represent a very homogenous group concerning their host specificity and effects on humans, non-target organisms and the environment, especially when compared with bacteria or fungi. Thus, all baculovirus species and all isolates within one species can be treated similarly if not equally in the assessment of risks for man or the environment. Regulation of further baculovirus species and isolates for the use in plant protection products can then be facilitated.

## 12.6 Proposal for Facilitated Regulation of Baculoviruses as Active Ingredients in Plant Protection Products

Based on the conclusions from the OECD consensus paper and on the expected results of the evaluation of dossiers submitted for the inclusion of isolates of CpGV, AoGV, and SeNPV, we propose that baculoviruses are not evaluated at strain level. The high similarity between baculoviruses justifies a general assessment at the level of the family Baculoviridae, considering species-specific information where necessary. Inclusion into Annex I shall be done at the level of individual species.

A facilitated procedure for the registration of new species or isolates could then perhaps be performed similarly to the procedure for "equivalence of technical material" as applied for chemically active substances for plant protection products. This would necessitate the submission of an application for national authorisation of a plant protection product containing the new species or the new isolate at member state level. After evaluation and approval of the application the member state then reports this to the Commission. Depending on the level of inclusion Annex I needs to be amended.

Formally, each data point for the active substance and the product has to be addressed. However, it is not necessary to submit isolate-specific information for many data points. Most of the data formally required are published and equal for all baculoviruses and already assessed by MS and EU authorities. Therefore, it is also possible to refer to already submitted own data or to relevant data already evaluated in other DARs. Species- or isolate-specific data have to be submitted for data points concerning the individual baculovirus species or isolate.

The following species/isolate-specific information – according to Annex II data requirements – has to be provided for the active substance:

- Origin of the isolate
- A molecular identification and characterisation, preferably by restriction length polymorphism (RFLP) analysis of DNA.
- Deposition of the new species/isolate in a recognized culture collection
- Biological properties, especially the host range

- The manufacturing process including threshold levels for contaminants
- Analytical methods for the detection of the new species/isolate as well as methods for the detection of microbial contaminants

Product-specific data, according to Annex III data requirements, have to be provided, including the production method (medium components, larvae hairy or not), information on the amount of non-pathogenic and pathogenic bacteria and fungi, and composition of the product. Changes when compared with methods already submitted for other products have to be declared. Data on toxicology and ecotoxicology should be based on the composition of the product. If the active substance is accepted to be safe without restrictions, risks can only result from other product components. The health and environmental hazards of a preparation should be assessed as described in article 6 and 7 of 99/45/EEC, hence by a conventional (calculation) method or by providing toxicological data on the preparation or its individual components. If the composition of the product is similar to an already evaluated product, information can be referred to this product (with appropriate justification and, if necessary, bridging studies). Efficacy data have to be submitted for a product containing a new species/isolate according to national regulations.

## 12.7 Data Protection

Data submitted for the inclusion of a baculovirus species in Annex I are protected. This means that all notifiers applying for national authorisation of a plant protection product containing an active substance, which was included in Annex I, must either prove access to all data that were necessary for the Annex I inclusion, or provide equivalent own data. This refers only to data still under data protection (i.e., literature that is not to be published). For submitted studies, for which the notifier claims data protection, the standard EU rules for data protection apply. Likewise, notifiers of products containing a new species have to provide own data or a letter of access to an already submitted dossier.

## 12.8 Remark on Genetically Modified Baculoviruses

This proposal explicitly does not include genetically modified baculoviruses.

## 12.9 Regulatory Situation

The proposal resulted in the "Guidance Document on the assessment of new isolates of baculovirus species already included in Annex I of Council Directive 91/414/EEC" (SANCO/0253/2008 rev. 2 from January 22, 2008). Until Summer 2010, three baculovirus isolates were included into Annex I using the procedure described in the SANCO document.

## 12.10 Proposal on Threshold Levels for Microbial Contaminations in Baculovirus Products

Baculoviruses for the use in plant protection products are multiplied in vivo using living host larvae. As these animals are not sterile, and separation of the virus from any contaminant is not feasible, microbial contaminations cannot be avoided and represent one risk associated with the use of products containing Baculoviruses. A draft OECD document prepared by Canada was discussed as the basis for threshold levels. The threshold levels listed in Table 12.1 were agreed between members of the working group and are proposed as general thresholds for microbial contaminants in plant protection end products containing baculoviruses.

*Bacillus cereus* represents a particular case for CpGV. *B. cereus* is a common spore forming, motile ubiquitous soil bacterium and an opportunistic human pathogen, causing diarrhoeal or emetic disease through the production of enterotoxins especially during inappropriate storage temperatures. *B. cereus* is frequently isolated as a contaminant of various foods. The consumption of foods that contain more than $10^5$ CFU *B. cereus* per gram may result in food poisoning. However, in some outbreaks, lower numbers in the food ($10^3 - 10^4$ CFU/g) were reported. As *B. cereus* is part of the intestinal flora of *Cydia pomonella* larvae, its presence in CpGV products cannot be avoided. CpGV products are highly diluted before application. As *B. cereus* is a soil bacterium, multiplication on fruit surfaces seems minimal due to lack of nutrients.

To estimate the populations of *B. cereus* on apples resulting from application of CpGV products, the following assumptions are made:

maximum accumulated application rate for CpGV products: 2.7 L/ha per season
maximum contamination *B. cereus*: $10^{10}$ CFU/L
apple yield: 28 t (average for Germany, in France 38–40 t)

$$2.7 \times 10^{10} \text{ CFU/28 t} = 1000 \text{ CFU/g or } 10^5 \text{ CFU/100 g.}$$

If a soil coverage of 60% is considered, maximum contamination levels are 600 CFU/g fruit or 60,000 CFU/100 g fruit. This calculation still does not take into account that the majority of *B. cereus* cells will end up on leaves and not on fruits, because fruit surface is still small when compared with leaves at the time of

**Table 12.1** Threshold levels for microbial contaminants (per gram or millilitre)

| Contaminant | Maximum content |
|---|---|
| Total mesophiles | $10^8$ CFU |
| *Bacillus cereus* | $10^7$ CFU |
| *Escherichia coli* | None in 1 g or mL |
| *Staphylococcus aureus* | None in 1 g or mL |
| *Salmonella spp.* | None in 25 g or mL |

Yeasts and moulds are visually checked during production

application of CpGV products. In addition, decrease of *B. cereus* between application of the product and harvest through UV radiation or washing of by rain is not considered. Further reduction of *B. cereus* on food can be achieved by washing or peeling.

The draft OECD document prepared by Canada on contaminants is currently out for comment by the REBECA participants. All participants agreed that animal testing is not required to guarantee absence of mammalian pathogens in Baculovirus products.

# Chapter 13
# Proposals for Bacterial and Fungal Biocontrol Agents

**Olaf Strauch, Hermann Strasser, Rüdiger Hauschild, and Ralf-Udo Ehlers**

**Abstract** Registration of biological control agents containing micro-organisms is a long-lasting and costly procedure and has discouraged companies investing in the development of microbial biological control products. Risk assessment and regulation of microbial biological plant protection products is reviewed to develop proposals for an improvement of the current system. Minimum data requirements to be presented during pre-submission meeting were defined in order to support decisions on data requirements for the dossier and possible waivers. Criteria for the possible inclusion of microbial biocontrol agents into the "low risk products" list and support for the discussion of a comparative risk analysis are discussed. Major problems are the lack of validated risk assessment methods for microbials, knowledge gaps on the natural distribution of the biocontrol micro-organisms and on natural exposure of humans and other non-target organisms, and missing definitions allowing the identification of low risk products. Potential alternative approaches for the assessment of infectivity, toxicity, identification of relevant metabolites and the risk assessment procedures for metabolites and sensitisation are discussed. Proposals for waivers for potentially obsolete data requirements related to infectivity, effect on soil biota, fate and behaviour in the environment and genetic stability are recommended.

## Contents



O. Strauch (✉)
Department for Biotechnology and Biological Control, Institute for Phytopathology, Christian-Albrechts-University Kiel, 24118 Kiel, Germany
e-mail: o.strauch@biotec.uni-kiel.de

R.-U. Ehlers (ed.), *Regulation of Biological Control Agents*,
DOI 10.1007/978-90-481-3664-3_13, 

## 13.1 Introduction

Microbial biological control agents (MBCAs) are living organisms, like bacteria and fungi, used to control insects, diseases or weeds. Insect viruses are included within this group. Proposals on how to improve registration of insect viruses are dealt with in Chapter 12.

MBCAs used in plant protection in the EU are regulated according to the EU Council Directive 91/414/EEC (EU 1991). This Directive was amended by the Commission Directive 2001/36/EC (EU 2001) regarding the data requirements for the Annex I inclusion of micro-organisms as active substances and national authorisation of products (Annex IIB and IIIB in the directive, respectively). The Uniform Principles for evaluation and authorisation of plant protection products containing micro-organisms are laid down in the Council Directive 2005/25/EC (EU 2005). Recently, the EU decided on the new Regulation (EC) 1107/2009 for placing plant protection products (PPPs) in the market (EU 2009).

In the past, the biocontrol industry continually complained about the registration system for MBCAs in the EU being too costly and time-consuming. Long registration periods of up to 10 years created a severe problem for the market access of their products (see also Chapter 2). In many cases, the related costs exceeded the expected profits. This situation clearly discouraged companies from investing in the development of new microbial products (see also Chapter 6).

The reasons for the long registration process for MBCAs are, among others, the regulation system in the EU now involves 27 Member States. Low level expertise in risk assessment of microbials among biocontrol industry and regulators and the lack of guideline contributed to the delay in authorisation. With the Regulation (EC) 1107/2009, time frames are defined for the registration process (EU 2009). If these time lines can be realized, an important hurdle in the registration of MBCAs will also be abolished. Additional proposals for improvements on administrative points are summarized in Chapter 17.

The data requirements and methodology for the risk assessment derived from the assessment system developed for synthetic chemical PPPs. Even though, the Commission has put effort into the development of better adapted data requirements for MBCAs (Directives 2001/36/EC and 2005/25/EC), the current procedure can still be judged as not appropriate. Some data requirements may be dispensable. The risk assessment methodology, based on the assessment of synthetic PPPs, is not properly adapted and has often not been validated for microbials.

Micro-organisms are known as pathogens and can produce toxins and antibiotics, which is why it is necessary to perform a risk assessment when these organisms are to be used in biocontrol. However, micro-organisms, in general, do not necessarily cause problems. Humans and animals are regularly exposed to micro-organisms. Most of them have beneficial functions in the environment and some might even function as symbionts. Those micro-organisms used as MBCAs have so far not been reported to cause any hazards. In contrast to chemicals, MBCAs have a history of safe use. A minority of micro-organisms are human, animal and plant pathogens. The society is spending considerable scientific resources to identify human and animal pathogens. Therefore, on the basis of a proper identification of the microbials, it should be possible to use this knowledge to a great extent to assess the risks related to the use of MBCAs. In consequence, a more balanced risk assessment for MBCAs is requested by the industry and supported by scientific experts in the field.

The European Union Policy Support Action REBECA organised several workshops, involving experts from science, regulatory authorities and industry, with the objective of first to identify possible risks posed by MBCAs and then to review data requirements and the risk assessment methodology and identify knowledge gaps. The potential risks of bacterial and fungal BCAs are discussed in Chapters 7 and 8 (bacteria) and 9 (fungi). The current data requirements for regulation of MBCAs are presented and compared with systems in non-EU countries in Chapter 2. Furthermore, rationales for the identification of low risk products have been delivered (see Chapter 17). Within the REBECA Action, proposals for a better adapted and more balanced risk assessment for MBCAs and the rationale for waivers on data requirements were also developed and discussed. These results are summarized in this chapter. Furthermore, research objectives are defined for projects to overcome major knowledge gaps hampering an adequate risk assessment of MBCAs.

## 13.2 Pre-submission Meeting

REBECA participants agreed that pre-submission meetings should be the rule in the registration process of BCAs in Europe (see also Chapter 17). Pre-submission meetings seem to be an indispensable tool to keep time-lines and to avoid unnecessary efforts and time for applicants and evaluators.

In several Member States of the EU informal meetings between regulation authorities and industry have been established as a useful tool for shortening the evaluation process and avoiding the preparation and submission of unnecessary data by the industry. Such pre-submission meetings between the applicant and the evaluating authorities are formally recommended and well established in the USA, Canada and Australia and have proved to be very effective. Applicants avoid producing unnecessary data, and regulators save time, because dossiers better address those points that the regulators consider important. Current practice in the EU demonstrated that these meetings were, in some cases, not offered by the authorities or not used by the applicants.

In order to simplify the registration procedure it is recommended to compile available data and to discuss relevant data requirements in a pre-submission meeting with the Rapporteur Member State (RMS) authorities prior to submission of the dossier. The decision on the relevant data to be provided can be based on the following information, which can originate from data produced by the applicant and/or from published scientific literature.

The correct identification is a prerequisite for the correct attribution of published scientific data to be presented during the pre-submission meeting. Information on the identity of the micro-organism is most important for a decision on data requirements for the subsequent risk assessment process and therefore indispensible for a pre-submission meeting. The identification of micro-organisms is not trivial, since the taxonomy is still in development and wrong synonymisations can easily occur (see also Chapter 8). Therefore, the identification methods used should always represent the state of the art and the identification should be approved by a specialist for the specific taxonomic group. A guidance document on the use of taxonomy in risk assessment of bacterial micro-organisms is available (OECD 2003).

In addition, the following data should be the minimum requirements to be available during a pre-submission meeting for a preliminary evaluation of possible hazards and their probability on the effects of non-targets:

- ☑ intended use of the product (target organisms)
- ☑ formulation of the product
- ☑ site and method of application
- ☑ mode of action
- ☑ listing in Dir. 2000/54/EC concerning worker's protection from micro-organisms or comparable MS documents
- ☐ host range
- ☐ relevant metabolites produced by the MBCA
- ☐ literature research on health and medical reports
- ☐ maximum growth temperature
- ☐ list of available effective antibiotics
- ☐ data on non-target toxicity and pathogenicity (virulence)
- ☐ natural distribution of the species in particular on food and feed and in agriculture environments; data on natural exposure of humans and animals

Dir. 2000/54 EC provides lists of micro-organisms that can be hazardous for workers when exposed during production or handling of these micro-organisms (EU 2000). Absence from this list, or comparable lists that exist in MS, would qualify a micro-organism as potentially non-pathogenic to humans. Further information is given at 13.5.1 in this chapter.

The ticked data points should always be available for a pre-submission meeting. The other points should be optional. However, a comprehensive literature research will simplify the definition of balanced data requirements. The data provided shall be the basis for a decision on the list of necessary additional data to be provided in the dossier, and this information can also be used to define waivers for data requirements.

A preliminary decision on data requirements during a pre-submission meeting will help to reduce efforts in investigations on the risks and will enable the applicants to review possible costs to take a decision whether or not a dossier for a particular MBCA should be submitted.

## 13.3 Identification of Low Risk Products

Among MBCAs many organisms pose remote risks to humans, non-targets and the environment. For that reason it would be possible, based on scientific knowledge, to award specific organisms the status of "low risk products". In the past, no definitions or criteria for low risk plant protection products or active substances existed in the legislative proposals from the European Commission, the US Environment Protection Agency (EPA) or the OECD BioPesticides Steering Group. The new regulation Regulation 1107/2009/EC contains paragraphs relating to "low risk substances", "basic substances", and "substances of concern". These categories, however, do not meet the standards for natural occurring substances and microorganisms. "Low risk" has been defined in the sense of natural origin, which is considered inherently unlikely to cause an adverse effect in humans, animals or the environment. This means that the products/substances must not be (i) carcinogenic, (ii) mutagenic, (iii) toxic to reproduction, (iv) very toxic or toxic, (v) sensitising or (vi) explosive (EU 2009). The criteria for low risk substances have been defined with chemically active substances in mind, but specific characters of MBCAs have not yet been taken into consideration.

During the last decades numerous attempts have been made to compare the risks associated with different pesticides, and to identify and replace hazardous pesticides. A list of low risk candidates of microbial plant protection products was presented by REBECA experts. The list contains MBCAs, which were classified as unlikely either to cause human disease or to pose even a remote risk to humans, animal health and the environment. Commercially used MBCAs with a long-term history of safe use are also on the list. This recommendation was based (i) on a case-by-case evaluation of MBCAs, (ii) the safety/data fact sheet published by the US EPA, and (iii) publications of the European Council reporting listing of MBCAs on Annex I. The following MBCAs were considered (product names in parenthesis): *Ampelomyces quisqualis* (AQ10), *Bacillus subtilis* (Serenade), *Bacillus thuringiensis kurstaki, tenebrionis and israelensis* (several products), *Beauveria bassiana* (BotaniGard), *Beauveria brongniartii* (Melocont), *Coniothyrium minitans* (Contans), *Gliocladium catenulatum* (Prestop), *Lecanicilium (Verticillium) lecanii* (Mycotal, Vertalec), *Metarhizium anisopliae* (GRANMET, BIO 1020), *Paecilomyces fumosoroseus* (Preferal), *Panthoea agglomerans* (BlightBan), *Pseudomonas chlororaphis* (Cedomon), *Serratia entomophila* (Invade), *Serratia plymuthica* (Rhizostar). The experts recommended to give these MBCAs a low risk status and suggested they might be exempt from further registration. As prior risk assessments were available for these MBCAs, it was possible to assign these to a low risk group. Such a decision is more difficult when only limited data are available.

Another approach to group MBCAs might be taken by a comparative analysis with products already on the market, including synthetic PPPs. Claims have often been made that MBCAs are much safer than conventional products. However, tools to back up such claims objectively were rare. During one REBECA workshop a risk index system was presented and evaluated for its possible suitability to define low risk products. *Bacillus thuringiensis*, *Beauveria* spp., *Metarhizium anisopliae*, *Coniothyrium minitans*, *Pantoea agglomerans*, *Pseudomonas fluorescens*, *Trichoderma harzianum*, Antrazine, Benomyl, Chlorpyrifos, DDT, Methyl bromide, Phorate, Pyrethrins and Streptomycin were compared. The calculation of the overall environmental risk score was based on criteria like (i) persistence of the substance, (ii) potential for dispersal, (iii) range of non-target organisms that are affected, and (iv) direct and indirect effects on the ecosystem. Each of the criteria values consisted of "likelihood" multiplied by a "magnitude" factor. Both values scored on a scale between 1 and 5. In addition, the direct effects were multiplied by a weighting factor if vertebrates or other groups of specific importance were affected. A minimum of 5 and maximum of 5,000 was possible.

A very low risk score of 24 was assessed for *Coniothyrium minitans*, which is applied to soil and a maximum risk score of 4,275 was assigned to the foliar application of DDT. The results of this evaluation were published by Laengle and Strasser (2010), who conclude that the score for low risk products should not exceed 100, whereas a threshold of 500 seems justified for the term "reduced risk". Cut-off criteria need further discussion on a broader basis. However, as the idea of a comparative risk assessment analysis finds it way into the discussion on risk assessment of PPPs, such comparative approaches might provide useful tools to define low risk categories and to decide on waivers for data requirements.

One problem with the definition of low risk products or the comparative analysis is that these decisions can only be made based on a broad basis of data on the safety of MBCAs or the products to be compared. These data will have to be provided by the applicants, can originate from existing Draft Assessment Reports (DARs) or might be part of the public domain as a result of scientific research supported with public funds. Although the definition of low risk groups and the comparative risk analysis will enable future applicants to refer to such data, it will have a limited potential to reduce data requirements for registration of MBCAs in the near future.

## 13.4 Risk Assessment Methodology

A major problem with evaluation of risks related to MBCAs is the lack of qualified guidelines to assess their risks. The more industry, science or regulators deal with the currently existing guidelines and directives the more it becomes obvious that most of the methods used in risk assessment for chemical PPPs cannot easily be conferred to MBCAs.

### *13.4.1 Infectivity*

Infectivity, the capability of entering, surviving and multiplying in a susceptible host, is a unique character of microbials. The related potential risks of microbial biocontrol products for humans and animals cannot be assessed by methods used for chemical products. Human pathogens are well known and for common micro-organisms the screening of the medical literature can be sufficient to assess the risks for infections (Möllby 1998). However, if the micro-organism is not well described and the natural exposure of humans and animals cannot be evaluated, the absence of clinical reports might preclude a presumption of safety. In such cases an assessment of the potential for infections is indispensable. Currently the infectivity is assessed by so called clearance investigations. In this investigation the clearance of the micro-organisms from the inner organs of rats or mice is assessed after an intratracheal instillation. This method has a number of disadvantages.

For several reasons, intratracheal instillation can cause high control mortalities as has been noted in the biopesticide registration action document for *Chondrostereum purpureum* strain PFC 2139 (EPA 2004) or the product monograph for *Pseudozyma flocculosa* strain PF-A22 UL (EPA 2002). The reason is high stress for the laboratory animals, and anesthetization and intoxication or irritation/sensitization of the lungs by the control substances. Control substances are the 'inert' additives, which are applied together with the micro-organisms in the treated group, or killed micro-organisms. Further complications can be caused by plugging of the trachea and bronchi by the test substances. This happens regularly if microbial products are applied with low cell densities. In order to instil sufficient amounts of micro-organisms these products needs to be applied in high volumes. Sometimes an intratracheal instillation is not feasible for that reason.

In several cases a slow clearance process from the lung and other organs could be observed, although no infections and no clinical manifestations were recorded. Any microscopic particles like micro-organisms can be transported from the application site to other organs by tissue fluids and circulate in the reticular connective tissue and the reticulo-endothelial system (Adlersberg et al. 1969). Therefore, relocation of micro-organisms from the injection site to other organs does not necessarily indicate an infection process. Regardless of a slow clearance process, MBCAs could be proven to be non-infective. This has, for example, been documented for *Bacillus sphaericus* strain 2362 (Health Canada 2006) or for *Trichoderma harzianum* Rifai strain T-39 (EPA 2000). Also for *B. thuringiensis* (Bt) it is known that clearance is not instantaneous (Siegel 2001). Bt persisted up to 49 days after intraperitoneal injection into mice (Siegel et al. 1987) and 21 days after intratracheal instillation into rats (Tsai et al. 1997) without evidence of infections. Therefore, if no short term clearance can be observed, clinical manifestations of the laboratory animals are more relevant for the risk assessment than the clearance process. Focussing from the beginning on clinical manifestations instead of clearance might be more reliable, while reducing costs for long term clearance investigations and the number of tested animals over time. The applicability of blood test systems might be reviewed for that purpose. A highly sensitive and cost efficient candidate test system is the quantification of C-reactive protein (CRP) (Pepys 1981; Volanakis 2001; Das et al. 2003). CRP is a member of the class of acute phase reactants as its levels rise dramatically during inflammatory processes occurring in the body. CRP binds to phosphorylcholine on microbes. Highly sensitive CRP test systems (e.g. ELISA) are standard methods in clinical diagnostics and can be used with rats and rabbits. Other available biomarkers for infections are, e.g., procalcitonin (Jones et al. 2007) and interleukin-6 (Heinrich et al. 2003), which would be also available for mice. In contrast to the clearance investigations, biomarker assessment enables the development of infections on single individuals over time to be monitored. However, all potential alternatives to clearance investigations need to be checked for the possibility of wrong negative and wrong positive results. Nevertheless, it should be taken into account that the current methodology also cannot avoid wrong results in regard to human risks.

Since the investigations based on whole animal testing are very expensive, the methodology can cause many complications, and in view of animal welfare, alternatives for the infectivity assessment methods should be investigated. Chicken embryo tissue assays are regularly used to compare the infectivity of micro-organisms. These assays are proved for several pathogens to be equivalent to whole animal testing (Ormsbee et al. 1978; Wooley et al. 2000; Olier et al. 2002; Gibb and Wooley 2003). Further candidates for vertebrate testing alternatives are tests developed for the investigation of human pathogens and opportunistic microbes. For such test systems the nematodes *Caenorhabditis elegans* or *Panagrellus redivivus* can be used (Kurz and Ewbank 2000; Cardona et al. 2005; Laws et al. 2005; Sifri et al. 2005, see also Chapter 8). Also, human cell lines (Caco-2 cells) were used for the evaluation of human pathogenicity of micro-organisms (Pine et al. 1991; Anderssona et al.1998). All these test systems are highly sensitive. However, the development of standard protocols and a critical validation for risk assessment purposes of these methods is still needed.

### *13.4.2 Toxicity/Toxigenicity*

Chemical pesticides are usually based on one active ingredient in a defined concentration. By contrast, microbials can produce a broad range of metabolites. Such metabolites can be the active ingredient, like the insecticidal crystalline protein of *Bacillus thuringiensis* products and be present in high concentrations. Others might be produced in micro-quantities, in situ, in direct contact with the target organism or at the site where the MBCAs has established after application. The metabolites can be an important factor involved in the mode of action and some might be toxic to non-target organisms. Furthermore, different metabolites and different amounts can be produced under different environmental conditions (Baker and Griffiths 1993; Kershaw et al. 1999; Amiri-Besheli et al. 2000; Quesada-Moraga and Vey 2003; Strasser et al. 2000; Vey et al. 2001; Wang et al. 2004; Dabrowski and Sikorski 2005). Therefore, beside the toxicity of the active ingredient in the product, the potential to produce toxins (toxigenicity) of the microorganism might be of interest in the risk assessment of microbials (see also Chapter 7).

In the Directive 91/414 toxicity assessments with purified relevant metabolites are required. However, it is not feasible to identify and quantify all metabolites produced or potentially produced by a micro-organism. Methods are needed to detect or exclude the occurrence of relevant metabolites in relevant amounts in a first instance. Most of these metabolites are produced in very small amounts. A purification of sufficient quantities allowing whole-animal testing can methodologically be impossible or extremely expensive, a situation that is discouraging biocontrol companies from developing microbial BCAs (Blum et al. 2003; Zimmermann 2004). Clear, sensible, simple, better adapted and cost-effective strategies are needed for the risk assessment of metabolites from microbial BCAs.

A way out from that dilemma can be the assessment of extracts from microbial cell cultures, produced under different conditions. REBECA proposed a tiered scheme for bacterial and fungal metabolites (Fig. 13.1), based on the assessment of supernatants and crude extracts from cultures of the micro-organisms in question (for fungal metabolites see also Chapter 11). Microbial metabolites may have additive or synergistic toxic effects. It is conceivable that the toxicological risk associated with a particular MBCA would be better foreseen by assaying mixtures of metabolites, like those in crude culture extracts, instead of assessing the toxicity of single metabolites. The use of crude extracts, however, might also have pitfalls. The crude extracts represent the "worst case" scenario as levels and spectrum of metabolites being assessed are far higher than occurring in nature. Crude extracts are hardly expected to show zero toxicity; therefore, it will be necessary to establish tolerance levels of toxicity in biological assays.

In Fig. 13.1, steps (1) and (2) indicates the situation when available (or public) information demonstrates the absence of relevant (toxic) metabolites. They are either not produced in relevant amounts or exposure to relevant metabolites will not occur. In this case, no data on metabolites should be required. Natural background levels and natural exposure to the micro-organism should always be taken into account. If no hazards are known from a regular exposure of humans and other non-targets to the micro-organism, no risks can be expected from the application of the same organism as a plant protection product in amounts comparable to background levels.

Step (3): If questions (1) and (2) cannot be answered with 'no', in the first instance it should be investigated whether the product contains relevant amounts of toxins and the toxigenicity of the micro-organism should be evaluated. At this stage, the toxicity assessment of defined toxins should be carried out using maximum doses (at least 10-fold higher than the maximum expected environmental exposure). The toxigenicity under different conditions should be evaluated by using culture supernatants and crude extracts, which is a mixture of all possible metabolites produced under different growth conditions. The growth conditions (temperature, substrate) after application of the MBCA should be taken into consideration. A prerequisite for these investigations will be the development and validation of sensitive, high-throughput and cost effective standard bioassays for the cyto- and genotoxicity

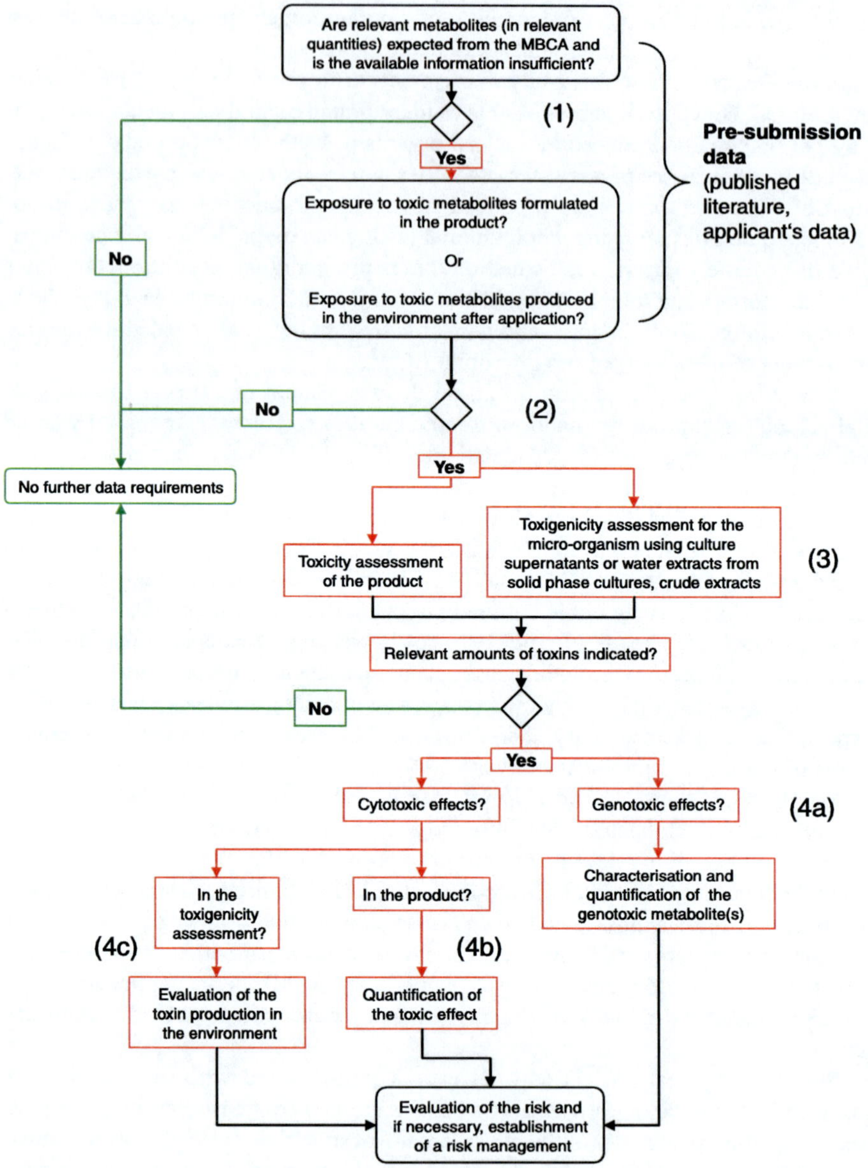


**Fig. 13.1** Scheme for assessment of potential relevant metabolites of bacterial and fungal biological control agents (MBCAs)

assessment. Whole animal (vertebrate) testing with single purified metabolites would not be feasible and not appropriate. Examples for bioassays, which potentially can be used, are given in the following publications: Walker et al. (1991), Guilhermino et al. (2000), Lagarto Parra et al. (2001), Sifri et al. (2005), Favilla et al. (2006), Skrobek et al. (2006). McLaughlin et al. (1993), Solis et al. (1993) and Logrieco et al. (1996) proposed human cell line tests. If no relevant amounts of toxins can be detected in the product or in the toxigenicity assessment, no further data requirements on metabolites would be necessary. For the identification of relevant amounts the application rate, growth conditions at the application site, natural background levels and the fact that metabolites are biodegradable should be taken into consideration.

Step (4a): If genotoxic effects were detected under (3), indicating relevant amounts of toxins in the product or potential production of such metabolites in the environment, hidden risks due to the application of the MBCA might exist. Consequently, the genotoxic metabolites need to be identified and characterised. In order to build up a risk management the amounts of these metabolites need to be quantified in the product and the production in the environment needs to be evaluated.

Step (4b): If cytotoxic effects were detected under (3), the toxic effect should be quantified using validated bioassays. Until now only a few relevant toxins have been purified and assessed (Boss et al. 2007; Favilla et al. 2006; Seger et al. 2005a; Skrobek and Butt 2005; Skrobek et al. 2006). The toxin contents were quantified in several batches in order to identify maximal exposure rates. However, it is conceivable that the toxicological risk associated to a particular MBCA would be better foreseen by assaying the mixtures of all metabolites/compounds in the product instead of assessing the toxicity of single metabolites (Favilla et al. 2006). Using validated sensible and low cost bioassay systems, a quantification of the toxicity of many product samples would be possible, less expensive and more reliable than the current practice.

Step (4c): If it was indicated under (3) that the micro-organism might produce cytotoxic compounds in relevant amounts in the environment, related exposure rates need to be evaluated. For this evaluation, data should be submitted allowing a worst-case estimation. These data are the application rate, persistence and growth rate and growth place of the micro-organism in the environment and the toxin production under different relevant environmental conditions (temperature, substrate). It should be determined if additional data to that collected under (3) and to public data are necessary for this assessment.

As a replacement of whole-animal (vertebrate) testing, it is proposed to use assays with cell lines, protozoa, arthropods or nematodes. Progress has been achieved in relating toxicity data for invertebrates to toxicity to vertebrates (Walker et al. 1991; Guilhermino et al. 2000; Lagarto Parra et al. 2001; Sifri et al. 2005; Favilla et al. 2006; Skrobek et al. 2006). Human cell lines have also been used (McLaughlin et al. 1993; Solis et al. 1993; Logrieco et al. 1996). Another well established alternative method to animal testing is the chicken embryo assay system, used already for the assessment of microbial toxin production (Griffin and Chu 1983, Veselý et al. 1984; Prelusky et al. 1987; Bacon et al. 1995; Sayers et al. 1995).

Bioassay systems may differ in the sensitivity to different chemicals and may represent different groups of non-target organisms. Therefore, different combinations of invertebrate cell and/or tissue culture bioassays may need to be evaluated for human, animal and ecotoxicological risk assessments, avoiding wrong negative and wrong positive results. Furthermore, standard protocols for sample (crude extract) preparations are necessary and need to be adapted for different groups of chemical/metabolites.

Higher throughput and low cost test systems have major advantages over animal testing and would allow the investigation of toxic metabolite production over a broader range of environmental conditions. Consequently, compared to the current methodology, improved risk assessment for microbials might be possible using innovative testing systems. However, research is needed to validate the methodology and to develop guidelines and protocols for the assessment of toxigenicity of the microorganism. Growth under different environmental conditions should be taken into consideration to understand the microbial activity under conditions closer to their practical application.

Genotoxicity assessment is a special part of the toxicity assessment, requiring a particular set of methodologies. Genotoxic effects are often cumulative in nature or can cause germline damage. Therefore, an acute toxicity assessment is usually not sufficient to detect hidden genotoxic effects. Genotoxicity assessment is based on in vitro assays, because whole animal testing is known to be inadequate. With conventional genotoxicity tests applied for small-molecule chemicals, appropriate protocols that avoid uninterpretable or misleading results when used with micro-organisms should be avoided (McGregor 2008). Mutagenic and carcinogenic secondary metabolites have been identified in micro-organisms, particularly in fungal species, using various methods of isolation and bioassay (e.g. Enomoto and Saito 1972; Steyn 1977; Tazima 1982; Rodericks et al. 1977). However, although specific products under consideration as microbial pesticides have been tested (e.g. Genthner et al. 1998), a general method of screening fungi or other micro-organisms for mutagenic activity has not been developed yet. Data requirements and assessment methods according to Directive 91/414 have been reviewed by McGregor (2008) and judged as less appropriate. Better adapted sample preparation protocols, guidelines on selection of the test system and test hierarchy in a tiered system need to be developed in relation to exposure scenarios. As a first step, the development of assays with crude extracts and culture supernatants should be developed for detecting or excluding the production of genotoxic metabolites in relevant quantities.

In the course of the toxicity and toxigenicty evaluation of a MBCA the biology of these micro-organisms should be always taken into account. In detail, the following general conclusions were drawn by REBECA experts based on the results of EU project RAFBCA (http://www.rafbca.com):

1. The biology of MBCAs should be more seriously taken into account when assessing the risks. For example, in most cases MBCAs are already in the environment. Although their density increases immediately after application, the density of MBCAs and the concentration of their metabolites decline over time returning to the naturally occurring background levels in the field.
2. Toxins are usually produced under inducible conditions within or in contact with the host or target. Their concentrations are low and they cannot be easily detected

in the crop or the environment in amounts sufficient to monitor their presence or fate. Therefore, such metabolities should be considered as of minor concern (Boss et al. 2007; Seger et al. 2005b, c; Strasser and Kirchmair 2006; Skrobek et al. 2007).

3. Fungal BCAs investigated produced metabolites in extremely small amounts both in vitro and in vivo and, therefore, the metabolites are unlikely to pose a threat to humans and the environment (Boss et al. 2007; Seger et al. 2004, 2005a, b; Shah et al. 2005; Skrobek and Butt 2005; Strasser et al. 2000).
4. None of the investigated fungal metabolites entered the food chain in quantifiable amounts, even when applied at rates ten times higher than the recommended application rate. Metabolite risks were assessed at all stages of the production and application cycle, i.e. in fermenters, unformulated inoculum, formulated product, on crops and in harvested crops (Skrobek and Butt 2005; Skrobek et al. 2006; Boss et al. 2007).
5. Purification of any metabolite is time consuming and requires the use of several analytical methods. Only few of the several possible metabolites produced by these organisms could be isolated. Therefore, a risk assessment investigation based on single metabolites is not feasible.
6. The action of microbial BCAs is in most cases related to the presence of an active living cell. Metabolites of microbial origin are biodegradable. They are produced in situ by the cell and are active within a limited time and space. They are not accumulated in the environment and consequently residues are not expected to be higher than the natural background levels.
7. Should pre-submission data already indicate that the MBCA is member of the microbial community at the application site, no major risk is expected because non-targets including consumers of the plants are and always have been naturally exposed to these organisms and their metabolites.

It is obvious that the assessment of potential risks related with the effect of microbial metabolites needs much more scientific investigation to come up with appropriate, better adapted and more-balanced test systems. The discussion on the potential risks related to the use of MBCAs in plant protection would also benefit from more scientific results on the occurrence of micro-organisms and the concentration of their metabolites in the natural environments. It should also be reviewed whether metabolites, which have not been identified in the acute toxicity testing, are produced in high enough amounts in the agriculture environment to be of concern.

### 13.4.3 Sensitisation/Irritation

Sensitisation or hypersensitivity is a delayed inflammatory reaction induced by a reaction of the immune system to a chemical compound. By contrast to irritation, which is a direct inflammatory response to a substance, sensitisation can only be routinely assessed by whole-animal testing (Chew and Maibach 2006; Simion 2006).

Data on sensitisation are required in Point IIM 5.3.1. of the OECD Series on Pesticides 23, Appendix 11 (OECD 2010). If no data are provided, the products are likely to be labelled as sensitizing with "Xi – R43 potentially sensitizing through skin contact" or "Xn – R42 potentially sensitizing through eye contact". If both classifications (R42 + R43) are given, the product is labelled as Xn. As the labels "Xi" or "Xn" exclude the use of the MBCA, e.g., in home gardening markets, applicants want to avoid this label. Sensitisation would currently also exclude the grouping as "low risk product".

The available methodology is based on assays developed for pure chemicals and is producing inconsistent results with microbials. Injective induction and challenge with foreign proteinaceous components into a laboratory animal regularly yields a positive response. On the other hand, topical induction and challenge with the active microbial ingredient would most probably lead to a negative response in most cases. No tests are available for assessing the potential sensitisation by inhalation of micro-organisms, most probably a greater problem compared with dermal exposure. The lack of suitable methods assessing the sensitising potential of microbials is acknowledged by the European and North American regulation authorities. As a consequence of the absence of proper test methods, the Directive 2001/36/EC (EU 2001) advises that all micro-organisms should be labelled as potential sensitizers, unless the applicant wants to demonstrate the non-sensitising potential by submitting data. The producers of microbial plant protection products demand applicable test methods, since the sensitizer label might unnecessarily restrict the use of their products, especially in organic farming and the amateur market. The same reservations apply to the irritation assessment. The methods used should be better adapted to products containing micro-organisms and they should substitute whole animal testing.

## 13.5 Proposed Waivers

### *13.5.1 Infectivity*

Humans are regularly exposed to a wide range of micro-organisms and the human community is spending a lot of resources to identify pathogens. Human pathogens are well described and documented in the relevant literature and databases (Möllby 1998). On the basis of this knowledge, microbes are categorised into four risk groups according to Directive 2000/54 EC (EU 2000). This Directive aims at protection of workers against risks to their health and safety, including the prevention of such risks, arising or likely to arise from exposure to biological agents at work. If a biological agent is included in risk group 1, it is unlikely to cause human diseases. In that case, no special measures are required according to the Directive to prevent or reduce the risk of exposure to such an organism (Article 4, Clause 1). Only general principles of good occupational safety and hygiene should be followed. Until today, all micro-organisms used in registered plant protection products are not listed in the risk groups 2–4.

Dir. 2000/54 only categorized organisms into groups 2–4. This means: "In line with the scope of the Directive, only agents, which are known to infect humans are to be included in the classified list. Animal and plant pathogens, which are known not to affect man are excluded". However, not explicitly listing the group 1 organisms is a drawback of this Directive, because non-listed micro-organisms might not have been categorized at all. By contrast, Germany includes the group 1 organisms in the so called "Technical Advice for Biological Substances" (Anonymus 2002 and 2006). In all EU Member States adaptations of Dir. 2000/54 EC exist. A quite similar categorisation of micro-organisms as used in the EU is used by the WHO (WHO 2003) and many non-European countries. Taking into consideration the vast amount on scientific information available on human pathogens, the risk for infection of humans by a micro-organisms is well known. EU and the Member States already performed risk assessment and made decisions concerning the risk regarding the exposure of workers. This information/classification should be used for the risk assessment of micro-organisms used in plant protection products.

Another possible source of information is available when screening clinical reports and scientific publications on adverse effects of species of micro-organisms. As mentioned earlier, a correct identification of the micro-organism by the applicant is an indispensable prerequisite. REBECA experts discussed whether the classification of a micro-organism into group 1 provides at least the rationale to waive the risk assessment requirements regarding extensive infectivity studies of the micro-organism or, in other words, to waive the clearance investigations in the Tier I assessment.

Despite the group 1 classification, further key indicators for the human (mammalian) safety of MBCAs are:

- no growth at temperatures >35°C
- no clinical reports and indications in relevant scientific literature or databases
- data on susceptibility of MBCA to antibiotics

In this aspect the potential of MBCAs to cause problems to immune-compromised patients was also discussed. Nosocomial infections of immune-compromised patients are a result of treatment in a hospital or a healthcare service unit, but secondary to the patient's original condition. Nosocomial infections are alarming as antibiotic resistance has spread widely. Data on the susceptibility of the MBCA to common antibiotics can minimize the risk of nosocomial infections. Reports on infections of immune suppressed patients, however, should not hamper registration of a micro-organism for use in PPP, since contact of immune-suppressed patients to PPP should be avoided in any case.

REBECA experts proposes that if all the following criteria are fulfilled, the data requirements for infectiveness in Directive 2001/36 EC point 5.2.2 (EU 2001) should be waived:

1. No (or few) clinical reports and indications of infectiveness in relevant scientific literature or databases. A low number is, in most cases, a wrong identification or

an indication for an opportunistic infection. This can be assessed from the data provided with the record.
2. Point 1 criteria should be cross-checked with Directive 2000/54 EC (EU 2000) or equivalent Member State documents.
3. Data on susceptibility of MBCA to antibiotics, indicating that the strain is susceptible to several available compounds.
4. Data on distribution and occurrence of species, which underpin the regular exposure of humans to the micro-organism in question (e.g. common on food and feed, common on food and feed plants foliage or roots, common in the soil etc.).

In other words, if humans are already regularly exposed to the micro-organism and no relevant clinical reports exist (risk group 1) and the micro-organism is susceptible to available antibiotics, the risks of infections are negligible.

### *13.5.2 Soil Biota*

REBECA experts proposed that data requirements on non-target effects on micro-organism in the soil should be waived. Soil seems to be characterised by a redundancy of functions (Nannipieri et al. 2003). The functional characteristics of component species are at least as important as the number of species for the maintenance of essential processes. Therefore, an expedient assessment of environmental risks caused by different agricultural practices should not be focused on possible changes of the abundances of particular species. Attention should be paid to preserve the functionality of the soil and keep the different functional groups of organisms in balance (see also Chapter 7). Directive 2005/25 EC mentions that micro-organisms may pose risks because of their potential to interfere with nitrogen and carbon mineralization in the soil (EU 2005). It is also mentioned that experimental data are not normally required (point 2.8.6.1). Carbon mineralization is the consequence of microbial activity in the soil. It was questioned whether the release of comparatively low numbers of additional micro-organisms poses a risk to the other soil micro-organism community responsible for carbon mineralization. Hazards have not been observed so far. Changes in the soil microbiota occur regularly, particularly in agricultural soil ecosystems after anthropogenic input. Severe impacts on the composition and quantities of soil micro-organisms have been observed during irrigation, tillage, application of organic or synthetic fertilizers or simply by crop rotation (Alabouvette 1998; Steenwerth et al. 2002; Buckley and Schmidt 2003; Clegg et al. 2003; Johnson et al. 2003; Garbeva et al. 2004; Grayston et al. 2004; Salles et al. 2006; Meyling and Eilenberg 2007). Agricultural measures with negative impacts on the functional soil characters are not regulated, but are always more severe than the release of comparatively few microbial plant protection organisms. Data on the effect of the release of MBCAs on other micro-organisms in the soil should therefore not be requested.

REBECA experts also proposed that the data requirements for effect on earthworms should be waived. Earthworms are well adapted to the broad spectrum of soil born pathogens. Therefore, real pathogens of earthworms are at least very rare and epizooties are not known. Thus, it is most improbable that effects on earthworms will be detected and any positive control for assays with earthworms can currently not be provided. As with the soil microbiota, earthworms will be more affected by agriculture measures than by the application of MBCAs.

### 13.5.3 Fate and Behaviour in the Environment

Experience with past dossiers for the registration of MBCAs indicates that data requirements on the fate and behaviour of MBCAs have been of minor concern in the risk assessment, and information from public data have often been accepted.

Persistence of an organism in the environment is an important factor in determining its risk because it strongly influences the likelihood for non-target organism exposure. Clearly, in the environment living organisms can have an entirely different behaviour from chemicals as they can proliferate. It is important to note that, from a risk assessment perspective, an organism or substance naturally present in the environment must be regarded differently from a new species or substance introduced to the ecosystem. Most MBCAs can be considered to be part of the "background population" (Annex II, 2001/36/EC). Natural habitat and application site for MBCAs are in most cases identical or similar (e.g., Damgaard 2000; Ramos 2004; Meyling and Eilenberg 2007). In these cases persistence should not be recognized as a risk. The density of microbiota fluctuates markedly depending on host, seasonal and micro-climatic conditions and agricultural measures. Therefore, the introduction of a relatively high and persistent population of an indigenous organism in the environment should not be a major concern. Application of microbial species to any particular environment usually results in a temporary increase of its population followed by a gradual decrease to background levels.

Most micro-organisms have a world-wide distribution. Among these are all registered MBCAs. However, some micro-organisms might be non-indigenous in a particular habitat. For those micro-organisms, release and persistence in the environment might pose a risk to non-target organisms that have never been exposed to the micro-organism before. Data on non-target effects will add to the assessment of potential risks due to persistence in a defined habitat. However, these data only need to be asked for if a MBCA is proven to be non-cosmopolitan, which is rarely the case.

### 13.5.4 Genetic Stability

The current data requirements demand, where appropriate, information on genetic stability (e.g., mutation rate of traits, related to the mode of action or uptake of exogenous genetic material). Environmental conditions of proposed use must be provided according to OECD Section 1, Point IIM 2.10 (OECD 2010). Information must also be provided on the micro-organism's capacity to transfer genetic material to other organisms, as well as its capacity to being pathogenic for plants, animals or man. If the micro-organism carries relevant additional genetic elements, the stability of the encoded traits should be indicated.

REBECA experts recommended that data requirements regarding the stability of genetic traits affecting the efficacy of the product should be waived or erased. Changes in the efficacy due to genetic instability will be detected during quality control procedures. Therefore, the applicant should demonstrate the use of proper quality control measures instead of demonstrating the genetic stability of the beneficial traits.

Genetic variations occur spontaneously. Statements on the stability of the MBCAs can only be based on investigations on their mutation rate, but the relevance of such studies for the assessment of risks is questioned. Results of mutations cannot be predicted. As MBCAs are not expected to be different from other micro-organisms in their capacity to transfer genetic information to other populations, data specific to the MBCA will not add more information on its safety.

## References

Adlersberg L, Singer JM, Ende E (1969) Redistribution and elimination of intravenously injected latex particles in mice. J Reticuloendothelial Soc 6:536–560

Alabouvette C (1998) Fate of micro-organisms introduced into soil, effects on autochthonous communities and activities In: KEMI (ed) Microbiological Plant Protection Products – Workshop on the Scientific Basis for Risk Assessment Proceedings Workshop Stockholm, Sweden, 26–28 October 1998, pp 57–64

Amiri-Besheli B, Khambay B, Cameron S, Deadman ML, Butt TM (2000) Inter- and intra-specific variation in destruxin production by insect pathogenic *Metarhizium* spp, and its significance to pathogenesis. Mycol Res 104:447–452

Anderssona A, Granumb PE, Rönnera U (1998) The adhesion of *Bacillus cereus* spores to epithelial cells might be an additional virulence mechanism. Int J Food Microbiol 39:93–99

Anonymus (2002) TRBA 460 Technische Regeln für biologische Arbeitstoffe: Einstufung von Pilzen in Risikogruppen Bundesarbeitsblatt 10–2002

Anonymus (2006) TRBA 466 Technische Regeln für biologische Arbeitstoffe: Einstufung von Bakterien (Bacteria) und Aracheabakterien (Archea) in Risikogruppen Bundesarbeitsblatt 7–2006

Bacon CW, Porter JK, Norred WR (1995) Toxic interaction of fumonisin B1 and fusaric acid measured by injection into fertile chicken egg. Mycopathologia 129:29–35

Baker JM, Griffiths MW (1993) Predictive modeling of psychrotrophic *Bacillus cereus*. J Food Protec 56:684–688

Blum B, Ehlers R-U, Haukeland-Salinas S, Hokkanen HMT, Jung K, Kuhlmann U, Menzler-Hokkanen I, Ravensberg W, Strasser H, Warrior P, Wilson M (2003) Letter to the editors – biological control agents: Safety and regulatory policy. BioControl 48:474–487

Boss D, Maurhofer M, Schläpfer E, Défago G (2007) Elsinochrome A production by the bindweed biocontrol fungus *Stagonospora convolvuli* LA39 does not pose a risk to the environment or the consumer of treated crops. FEMS Microbiol Ecol 59:194–205

Buckley DH, Schmidt TM (2003) Diversity and dynamics of microbial communities in soils from agro-ecosystems. Environ Microbiol 5:441–452

Cardona ST, Wopperer J, Eberl L, Valvano MA (2005) Diverse pathogenicity of *Burkholderia cepacia* complex strains in the *Caenorhabditis elegans* host model. FEMS Microbiol Lett 250:97–104

Chew A-L, Maibach HI (2006) In vitro methods to predict skin irritation. In: Chew A-L, Maibach HI (eds) Irritant Dermatitis XI. Springer, Berlin Heidelberg, pp 501–508

Clegg CD, Lovell RDL, Hobbs PJ (2003) The impact of grassland management regime on the community structure of selected bacterial groups in soils. FEMS Microbiol Ecol 43:263–270

Dabrowski WM, Sikorski ZE (2005) Toxins in food. CRC Press, Boca Raton

Damgaard PH (2000) Natural occurrence and dispersal of *Bacillus thuringiensis* in the environment. In: Charles J-F, Delécluse A, Nielsen-Le Roux C (eds) Entomopathogenic bacteria from laboratory to field application. Springer, New York, NY, pp 23–40

Das T, Sen A, Kempf T, Pramanik SR, Mandal C, Mandal C (2003) Induction of glycosylation in human C-reactive protein under different pathological conditions. Biochem J 373:345–355

Enomoto M, Saito M (1972) Carcinogens produced by fungi. Ann Rev Microbiol 26:279–312

EPA (2000) Technical Document: *Trichoderma harzianum* Rifai Strain T-39 (PC Code 119200) US Environmental Protection Agency http://www.epagov/oppbppd1/biopesticides/ingredients/factsheets/factsheet.119200.htm

EPA (2002) Product monograph: *Pseudozyma flocculosa* strain PF-A22 UL (PC Code 119196). US Environmental Protection Agency, September 2002 http://www.epa.gov/oppbppd1/biopesticides/ingredients/factsheets/factsheet_119196.pdf

EPA (2004) Biopesticides registration action document: *Chondrostereum purpureum* strain PFC 2139 (PC Code 081308). US Environmental Protection Agency, September 2004 http://www.epa.gov/oppbppd1/biopesticides/ingredients/factsheets/factsheet_081308.pdf

EU (1991) 1991/414/EEC: Council Directive of 15 July 1991 concerning the placing of plant protection products on the market Official Journal of the European Communities L 230

EU (2000) 2000/54/EC: Directive of the European Parliament and of the Council of 18 September 2000 on the protection of workers from risks related to exposure to biological agents at work Official Journal of the European Communities L 262/21

EU (2001) 2001/36/EC: Commission Directive of 16 May 2001 amending Council Directive 91/414/EEC concerning the placing of plant protection products on the market Official Journal of the European Communities L 164/1

EU (2005) 2005/25/EC: Commission Directive of 14 March 2005 amending Annex VI to Directive 91/414/EEC as regards plant protection products containing micro-organisms Official Journal of the European Union L 90/1

EU (2009) Regulation (EC) No 1107/2009 of the European Parliament and of the Council of 21 October 2009 concerning the placing of plant protection products on the market and repealing Council Directives 79/117/EEC and 91/414/EEC Official Journal of the European Union L 309/1

Favilla M, Macchia L, Gallo A, Altomare C (2006) Toxicity assessment of metabolites of fungal biocontrol agents using two different (*Artemia salina* and *Daphnia magna*) invertebrate bioassays. Food Chem Toxicol 44:1922–1931

Garbeva P, van Veen JA, van Elsas JD (2004) Assessment of the diversity and antagonism towards *Rhizoctonia solani* AG3 of *Pseudomonas* species in soil from different agricultural regimes. FEMS Microbiol Ecol 47:51–64

Genthner FJ, Chancy CA, Couch JA, Foss SS, Middaugh DP, George SE, Warren MA, Bantle JA (1998) Toxicity and pathogenicity testing of the insect pest control fungus *Metarhizium anisopliae*. Arch Environ Contam Toxicol 35:317–324

Gibbs PS, Wooley RE (2003) Comparison of the intravenous chicken challenge method with the embryo lethality assay for studies in avian colibacillosis. Avian Dis 47:672–680

Grayston SJ, Campbell CD, Bardgett RD, Mawdsley JL, Clegg CD, Ritz K, Griffiths BS, Rodwell JS, Edwards SJ, Davies WJ (2004) Assessing shifts in microbial community structure across a range of grasslands of differing management intensity using CLPP, PLFA and community DNA techniques. Appl Soil Ecol 25:63–84

Griffin GF, Chu FS (1983) Toxicity of the *Alternaria* metabolites alternariol, alternariol methyl ether, altenuene, and tenuazonic acid in the chicken embryo assay. Appl Environ Microbiol 46:1420–1422

Guilhermino L, Diamantino T, Carolina Silva M, Soares AMVM (2000) Acute toxicity test with *Daphnia magna*: an alternative to mammals in the pre-screening of chemical toxicity? Ecotoxicol Environ Saf 46:357–362

Health Canada (2006) Regulatory note *Bacillus sphaericus* Strain 2362 (REG2006-02). Pest Management Regulatory Agency. April 7, 2006

Heinrich PC, Behrmann I, Haan S, Hermanns HM, Müller-Newen G, Schaper F (2003) Principles of interleukin (IL)-6-type cytokine signalling and its regulation. Biochem J 374:1–20

Johnson MJ, Lee KY, Scow KM (2003) DNA fingerprinting reveals links among agricultural crops, soil properties, and the composition of soil microbial communities. Geoderma 114: 279–303

Jones AE, Fiechtl JF, Brown MD, Ballew JJ, Kline JA (2007) Procalcitonin test in the diagnosis of bacteremia: a meta-analysis. Ann Emerg Med 50:34–41

Kershaw MJ, Moorhouse ER, Bateman RP, Reynolds SE, Charnley AK (1999) The role of destruxins in the pathogenicity of *Metarhizium anisopliae* for three species of insect. J Invertebr Pathol 74:213–223

Kurz CL, Ewbank JJ (2000) *Caenorhabditis elegans* for the study of host–pathogen interactions. Trends Microbiol 8:142–144

Laengle T, Strasser H (2010) Developing a risk index to comparatively assess environmental risks posed by microbial and conventional pest control agents. Biocont Sci Technol 20:659 – 681

Lagarto Parra A, Silva Yhebra R, Guerra Sardiñas I, Iglesias Buela L (2001) Comparative study of the assay of *Artemia salina* L and the estimate of the medium lethal dose (LD50) in mice in order to determine oral acute toxicity of plant extracts. Phytomed 8:395–400

Laws TR, Smith SA, Smith MP, Harding SV, Atkinsaa TP, Titball RW (2005) The nematode *Panagrellus redivivus* is susceptible to killing by human pathogens at 37°C. FEMS Microbiol Let 250:77–83

Logrieco A, Moretti A, Fornelli F, Fogliano V, Ritieni A, Caraffa MF, Randazzo G, Bottalico A, Macchia L (1996) Fusaproliferin production by *Fusarium subglutinans* and its toxicity to *Artemia salina*, SF-9 insect cells, and IARC/LCL 171 human B lymphocytes. Appl Environ Microbiol 62:3378–3384

McGregor JT (2008) Genetic toxicity assessment of microbial pesticides. In: OECD (ed) Working document on the evaluation of microbials for pest control. OECD environment, health and safety publications series on pesticides No. 43, Paris, France, pp 27–37. http://www.oecd.org/dataoecd/45/46/41946259.pdf

McLaughlin JL, Chang C, Smith DL (1993) Simple bench-top bioassays (brine shrimp and potato discs) for the discovery of plant antitumor compounds. In: Kinghorn AD, Balandrin MF (eds) Human medicinal agents from plants, Vol 534. ACS Symposium, Washington DC, 112–134

Meyling NV, Eilenberg J (2007) Ecology of the entomopathogenic fungi *Beauveria bassiana* and *Metarhizium anisopliae* in temperate agroecosystems: Potential for conservation biological control. Biological Control 43:145–155

Möllby R (1998) Pathogenic properties of micro-organisms – human hazard identification In: KEMI (ed) Microbiological Plant Protection Products – Workshop on the Scientific Basis for Risk Assessment. Proceedings Workshop Stockholm, Sweden, 26–28 October 1998, pp 49–56

Nannipieri P, Ascher J, Ceccherini MT, Landi L, Pietramellara G, Renella G (2003) Microbial diversity and soil functions. Europ J Soil Sci 54:655–670

OECD (2003) Guidance document on the use of taxonomy in risk assessment of micro-organisms: Bacteria. OECD Environment, Health and Safety Publications Series on Harmonisation of Regulatory Oversight in Biotechnology No29

OECD (2010) OECD guidance documents for pesticide registration http://www.oecd.org/document/48/0,2340,en_2649_34383_2085104_1_1_1_1,00.html

Olier M, Pierrea F, Lemaître J-P, Divies C, Rousset A, Guzzo J (2002) Assessment of the pathogenic potential of two *Listeria monocytogenes* human faecal carriage isolates. Microbiol 148:1855–1862

Ormsbee R, Peacock M, Gerloff R, Tallent G, Wike D (1978) Limits of rickettsial infectivity. Infect Immun 19:239–245

Pepys MB (1981) C-reactive protein fifty years on. Lancet 317:653–657

Pine L, Kathariou S, Quinn F, George V, Wenger JD, Weaver RE (1991) Cytopathogenic effects in enterocytelike Caco-2 cells differentiate virulent from avirulent *Listeria* strains. J Clin Microbiol 29:990–996

Prelusky DB, Hamilton RMG, Forest BC, Trenholm HL, Thompson BK (1987) Optimization of chick embryotoxicity bioassay for testing toxicity potential fungal metabolites. J Ass Off Anal Chem 70:1049–1055

Quesada-Moraga E, Vey A (2003) Intra-specific variation in virulence and in vitro production of macromolecular toxins active against locust among *Beauveria bassiana* strains and effects of in vivo and in vitro passage on these factors. Biocontr Sci Technol 13:323–340

Ramos J-L (2004) Pseudomonas. Kluwer, Dortrecht

Rodericks JV, Hesseltine CW, Mehlman MA (1977) Mycotoxins in human and animal health. Pathotox Publishers, Park Forest South IL

Salles J, van Elsas J, van Veen J (2006) Effect of agricultural management regime on *Burkholderia* community structure in soil. Microb Ecol 52:267–279

Sayers NM, Drucker DB, Morris JA, Telford DR (1995) Lethal synergy between toxins of staphylococci and enterobacteria: implications for sudden infant death syndrome. J Clin Pathol 48:929–932

Seger C, Erlebach D, Stuppner H, Griesser UJ, Strasser H (2005a) Physicochemical properties of oosporein, the major secreted metabolite of the entomopathogenic fungus *Beauveria brongniartii*. Helvet Chim Acta 88:802–810

Seger C, Längle T, Pernfuss B, Stuppner H, Strasser H (2005b) High-performance liquid chromatography-diode array detection assay for the detection and quantification of the *Beauveria* metabolite oosporein from potato tubers. J Chromatogr A 1092:254–257

Seger C, Sturm S, Längle T, Wimmer W, Stuppner H, Strasser H (2005c) Development of a sensitive HPLC-DAD assay for the detection and quantification of the fungal metabolite oosporein from fungal culture broth and biological pest control formulations. J Agri Food Chem 53:1364–1369

Seger C, Sturm S, Stuppner H, Butt TM, Strasser H (2004) Combination of a new sample preparation strategy and an accelerated high-performance liquid chromatography assay with photodiode array and mass spectrometric detection for the determination of destruxins from *Metarhizium anisopliae* culture broth. J Chromatogr A 1061:35–43

Siegel JP (2001) The mammalian safety of *Bacillus thuringiensis* based insecticides. J Invertebr Pathol 77:13–21

Siegel JP, Shadduck JA, Szabo J (1987) Safety of the entomopathogen *Bacillus thuringiensis* var *israelensis* for mammals. J Econ Entomol 80:717–723

Sifri CD, Begun J, Ausubel FM (2005) The worm has turned – microbial virulence modelled in *Caenorhabditis elegans*. Trends Microbiol 13:119–127

Simion F (2006) In vivo models of skin irritation. In: Chew A-L, Maibach HI (eds) Irritant dermatitis XI". Springer, Heidelberg, pp 489–499

Skrobek A, Boss D, Defago G, Butt TM, Maurhofer M (2006) Evaluation of different biological test systems to assess the toxicity of metabolites from fungal biocontrol agents. Toxicol Lett 161:43–52

Skrobek A, Butt TM (2005) Toxicity testing of destruxins and crude extracts from the insect-pathogenic fungus *Metarhizium anisopliae*. FEMS Microbiol Lett 251:23–28

Skrobek A, Shah FA, Butt TM (2007) Destruxin production by the entomogenous fungus *Metarhizium anisopliae* in insects and factors influencing their degradation. BioControl 53:361–373

Solis PN, Wright CW, Anderson MM, Gupta MP, Phillipson JD (1993) A microwell cytotoxicity assay using *Artemia salina* (brine shrimp). Planta Med 59:250–252

Steenwerth KL, Jackson LE, Calderon FJ, Stromberg MR, Scow KM (2002) Soil microbial community composition and land use history in cultivated and grassland ecosystems of coastal California. Soil Biol Biochem 34:1599–1611

Steyn PS (1977) Mycotoxins, excluding aflatoxin, zearalenone, and the trichothecenes. In: Rodericks JV, Hesseltine CW, Mehlman MA (eds) Mycotoxins in human and animal health. Pathotox Publishers, Park Forest South, pp 419–467

Strasser H, Abendstein D, Stuppner H, Butt TM (2000) Monitoring the spatial-temporal distribution of secondary metabolites produced by the entomogenous fungus *Beauveria brongniartii* with particular reference to oosporein. Mycolog Res 104:1227–1233

Strasser H, Kirchmair M (2006) Potential health problems due to exposure in handling and using biological control agents. In: Eilenberg J, Hokkanen HMT (eds) An ecological and societal approach to biological control. Springer, Dordrecht, pp 275–293

Tazima Y (1982) Mutagenic and carcinogenic mycotoxins. In: Klekowski EJ Jr (eds) Environmental mutagenesis, carcinogenesis and plant biology. Praeger Scientific Greenwood Publishing Group Inc, Westport, CT, pp 67–95

Tsai S, Liao J, Wang S (1997) Clearance and effects of intratracheal instillation of spores of *Bacillus thuringiensis* or *Metarhizium anisopliae* to rats. J Chin Soc Vet Sci 23:515–522

Veselý D, Veselá Jelínek R (1984) Use of chick embryo in screening for toxin-producing fungi. Mycopathologia 88:135–140

Vey A, Hoagland R, Butt TM (2001) Toxic metabolites of fungal biocontrol agents. In: Butt TM, Jackson CW, Magan N (eds) Fungi as biocontrol agents. CAB International, Wallingford, pp 311–346

Volanakis JE (2001) Human C-reactive protein: expression, structure, and function. Mol Immunol 38:189–197

Walker CH, Greig-Smith PW, Crossland NO, Brown R (1991) Ecotoxicology. In: Balls M, Bridges J, Southee J (eds) Animals and alternatives in toxicology. Macmillan, Basingstoke, pp 223–252

Wang CS, Skrobek A, Butt TM (2004) Investigations on the destruxin production of the entomopathogenic fungus *Metarhizium anisopliae* in liquid and solid media. J Invertebr Pathol 85:168–174

WHO (2003) Laboratory Biosafety Manual WHO/CDS/CSR/LYO/20034 Geneva

Wooley RE, Gibbs PS, Brown TP, Maurer JJ (2000) Chicken embryo lethality assay for determining the virulence of avian *Escherichia coli* isolates. Avian Diseases 44:318–324

Zimmermann O (2004) The use of beneficials for biological control in Germany – notes on the present status. Gesunde Pflanze 56:151–156

# Chapter 14
# Proposals for Regulation of Botanicals

**Lucius Tamm, Bernhard Speiser, and Thierry Mercier**

**Abstract** Plants and plant extracts, here called 'botanicals', have been used for plant protection for a long time. Quantitatively, the most important botanical is pyrethrum, followed by azadirachtin, rotenone and essential oils. The current regulatory system for pesticides is often seen as a major hurdle for the market introduction of new botanicals. The EU-funded Specific Support Action project 'REBECA' has held a series of workshops with stakeholder representatives. The following proposals for improvement were elaborated: (1) development of a specific guidance document for botanicals; (2) adapted requirements concerning the characterisation of the active substance(s); (3) relaxations concerning identification and analytical methods for 'impurities'; (4) adapted requirements concerning the description of manufacturing methods; (5) adapted requirements for risk assessment, taking into account the history of safe use of the substance; (6) adapted requirements concerning efficacy evaluation. During the final conference of the REBECA project, it was evaluated which proposals can be implemented easily (and therefore in a short timespan). Also, the impact on the duration of the registration process and on the costs of registration (for the applicant) were assessed for each proposal. Fenugreek, neem extract/Quassia, lecithine and laminarine were selected as representative botanicals. For these substances, the REBECA proposals would decrease registration costs substantially.

## Contents



---

L. Tamm (✉)
Research Institute of Organic Agriculture (FiBL), Ackerstrasse, 5070 Frick, Switzerland
e-mail: lucius.tamm@fibl.org

R.-U. Ehlers (ed.), *Regulation of Biological Control Agents*,
DOI 10.1007/978-90-481-3664-3_14, 

## 14.1 Introduction

In this chapter, the term 'botanicals' is employed for plant extracts used in plant protection. At the heart of this chapter are the proposals for improvement of the current regulatory situation with respect to botanicals. These proposals were developed in a series of workshops held in the framework of the REBECA project, leading to a final draft at the REBECA Final Conference on 20–21 September 2007. Representatives of all stakeholder groups showed a high degree of consensus with these proposals.

The document is intended as a proposal to the Commission and EU Member States on how to facilitate the registration of plant protection products containing botanicals. Although we use the term 'registration', the intention is to cover all regulatory processes which affect the commercialisation of botanicals. This covers the procedure for Annex I inclusion of botanical active substances and the subsequent national registration of plant protection products containing these botanicals.

These proposals represent the state-of-the-art at the time of the REBECA project and can be improved by the insights and experience gained during the evaluation of those botanical active substances on the 4th list of the 91/414 review programme.

The proposals are aimed at the European process, but they consider also non-European systems. Based on extensive experience from registration, the USA has recently relaxed registration requirements for 'biopesticides', which includes many botanicals. In the opinion of the REBECA project partners[1], the EU should envisage

---

[1]REBECA aims to find proposals which are based on a broad consensus in the workshops, but does not claim that all participants agree with all proposals. In case of major disagreements, REBECA will describe the advantages and disadvantages of a proposal.

similar relaxations, because potential risks are similar in the EU as in the USA. Over extensive and non-harmonised data requirements are a substantial economic hurdle for the manufacturers of botanicals, with very little benefit for the European market and the environment.

## 14.2 Widely Used Botanicals

Materials of plant origin have a very long history of use by mankind. The uses include plant protection and other agricultural uses, as well as non-agricultural uses.

### *14.2.1 Botanicals in Plant Protection*

Plants, and particularly plant extracts, have been used for plant protection for a long time (Philogène et al. 2005). Extracts can range from crude to highly purified. In this document, the term 'botanical' is used to describe active substances made from plants, as defined in the draft working document SANCO/10472 (Anonymous 2004a). However, in contrast to SANCO/10472, the range of plants and of solvents is not restricted in this chapter. Quantitatively, the most important botanical is pyrethrum, followed by azadirachtin, rotenone and essential oils. Ryania, nicotine, sabadilla, garlic oil and *Capsicum* oleoresin have limited use (Isman 2006). Typical uses are:

- insecticides (e.g. pyrethrum, rotenone, rape seed oil, quassia extract, azadirachtin, nicotine),
- repellents or antifeedants (e.g. neem),
- fungicides and inducers of resistance (e.g. laminarine, fennel oil, lecithine),
- herbicides (e.g. pine oil),
- nematicides (e.g. neem),
- sprouting inhibitors (e.g. caraway seed oil) and
- adjuvants such as stickers and spreaders (e.g. pine oil).

Some botanicals may have more than one use.

### *14.2.2 Botanicals Used for Other Purposes*

Plant extracts and other materials of plant origin are also used for purposes other than plant protection. In this case, their use is not subject to Dir. 91/414. Examples of such uses are:

- fertilizers and soil conditioners (e.g. green waste compost, seaweeds and seaweed extracts, sawdust, wood chips, composted bark, bark of hemlock pine [iron micronutrient fertiliser]),
- biocides (e.g. pyrethrum and azadirachtin as insecticides, citronella as repellent),

- foods and spices. Some plant products which are mainly used as foods or spices also have a secondary use in plant protection, e.g. rape seed oil, lecithine, garlic, mustard powder, fennel and caraway. In these cases, use for plant protection is quantitatively minimal in comparison to use for human consumption.
- Perfumes, cosmetics and medicine (e.g. limonene).

## 14.3 Environmental Impact and Human Health Risks of Botanicals

'Botanicals' covers an extremely heterogeneous group of substances, i.e. unprocessed and processed plant extracts. Furthermore, plant extracts may be highly refined (i.e. one single active substance) or represent a 'cluster' of substances present in an extract. Characteristics of substances may be very well known or virtually unknown a priori. Extracts of plants may vary due to variability in the composition of the raw material (see below) and/or due to processing/storage conditions. In conclusion, risks associated with the use of 'botanicals' may vary between very low and very high. Thus, it is extremely difficult to compile a definitive set of data requirements which would be equally applicable to all botanicals. Instead, it is proposed to compile a guidance document which explains when the data requirements of Dir. 91/414/EEC will need to be addressed by specific studies and when data requirements can be met with reasoned scientific cases (often referred to as 'waivers').

## 14.4 Overview of Regulation and Regulatory Efforts for Botanicals

### *14.4.1 Directive 91/414/EEC*

Plant extracts or 'botanicals' are not defined in the EU legislation, and no separate data requirements exist in Directive 91/414. Thus, the full data requirements given in Dir. 91/414 must be fulfilled. However, it can always be negotiated between the applicant and the evaluators whether a specific data requirement must be met with a study, or whether a reasoned case (waiver) is accepted. A detailed description of the data requirements for botanicals is given in Hauschild et al. (2008). The first botanical authorized under Dir. 91/414 is laminarin. For the review report, see Anonymous (2004b).

### *14.4.2 Draft Working Document SANCO/10472*

Reduced data requirements are described in the SANCO draft working document 10472 (Anonymous 2004a). However, this document is not legally binding. Regulators and applicants have gained experience with this document, applying it

**Table 14.1** Examples of (low risk/concern) candidate botanicals

| | |
|---|---|
| (A) Edible parts of plants used for human nutrition or animal feed | |
| | Artichoke (edible parts), basil (whole plant), black pepper (fruit), carvi (fruit), chives (clove), coriander (fruit), elder (bark, flower, fruit), garden sage (whole plant), garlic (clove), horse tail (leaf), laurel (leaf), mint (whole plant), olive (oil), onion (bulb), oil seed-rape (oil), sesame (seed), soybean (oil), squash (seed), sunflower (oil), tomato (fruit) |
| (B) Parts of plants authorized as herbal drugs | |
| | Bladder wrack (thallus), feverfew (whole plant), lavander (whole plant), nettle (whole plant), rhubarb (rhizome only), sweet chamomille (whole plant) |
| (C) Plant extracts classified as minimal risk pesticides in the USA | |
| | Castor oil, cedar oil, cinnamon and cinnamon oil, citric acid, citronella and citronella oil, cloves and clove oil, corn gluten meal, corn oil, cottonseed oil, eugenol, garlic and garlic oil, geraniol, gernanium oil, lauryl sulfate, lemongrass oil, linseed oil, malic acid, mint and mint oil, peppermint and peppermint oil, rosemary and rosemary oil, sesame (includes ground sesame plant) and sesame oil, sodium lauryl sulfate, soybean oil, thyme and thyme oil and white pepper |
| (D) Example of a plant extract classified as GRAS | |
| | Lecithine |
| (E) Plant extracts excluded from registration requirements in Australia | |
| | Cabbage extract, canola (=rapeseed) oil, capsicum oleoresin, chilli extract, citronella oil, cypress wood oil, *Derris* dust, *Eucalyptus* oil, garlic extract, garlic oil, lanolin oil, lavender fragrance, lime oil, orange oil, pine oil, pyrethrins, pyrethrin I, pyrethrin II, *Quassia*, rotenone, salicylic acid, sesame, tea tree oil and thymol |

Sources of information: (A) and (B) SANCO/10472; (C) '25b list' of the US EPA; (D) 21 CFR 184.1400; (E) AVPMA (Australian Veterinary Practice Management Association) registration requirements

to active substances of the fourth stage review programme for existing substances. The data requirements in SANCO/10472 apply to plant protection products made from all edible parts of plants used for animal or human feed, and in addition to other parts of plants which are listed explicitly (see Table 14.1 A). It also applies to parts of plants currently authorised as herbal drugs in European pharmacopoeia and known traditionally for plant protection properties, which are listed explicitly (see Table 14.1 B). Further, the data requirements are restricted to plant extracts made with water and/or ethanol. These restrictions allow the data requirements of Directive 91/414 to be adapted, as there is no evidence that the manufacturing process (e.g. crushing, drying, water and/or ethanol extraction) could considerably modify the toxicity or ecotoxicity profiles. For plant protection products made from other plants or plant parts or with other solvents, data requirements will be established case-by-case in a pre-submission meeting, based on the available information. It is not exceptional that plant extracts are used as such. For registration purposes, the extract is considered as the active ingredient and also as the plant protection product. Therefore, the guidance has not been separated in requirements for Annex II (active substance) and Annex III (plant protection product), but has to be interpreted on a case-by-case basis. The document identifies two categories: Category 1: Plant protection products made from one or several plants; category 2: Plant protection products made from one or several plant extracts. A tiered approach is taken, where the tier 1 data requirements are explicitly described and tier 2 data can be requested on a case-by-case basis.

### 14.4.3 Low-Risk and Basic Substances

Since this chapter was written (June 2008), the EU 'Pesticides Directive' 91/414/EEC has been replaced by regulation 1107/2009 (EC 2009). This regulation contains some facilitation for 'low risk active substances' and for 'basic substances'. Low risk active substances are dealt with in Article 22 and 47. Low risk active substances shall be approved for 15 years (as opposed to 10 years for other active substances), and plant protection products containing low risk active substances shall be authorised within a short period.

Basic substances are defined in Article 23 as active substances which (i) are not a substance of concern and no not have an inherent capacity to cause endocrine disrupting, neurotoxic or immunotoxic effects, (ii) are not predominantly used for plant protection purposes but (iii) nevertheless have some are useful in plant protection, either directly or in a product consisting of the substance and a simple diluent, and (iv) are not directly marketed for use as plant protection products. Basic substances shall be approved as active substances for an unlimited period of time. Plant protection products containing exclusively basic substances do not need to be authorised. Examples of basic substances could be rape seed oil, garlic oil, fennel oil, caraway seed oil, lecithine or essential oils.

### 14.4.4 Plant Strengtheners

Plant strengtheners are not specifically regulated at EU level. There have been attempts to define data requirements for plant strengtheners with low risk profile (Anonymous 2001). However, these activities have been discontinued.

### 14.4.5 Various Regulations Which Indicate Low Risk

In the USA, there is a list of substances that can be used as pesticides without any registration. These substances are called Minimal Risk Pesticides, the list is known as '25b list'. The list contains many essential oils. At the time of writing, all inerts must be on EPA's 4A inert list, all ingredients must be identified on the label, and the label may not contain false or misleading claims. The botanicals of the 25b list are given in Table 14.1 C.

The US FDA (Food and Drug Administration) has a list of substances which are considered to be safe for use as food additives ('GRAS' substances; Generally

Recognized As Safe). Lecithine, which can also be used as a fungicide, has GRAS status (see Table 14.1 D).

In Australia, a number of plant extracts are excluded from the requirements of AVPMA (Australian Veterinary Practice Management Association) approval as constituents in plant protection products. These are given in Table 14.1 E.

### *14.4.6 Fourth Stage Review of Botanicals in the EU*

The 4th list of the review programme under Directive 91/414 contains a diverse range of products and uses, including other substances considered under the REBECA project. One of the principles of the fourth stage review is to employ a lighter regime which, whilst ensuring appropriate health and environmental safeguards, recognises the economic impact of generating significant amounts of data for what are often specialised niche products.

Many botanicals are subject to re-evaluation under the 4th stage of the review programme under Directive 91/414, for example azadirachtin, pyrethrins, rape seed oil and quassia. Many of these botanicals have been used in plant protection for many years, without evidence of adverse effects on human health. In most cases, there are limited experimental data on human health effects available. At the time of writing, most of the stage 4 botanicals were expected to be included in annex I. At the time of proofreading, pyrethrins and rape seed oil were included, but rotenone was not included. Azadirachtine and Quassia were pending (original dossier withdrawn, and amended dossier re-submitted).

## 14.5 Bottlenecks Under the Current System

The current regulatory system is often seen as a major hurdle for the market introduction of new botanicals. From an applicant's perspective, registration costs are not well predictable and often high in relation to the expected sales. Furthermore, the duration of the registration process is also difficult to predict. For an in-depth analysis, see Chapter 2. The REBECA project has identified specific bottlenecks and elaborated proposals for improvement. These are described below.

Botanicals present different regulatory challenges from the other substances studied in the REBECA project (semiochemicals, micro-organisms and invertebrate biocontrol agents). The major challenges for botanicals are:

- they are often complex mixtures which are difficult to characterise and standardise;
- they comprise both harmless and potentially harmful substances;
- there is uncertainty regarding the data requirements, because SANCO/10472 is not legally binding and does not cover extracts from all plants and with all solvents.

Accordingly, the REBECA proposals given below are specific for botanicals.

## 14.6 Proposals of the REBECA Project

The REBECA project has developed several proposals for improvement of the current situation for botanicals. These proposals aim mainly at simplifying the data requirements for evaluation of the active substance. This reduces the costs for dossier preparation (for the applicant). It may also reduce the duration of the registration process and the efforts needed for dossier evaluation (which might lead to reduced registration fees). The proposals concerning efficacy evaluation also aim at national product registration. An overview of the proposals in relation to the regulatory processes is given in Fig. 14.1.

The proposals described in this chapter are specific for botanicals and provide technical solutions for selected problems. In the botanicals working group, they were therefore called 'pragmatic' proposals. In addition, the working group has also discussed 'visionary' approaches, which aim at the current regulatory processes in general and would affect other natural substances as well. The visionary proposals are much more difficult to implement, and there was considerably more disagreement regarding their usefulness. Such proposals are described in Chapters 6 and 17 of this volume.

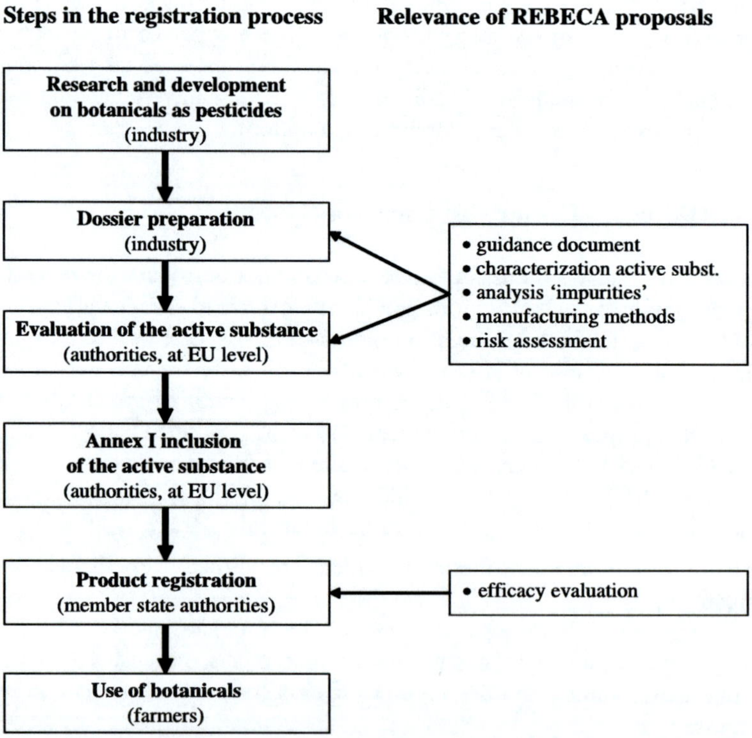


**Fig. 14.1** Schematic representation of the step in the registration process for which each REBECA proposal is relevant. The proposals are indicated by an abbreviation of their title in the text

### 14.6.1 Proposal to Develop a Guidance Document for Botanicals

The lack of specific data requirements for botanicals causes uncertainty regarding the data requirements. The REBECA project therefore recommends that a comprehensive guidance document should be formally adopted for botanicals. This could be based on SANCO/10472 (Anonymous 2004a), with some amendments. Certain amendments can already be proposed at this stage. After completion of the 4th stage review, further amendments may seem adequate. The REBECA project proposes the following amendments to SANCO/10472:

- The scope of extraction methods should be broadened. Currently, SANCO/10472 covers only water and ethanol extracts. Its scope should be broadened to cover *all* extraction methods.
- The scope of plants should be broadened. Currently, SANCO/10472 lists only a limited number of plant parts in the annex. Its scope should be broadened to cover *all* plants and plant parts.
- The document should contain a list of plants and/or combinations of plants and extraction methods which are recognised as of low risk/concern. This should be an open list that can be amended when new botanicals have been evaluated (taking into account issues of data protection). As a starting point, all substances which are currently listed in SANCO/10472, all substances on the '25b list' of the US EPA and all substances with GRAS status should be considered for such a list. These substances are given in Table 14.1. For this task, support by an EU funded research project would be useful.
- As a result of the broadened scope of plants and extraction methods, a tiered system will be needed. It is desirable to establish a system to identify substances/extracts of low risk/concern at an early stage of the process. For these substances, only tier I data requirements apply.

### 14.6.2 Proposals Concerning Characterization of the Active Substance(s)

Often, the 'active substance' is a cluster of very similar substances. For example, pyrethrum contains three esters of chrysanthemic acid and three esters of pyrethric acid (Isman 2006). Of these, pyrethrin I and II are the most abundant and account for most of the insecticidal activity. Neem contains more than a dozen azadirachtin analogues, but the major form is azadirachtin, and the other analogues contribute little to overall efficacy (Isman 2006). Neem also contains other triterpenoids such as salannin, nimbin, and derivatives thereof. Their role has been controversial, but seems to be minor in comparison to azadirachtin. In conclusion, it is often not possible in botanicals to draw a clear line between active and inactive substances. However, it is usually possible to identify one or a few substances which are responsible for most of the activity of the extract.

REBECA proposes that identification and analytical methods shall be required for the active substance(s), or for those substances which are mainly responsible for the effects on the target pest. If these are not identified, it should be determined case-by-case whether one or several representative lead substances (markers) may be used instead.

### *14.6.3 Proposals Concerning Identification and Analytical Methods for 'Impurities'*

The content of metabolites in plants (naturally occuring compounds from plant metabolism) is subject to great quantitative and sometimes also qualitative variation. Variation occurs between different plant parts, different physiological ages, different harvesting times, different growing conditions (e.g. nutrient, water or light availability), different regions and different genotypes. Due to this variability in the material of origin, the composition of plant extracts is inevitably variable.

In the regulatory context, *all* substances which are not responsible for the effect on the target pest are considered to be 'by-products' or 'impurities'. If such substances are present in quantities $\geq 1$ g/kg in the active substance as manufactured, they are considered to be 'significant impurities'. If they are of toxicological and/or ecotoxicological or environmental concern, they are considered to be 'relevant impurities' (see Annex IIA, Point 4.). Currently, all components present in quantities $\geq 1$ g/kg must be identified with validated methods. This causes very high costs.

The alternative approach proposed here for botanicals reduces the need for validation of all significant/non-relevant 'impurities'. The aim of this proposal is to exclude toxicological risks and to ensure constant quality and composition of plant extracts, while keeping the analytical efforts reasonable. The substances present in a plant extract can be categorized as follows:

- Active substances: see separate section above.
- Plant constituents of concern: certain plant metabolites with high toxicity (e.g. alkaloids) may present a hazard to human or animal health or the environment.
- Impurities: microbial metabolites or decay products which may be formed before and during manufacture, as well as process impurities may also present a hazard to human or animal health or the environment. Microbial contaminants and process impurities can be avoided with appropriate quality management.
- Other plant constituents: by definition, this category contains only those plant constituents that do not present a hazard to human or animal health or the environment, such as sugars and fatty acids. The presence of such substances is unavoidable, except if they can be eliminated by purification. Although the composition of other plant constituents is not toxicologically relevant, it mirrors the production and extraction process. Thus, it adequately reflects the constant quality of the production process and indicates deviations that may lead to unforeseen impurities in the technical active substance, as well as to variability in efficacy.

REBECA proposes that identification and analytical methods shall be required for all plant constituents of concern and for all impurities of concern (='relevant impurities' *sensu* Dir. 91/414, Annex IIA, Point 4).

REBECA proposes that other plant constituents should be characterised with sum parameters (e.g. by group analysis for sugars, fatty acids, terpenoids, or by measurement of pH). However, identification and validated analytical methods should not be required for each component present in quantities ≥1 g/kg (='significant impurities' *sensu* Dir. 91/414, Annex IIA, Point 4). To achieve this, the threshold for 'significance' (currently ≥1 g/kg for all substances) could be raised for botanicals.

In conclusion, it is not feasible for botanicals to characterise at least 98% of their composition, as required in conventional pesticides. However, the relevant components should be identified as far as necessary in order to ensure reproducibility of the product (on a case-by-case base). Where possible and necessary, purification of plant extracts should be encouraged.

### *14.6.4 Proposals Concerning the Description of Manufacturing Methods*

Botanicals are manufactured in a completely different way from synthetic pesticides. Therefore, the description of the manufacturing method must focus on different aspects. The description should include information on the plant material of origin, and should indicate the range of materials used. Greater variation in the composition should be acceptable for botanicals than for synthetically produced substances. The plant material should be produced with sustainable methods and the Rio convention on biological diversity must be respected.

In view of the great variability in plants, the applicant is free to define terms such as 'growing conditions' and 'region', based on the biology and distribution of each plant species. It is advised that applicant and evaluator agree on this in a pre-submission meeting. If other plant material is used in the future, the applicant has to demonstrate equivalency of the technical material with the criteria outlined here. If the supplier of the plant extract does not want to disclose this information to the manufacturer of a plant protection product, the regulatory authority may obtain the information directly from the supplier.

REBECA proposes that the description of the plant material used should include the plant parts used, the physiological ages, the harvesting time, the growing conditions (e.g. nutrient, water or light availability), the regions and the genotypes/chemotypes (if known). The samples analysed should cover the full range of plant materials used for production of the extract (all plant parts, physiological ages, harvesting times, growing conditions, regions and genotypes/chemotypes). As a minimum, the analytical profile of batches must be based on five samples and cover the harvest from at least 2 years. The range of samples to be analysed should be agreed in a pre-submission meeting.

If plant material or extracts are stored inappropriately, hazardous microbial decay products may be formed during manufacture, e.g. mycotoxins. REBECA proposes that the description of the method of manufacture should include all measures taken to prevent the formation of hazardous microbial decay products during manufacture, according to HACCP procedures. The description should cover harvesting, drying, storage and transport of plant material, as well as manufacture and storage of the extract and the final plant protection product. If the formation of hazardous decay products or microbial contamination is expected to occur, analytical or microbiological data for these substances have to be provided.

### *14.6.5 Proposals Concerning Risk Assessment*

Botanicals differ from synthetic plant protection products in the following ways:

- They have always been present in natural environments, although not necessarily in the same quantities and purity.
- Some are rapidly and completely metabolised in the environment. For example, pyrethrins have an outdoor half-life of 2 h or less, and azadirachtin has a half-life of a few days or less (Isman 2006). Rapid and complete degradation into non-active degradation products is a very good argument to 'waive' residue and environmental fate and behaviour studies.
- Many plants and plant extracts have extensively been used for a long time, without evidence of adverse effects ('history of safe use'). Note: The legislation relating to medicinal products for human use establishes a simplified registration procedure ('traditional-use registration') for herbal medicinal products that fulfil certain criteria. The substance must have been in use for at least 30 years, and at least 15 years within the European Community.

REBECA proposes to establish a system in which botanicals of low risk/concern are identified early in the process, and are subject to reduced data requirements for risk assessment. This system should be part of the guidance document described above. The section on a guidance document explains how a list of candidate substances could be compiled and reviewed. Examples of candidate substances are given in Table 14.1.

REBECA proposes that the history of safe use of a botanical in plant protection or for other purposes shall be adequately taken into account. This includes the use of information from the literature and from other public sources. Details of a 'safe use' such as the concentration and level of exposure have to be considered. Bridging of information from similar extracts should be encouraged, but the relevance must be justified by the applicant in each case. Reasoned cases should be based on exposure, dose, natural background levels, and application pattern. Which data requirements can exactly be fulfilled by such data should be determined in a pre-submission meeting. The following provides some guidance how safe use could be considered in risk assessment:

- Safe use in *human nutrition* (e.g. lecithine, rapeseed oil) may provide justifications to replace some or all oral toxicity and residue studies.
- Safe use in *animal feeding* may provide justifications to replace some or all oral toxicity and residue studies.

- Safe use in *cosmetics* may provide justifications to replace some or all dermal irritation/sensitization and oral toxicity studies.
- Safe use as a *fertiliser* may provide justifications to replace some or all ecotoxicology and environmental fate studies, but this needs to be verified case-by-case.
- Safe use in *pharmacopoeia* (called 'traditional use' in this context) may provide justifications to replace studies on a case-by-case basis.
- Safe use as a *biocide* may provide justifications to replace studies on a case-by-case basis.
- Safe *technical use* (e.g. oils) may provide justifications to replace studies on a case-by-case basis.
- *Natural occurrence* (e.g. nettle) may provide justifications to replace some or all ecotoxicology and environmental fate studies, but this needs to be verified case-by-case.

### 14.6.6 Proposals Concerning Efficacy Evaluation

Botanicals may have a repellent and/or toxic mode of action on pests, and the use of botanicals can involve specialized techniques, which require adapted trial protocols (e.g. plot size, replicates). REBECA proposes that requirements for efficacy data should be flexible. Authorities should accept modified trial protocols, if the rationale for modification can be justified by the applicant. Selectivity tests should be included in efficacy tests. Recently, specific protocols have been agreed for regulatory purposes in France (AFPP 2010).

Botanicals may be less efficient and/or more variable in their performance than conventional chemical pesticides. REBECA proposes that even products with only minor beneficial effects should be acceptable, provided that a demonstrable and consistent benefit is achieved. The product label should accurately reflect the levels of performance that may be expected, as well as provide guidance on how to achieve these.

REBECA recommends that a possible introduction of efficacy evaluation at EU level should be accompanied by appropriate guidance on evaluation criteria.

## 14.7 Evaluation of the REBECA Proposals

During the final conference of the REBECA project, the botanicals working group has evaluated the proposals described above. On one hand, it was evaluated whether a proposal could be implemented easily (and therefore in a short time-span). This evaluation was mainly based on the expert judgement by the regulators. On the other hand, the potential impact of the proposals was estimated. The impact on the duration of the registration process and on the costs of registration (for the applicant) was assessed separately. This evaluation was mainly based on the expert judgement by the industry. In these assessments, it was only discriminated between small, intermediate and large effects, to reflect the uncertainty associated with these processes. Despite this, the comparison between different proposals gives a clear indication whether or not a proposal is easy to implement, and whether its potential impact is large or small.

**Table 14.2** Evaluation of the REBECA proposals

| REBECA proposal concerning | Implementation | Potential impact On duration | Potential impact On costs |
|---|---|---|---|
| Guidance document for botanicals | ** | ++ | ++ |
| Characterisation of active substance | *** | +++ | ++ |
| Identification of imputrities | *** | +++ | ++ |
| Description of plant material | *** | ++ | ++ |
| Description of manufacturing methods (HACCP) | *** | +? | + |
| Early identification of low risk/concern Substances | * | +++ | +++ |
| Risk assessment based on history of safe use | *** | ++? | ++ |
| Flexible trial protocols for efficacy evaluation | ** | ++ | + |
| Acceptability of minor beneficial effects | ** | ++ | + |
| Guidance for efficacy evaluation | *** | ? | ? |
| SELECTIVITY data | *** | + | ++ |

Proposals were evaluated with respect to their implementation and their potential impact. For details, see Section 14.7
*** Implementation of the proposal is easy and/or fast
** Intermediate estimate for implementation
* Implementation of the proposal is difficult and/or slow
+++ The proposal has a large impact (greatly reduces duration of the process or costs for the applicant)
++ The proposal has an intermediate impact
+ The proposal has a small impact (slightly reduces duration of the process/costs for the applicant)
? no estimate was made, or estimate is uncertain

Early identification of low risk/concern substances was evaluated as relatively difficult to implement (score in Table 14.2: *), but would have the greatest potential impact. All other proposals were evaluated as relatively easy to implement (score in Table 14.2: ** or ***). Among these, characterisation of the active substance and identification of impurities have the greatest potential effect (score in Table 14.2: +++ for duration and/or costs). The other proposals have a lower potential effect (score in Table 14.2: never above ++).

The working group has then selected several botanicals as case studies, and has estimated the percentage of the total registration costs (including costs from prolonged duration of the registration process) which could be saved by each REBECA proposal (Table 14.3). Fenugreek (*Trigonella foenum-graecum*) is a plant of the family Fabaceae, which is cultivated worldwide in semi-arid regions. Its leaves can be used as herbs, and its seeds are a frequent component of curries. Fenugreek has insecticidal properties (e.g. El-Lakwah et al. 1993). Neem extract is made from the seeds or other plant parts of the neem tree, *Azadirachta indica*. It is a widely used insecticide that interferes with insect molting hormones, and also has antifeedant properties (Isman 2006). Quassia is an extract of the bitterwood, *Quassia amara*. In plant protection, it can be used as an antifeedant. Both neem extract and quassia are subject to the 4th stage re-evaluation, and were pooled for this evaluation. Lecithine is a phospholipid extracted from soybeans. It is primarily used as a food additive (E322, e.g. emulsifier in chocolate) and is virtually non-toxic to humans.

**Table 14.3** Estimates of potential cost reductions for selected botanicals

| REBECA proposal concerning | Fenugreek | Neem extract/ Quassia | Lecithine | Laminarine |
|---|---|---|---|---|
| Guidance document for botanicals | 20% | 0% | 0% | 0% |
| Characterisation of active substance | (20%) | 20% | 0% | 10–20% |
| Identification of imputrities | (20%) | 20% | 0% | 10–20% |
| Description of plant material | 10% | 5% | 0% | 10% |
| Description of manufacturing methods (HACCP) | b | 5–10% | 0% | 15% |
| Early identification of low risk/concern substances | 20% | 0–50% | 20% | 0% |
| Risk assessment based on history of safe use | >50% | 0–50% | 10%[a] | 50% |
| Flexible trial protocols for efficacy evaluation | 0–5% | 5% | 5% | 0% |
| Acceptability of minor beneficial effects | 0–5% | 5% | 5% | 0% |
| Guidance for efficacy evaluation | 0% | 0% | 0% | 0% |
| Selectivity data | b | b | b | b |

For details, see Section 14.7. Not all proposals have an additive economic effect
[a]In the case of lecithine, 'waivers' were already granted in the risk assessment.
[b]Not evaluated.

Lecithine can also be used as a fungicide against Powdery Mildew in various crops. Lecithine was also subject to the 4th stage re-evaluation, but was not defended by the notifier for economic considerations. It is still on the European market, because its use was declared essential. Laminarine is a frequently occurring polysaccharide. It is extracted from kelp (brown algae), *Laminaria digitata*. Laminarine was the first botanical listed in Annex I of Dir. 91/414. It stimulates the resistance of crops against various diseases.

The assessment of these case studies indicates that the REBECA proposals have the potential to decrease registration costs substantially. It must be noted that not all proposals have an additive economic effect.

**Acknowledgements** We gratefully acknowledge the contribution from many colleagues who provided their expertise within the 'botanicals working group' of the REBECA project. The Specific Support to Policy Action REBECA ('Regulation of Biological Control Agents') was funded by the European Commission (SSPE-CT-2005-022709). For details see project website www.rebeca-net.de.

## References

AFPP (2010) Document Technique N°18 de la CEB : Expérimentation des préparations naturelles à activité directe ou indirecte contre les champignons phytopathogénes ou les insectes ravageurs. Association Française de Protection des Plantes. www.afpp.net

Anonymous (2001) Draft working document concerning data requirements for plant strengtheners with low risk profile, European Commission, Health and Consumer Protection Directorate-General. Sanco/1003/2000 – rev. 3

Anonymous (2004a) Draft working document concerning the data requirements for active substances of plant protection products made from plants or plant extracts, European Commission, Health and Consumer Protection Directorate-General. Sanco/10472/2003 – rev. 5

Anonymous (2004b) Review report for the active substance laminarin, European Commission, Health and Consumer Protection Directorate-General. Sanco/10488/04 – rev. 3

EC (2009) Regulation (EC) No 1107/2009 of the European Parliament and of the Council of 21 October 2009 concerning the placing of plant protection products on the market and repealing Council Directives 79/117/EEC and 91/414/EEC. Off J Eur Union L 309: 1–50

El-Lakwah FA, Khaled OM, Darwish AA (1993) Toxic effect of fenugreek (*Trigonella foenum-graecum* L.) and fennel (*Foeniculum vulgare* L.) extracts alone and in mixtures with two organophosphorus insecticides against *Rhizopertha dominica* (F.). Ann Agric Sci 31:605–615

Hauschild R, Speiser B, Tamm L, Ehlers RU (2008). Réglementation et homologation des produits phytopharmaceutiques à base d'extraits végétaux dans la législation de l'Union Européenne (UE): présent et perspectives d'évolution. Biopesticides d'origine végétale, 2e édition. C. Regnault-Roger, B. J. R. Philogène and C. Vincent. Paris, Lavoisier, pp 477–496

Isman MB (2006) Botanical insecticides, deterrents and repellents in modern agriculture and an increasingly regulated world. Annu Rev Entomol 51:45–66

Philogène BJR, Regnault-Roger C, Vincent C (2005). Botanicals: yesterday's and today's promises. Biopesticides of plant origin. C. Regnault-Roger, B. J. R. Philogène and C. Vincent. Paris, Lavoisier, pp 1–15

# Chapter 15
# Proposals for Regulation of Semiochemicals

**Bernhard Speiser, Lucius Tamm, and Susan Mattock**

**Abstract** Semiochemicals are substances that evoke behavioural or physiological responses. Pheromones modify the behaviour of other individuals of the same species, while allelochemicals act on individuals of different species. Pheromones are used in plant protection (i) to interfere with the mating behaviour of pests ('mating disruption'), (ii) to attract pests to insecticidal traps or baits ('attract and kill'), (iii) to trap pests ('mass trapping'), and (iv) to monitor pest populations. Semiochemicals present a particular case among active ingredients used in plant protection products, as they are the only substances not intended to kill the pest organism directly. The current regulatory system for pesticides is often seen as a major hurdle for the market introduction of new semiochemicals. The EU-funded Specific Support Action project 'REBECA' has held a series of workshops with stakeholder representatives. The following proposals for improvement were elaborated: (1) collective listing of the 'straight-chained lepidopteran pheromones' (SCLPs) in Annex I of directive 91/414/EEC; (2) treating SCLPs as 'low risk' substances under the new pesticides legislation; (3) relaxations concerning the identification of impurities; (4) more flexibility in the number of samples to be analysed; (5) facilitations in the risk assessment of SCLPs; (6) a procedure which will lead to facilitations in the risk assessment of other semiochemicals in the long term; (7) flexibility in efficacy evaluation; (8) harmonisation of registration requirements. During the final conference of the REBECA project, it was discussed whether the REBECA proposals can be implemented easily, and therefore in a short time-span. Also, the impact on the duration of the registration process and on the costs of registration (for the applicant) was assessed for each proposal. Among the proposals that were evaluated as relatively easy to implement, collective listing of SCLPs, relaxations concerning the identification of impurities and flexibility in efficacy evaluation have the greatest potential impact. All proposals which were evaluated as more difficult to implement have a great potential impact, but harmonisation of registration would reduce costs much more than any other proposal.

---

B. Speiser (✉)
Research Institute of Organic Agriculture (FiBL), 5070 Frick, Switzerland
e-mail: bernhard.speiser@fibl.org

R.-U. Ehlers (ed.), *Regulation of Biological Control Agents*,
DOI 10.1007/978-90-481-3664-3_15, © Springer Science+Business Media B.V. 2011

## Contents



## 15.1 Introduction

At the heart of this chapter are the proposals for improvement of the current regulatory situation with respect to semiochemicals. These proposals were developed in a series of workshops held in the framework of the REBECA project, leading to a final draft at the REBECA Final Conference on 20–21 September 2007. Representatives of all stakeholder groups showed a high degree of consensus with these proposals. The proposals on efficacy evaluation were further elaborated during the preparation of the manuscript.

The document is intended as a proposal to the Commission and EU Member States on how to facilitate the registration of plant protection products containing semiochemicals. Although we use the term 'registration', the intention is to cover all regulatory processes which affect the commercialisation of semiochemicals. This covers the procedure for Annex I inclusion of semiochemical active substances and the subsequent national registration of the plant protection products containing these semiochemicals. The REBECA project recommends that similar facilitations are considered in the registration process for semiochemical biocides.

These proposals represent the current state-of-the-art and can be improved by the insights and experience gained during the evaluation of those semiochemical active substances on the 4th list of the 91/414 review programme.

The proposals are aimed at the European process, but they consider also non-European systems. Based on extensive experience from registration, the USA has recently relaxed registration requirements for a group of semiochemicals. In the opinion of the REBECA project partners,[1] the EU should envisage similar relaxations, because potential risks are similar in the EU as in the USA. Over extensive and non-harmonized data requirements are a substantial economic hurdle for the manufacturers of semiochemicals, with very little benefit for the European market and the environment.

## 15.2 Semiochemicals in Plant Protection

'Semiochemicals' are chemicals emitted by plants, animals, and other organisms – and synthetic analogues of such substances – that evoke a behavioural or physiological response in individuals of the same or other species. They include pheromones and allelochemicals and have a high specificity for the target species.

'Pheromones' modify the behaviour of other individuals of the same species, while 'allelochemicals' act on individuals of different species (definitions taken from OECD (2001)). Pheromones are involved in various behaviours such as social communication, flight response, sexual activity and aggregation. Allelochemicals can be further defined depending on which organism benefits. 'Allomones' are used to the advantage of the producer e.g. plants producing feeding repellents. 'Kairomones' are produced by one species but are taken advantage of by another e.g. insects using plant volatiles to locate food sources. 'Synomones' are advantageous to both species e.g. floral scents that attract insects, allowing pollination.

Semiochemicals present a particular case among active ingredients used in plant protection products, as they are the only substances not intended to kill the pest organism directly. This has to be reflected both in trials design to demonstrate effectiveness, and also in the claims made on the product label.

Most semiochemicals currently used commercially in plant protection products are 'straight-chained lepidopteran pheromones' (SCLPs). Their natural function is as sex pheromones produced and released by female Lepidoptera to attract ('call') males for mating. Males are able to find females over relatively long distances, using the concentration gradient of the pheromone in the air. In plant protection, pheromones can be used in different ways:

- In the 'mating disruption' technique, the pheromone is artificially applied in excess, so that no gradient from a calling female can be built up. Therefore, males are no longer able to find females, resulting in unfertilized females and a reduction in offspring.
- In the 'attract and kill' strategy, pheromones attract the pest species to traps or baits impregnated with lethal doses of insecticides.
- In 'mass trapping', pests are attracted to sticky traps, which are distributed in large numbers with the aim of significantly reducing the pest population.

---

[1] REBECA aims to find proposals which are based on a broad consensus in the workshops, but does not claim that all participants agree with all proposals.

- In 'monitoring', pests are also attracted to sticky traps (often identical in design to those employed in mass trapping techniques). In this case, however, only a few traps are deployed. These are regularly inspected to determine the start and peak of pest flight activity. This information is used in conjunction with local treatment threshold levels in order to apply insecticides at optimum timings.

From a regulatory point of view within the EU, semiochemicals are considered to be active ingredients of plant protection products, when they are used for mating disruption or mass trapping. If they are used in formulated 'attract and kill' products, they are considered to be co-formulants, not active substances. If they are used for monitoring purposes, they do not need to be registered.

## 15.3 Environmental Impact and Human Health Risks of Semiochemicals

The OECD 12 consensus document on semiochemicals (OECD 2001) summarises environmental impact and human health risks of semiochemicals. This concludes that reduced data requirements, and the use of reasoned cases in lieu of data, are appropriate for semiochemicals. The major arguments are based around the fact that semiochemicals are target specific and generally effective at very low rates, and applied at levels comparable to those occurring naturally. They are generally volatile and usually dissipate rapidly in the environment. In addition, many end use products are formulated in passive dispensers (hollow fibres, tapes) that present little direct exposure to humans and non-target organisms. Furthermore, they are usually not directly applied to the crop. In this case, exposure is limited to localised areas where the dispensers are placed (note: exposure may be higher in the case of direct application). All these factors minimise the risk of adverse effects from the use of semiochemicals.

For SCLPs, the most frequently used semiochemicals, the following conclusions are drawn:

- SCLPs are of low toxicity to mammals.
- The application rate is typically low and probably comparable to natural emissions over the course of a season.
- Volatility and rapid environmental transformation minimise residues in crops and exposure of non-target organisms.

These findings have been supported by experience of the US EPA, resulting in a recent proposal for relaxed registration requirements for semiochemicals.[2]

[2]Federal Register, Part III, Environmental Protection Agency, 40 CFR Parts 158 and 172, Pesticides; Data Requirements for Biochemical and Microbial Pesticides; Proposed Rule, March 8, 2006

## 15.4 Fourth Stage Review of Semiochemicals in the EU

The 4th list of the review programme under Directive 91/414 contains a diverse range of products and uses, including other substances considered under the REBECA project. One of the principles of the fourth stage review is to employ a lighter regime which, whilst ensuring appropriate health and environmental safeguards, recognises the economic impact of generating significant amounts of data for what are often specialised niche products.

A number of semiochemicals are subject to evaluation under the 4th stage. The great majority of these semiochemicals are SCLPs, which make up a very homogenous group of substances. Not only single substances were notified, but also blends of substances. Most of the SCLPs were notified with the 'single evaluation dossier' prepared by the IBMA[3] task force. The semiochemicals contained in the single evaluation dossier are given in Table 15.1 as examples. The rapporteur Member State is Austria.

At the time of writing, the final decisions for inclusion or non-inclusion of active substances on Annex I for the 4th list were not yet taken. In order to speed up the process, the Commission has adopted an amending Regulation (EC 2007). This has new procedures for handling the remaining 3rd list and all of the 4th list substances. There will continue to be an evaluation and risk assessment by a 'rapporteur' Member State, presented in a draft assessment report (DAR). However, instead of all substances going through the lengthy, normal process of Annex I decision (called 'amber route'), provisions are made for several forms of 'short cut' to a decision on inclusion or non-inclusion. The 'short cuts' are made possible by the establishment of criteria for 'clear indications of harmful effects', and for 'clear indications of no harmful effects'. Active substance risk assessments will be considered against these criteria by the Commission, with the involvement of Member States and the European Food Safety Authority (EFSA). For those substances considered to meet the criteria for 'no harmful effects' (or 'green route'), the Commission will present a draft Directive for the inclusion of the substance in Annex I. After Annex I listing, EFSA will conduct a peer review to confirm the 'end points' which Member States will use when re-registering products nationally. As it was anticipated the SCLPs were able to meet the criteria for 'no harmful effects' ('green route').

## 15.5 Bottlenecks Under the Current System

The current regulatory system is often seen as a major hurdle for the market introduction of new semiochemicals. From an applicant's perspective, registration costs are not easily predictable and are often high in relation to the expected sales. Furthermore, the duration of the registration process is also difficult to predict (as can also apply to conventional chemicals). For an in-depth analysis, see Chapter 17.

---

[3]IBMA = International Biocontrol Manufacturer's Association

**Table 15.1** Semiochemicals notified in the 4th stage review with the 'single evaluation dossier' by IBMA

1.A SCLP acetates notified as single substances

- 5-Decen-1-yl acetate
- (E)-8-dodecenyl acetate
- (E/Z)-8-dodecenyl acetate
- (Z)-8-dodecenyl acetate
- (Z)-9-dodecenyl acetate
- (7E, 9Z)-dodecadien-1-yl acetate
- (E)-11-tetradecenyl acetate
- (Z)-9-tetradecenyl acetate
- (Z)-11-tetradecenyl acetate
- (9Z, 12E)-tetradecadien-1-yl acetate
- (Z)-11-hexadecen-1-yl acetate
- (Z)-13-hexadecen-11-ynyl acetate
- (7Z, 11E)-hexadecadien-1-yl acetate
- (2E,13Z)-octadecadien-1-yl acetate
- (Z,Z,Z,Z)-7,13,16,19-docosatetraen-1-yl isobutyrate

1.B SCLP alcohols notified as single substances

- 5-decenol
- (Z)-8-dodecenol
- (E, E)-8,10-dodecadien-1-ol
- 1-tetradecanol
- (Z)-11-hexadecen-1-ol

1.C SCLP aldehydes notified as single substances

- (Z)-7-tetradecenal
- (Z)-9-hexadecenal
- (Z)-11-hexadecenal
- (Z)-13-octadecenal

1.D Blends of SCLP acetates

- (Z)-8-dodecenyl acetate and dodecan-1-yl acetate[a]
- (Z)-9-dodecenyl acetate and dodecan-1-yl acetate[a]
- (7E, 9Z)-dodecadienyl acetate and (7E, 9E)-dodecadienyl acetate[a]
- (7Z, 11Z)-hexadecadien-1-yl acetate[a] and (7Z, 11E)-hexadecadien-1-yl acetate

1.E Blends of SCLP aldehydes

- (Z)-9-hexadecenal; (Z)-11-hexadecenal and (Z)-13-octadecenal

1.F Blends of different SCLPs

- 5-decen-1-yl acetate and 5-decen-1-ol
- (E/Z)-8-dodecenyl acetate and (Z)-8-dodecenol
- (Z)-11-hexadecenal and (Z)-11-hexadecen-1-yl acetate

1.G Other semiochemicals

- 1,4-Diaminobutane (putrescine)

The dossier contains SCLP acetates, alcohols and aldehydes (as single substances and as blends), and one substance which is not an SCLP

[a]Notified as components of blends

The REBECA project has identified specific bottlenecks and elaborated proposals for improvement. These are described below.

Semiochemicals present different regulatory challenges from the other substances studied in the REBECA project (botanicals, micro-organisms and invertebrate biocontrol agents). The major challenges for semiochemicals are:

- They are the only pesticides not intended to kill the target pests.
- They are used in very low quantities that are often comparable with natural levels
- They may not be applied directly onto crops (particularly pheromones);
- The use of semiochemicals often requires specialist techniques, which requires adapted efficacy trial protocols;
- Particularly for SCLPs, the risk assessment should reflect the low risk.

Accordingly, the REBECA proposals given below are specific for semiochemicals.

## 15.6 Proposals of the REBECA Project

The REBECA project has developed several proposals for improvement of the current situation for semiochemicals. These proposals are aimed at various regulatory processes which affect the commercialisation of semiochemicals: (i) evaluation for Annex I inclusion, (ii) the procedure of Annex I inclusion, (iii) national registration of plant protection products, (iv) similar processes for biocides and for products for human and veterinary medicine. An overview of the proposals in relation to the regulatory processes is given in Fig. 15.1.

The proposals described in this chapter are specific for semiochemicals and provide technical solutions for selected problems. In the semiochemicals working group, they were therefore called 'pragmatic' proposals. In addition, the working group has also discussed 'visionary' approaches, which are aimed at the current regulatory processes in general and would affect other natural substances as well. The visionary proposals are much more difficult to implement, and there was considerably more disagreement regarding their usefulness. Such proposals are described in Chapters 6 and 17 in this volume.

### *15.6.1 Proposal for Collective Listing of SCLPs in Annex I*

The REBECA project anticipates that the 4th stage of the review will demonstrate that all evaluated SCLPs have similar, very low environmental impact and human health risks, and can therefore be included in Annex I of Dir. 91/414. In this case, the REBECA project proposes to list 'straight-chained lepidopteran pheromones' (SCLPs) collectively. The proposal should be limited to SCLPs applied at <375 g/ha of active substance. Such application rates are low and probably comparable to natural emissions.

This proposal facilitates the registration of products based on SCLPs which are already listed, as well as new SCLPs, which could be registered according to the

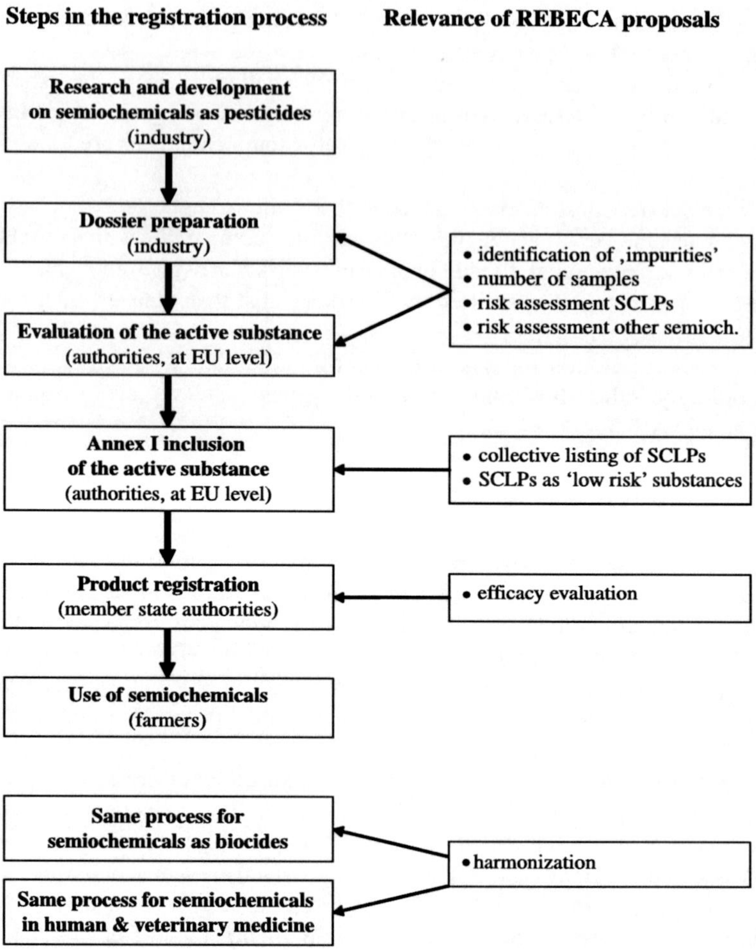


**Fig. 15.1** Schematic representation of the step in the registration process for which each REBECA proposal is relevant. The proposals are indicated by an abbreviation of their title in the text

rules for 'equivalence of technical material' (see Dir. 91/414). The main arguments for collective listing of all SCLPs are:

- Because of structural similarity, new SCLPs are likely to be of similarly low toxicity to mammals as SCLPs which are already registered.
- Volatility and rapid environmental transformation minimise residues in crops and exposure of non-target organisms.
- These proposals would harmonise EU regulation with the US EPA's proposed new rule.

In the discussions, some issues were raised which must be addressed before this proposal can be implemented: (i) data protection must be ensured; (ii) some regulators feared that this might set an unwanted precedent for certain groups of chemical substances such as pyrethroids, triazoles or sulfonylureas; (iii) under the new pesticides legislation, active substances have to be re-evaluated 10 years after Annex I inclusion (15 years for low risk substances). In the case of collective listing, it would need to be clarified how this provision is implemented. The approach of collective listing has recently been recommended for baculoviruses in an EU guidance document (Anonymous 2008a). Thus, it will be necessary to clarify the above mentioned critical issues for baculoviruses. This should facilitate collective listing of SCLPs.

Alternatively to the REBECA proposal, each substance could be listed separately. In comparison with the REBECA proposal, this possibility would increase the burden for registration of new SCLPs. In the case of individual listing, it must be determined how blends of SCLPs are listed: each component of the blend could be listed separately, or the blend could be listed as a whole. The listing of individual components would allow new blends of the same components to be registered more easily, while the second possibility would necessitate each new blend to undergo the whole procedure of Annex I-listing.

Meanwhile, SCLPs were included collectively in Annex I, with each specific substance evaluated under the review referenced (EC 2008). If new SCLPs are considered comparable to the already listed substances, they can be added to the review report and will not require separate Annex I listing (SANCO 2009).

### *15.6.2 Proposal to Treat SCLPs as 'Low Risk' Substances*

Since this chapter was written (June 2008), the EU 'Pesticides Directive' 91/414/EEC has been replaced by Regulation 1107/2009 (EC 2009). This Regulation contains some facilitation for 'low risk active substances' in Article 22 and 47. Low risk active substances shall be authorised for 15 years (as opposed to 10 years for other active substances), and plant protection products containing low risk active substances shall be authorised within a short period.

The REBECA project proposes that all SCLPs are treated as 'low risk' active substances, and are therefore subject to this facilitation. Clarity and predictability of low risk status would benefit all sides and would lead to cost and time reductions for all parties. Note: this proposal reflects the opinions of the participants in the semiochemicals working group, when the finalized definition of 'low risk' and the outcomes of the re-evaluation of the SCLPs were not yet known.

### *15.6.3 Proposals Concerning Identification of Impurities*

In the case of SCLPs, two types of impurities occur, (i) other SCLPs and (ii) unrelated contaminants. Other SCLPs are typically stereo-isomers of the active substance, or closely related molecules that differ from the active substance in the position/orientation of a double bond, alcohol, acetate or aldehyde group. These result mainly because a small proportion of the material does not undergo the synthetic pathway completely. There is no indication that these substances are of toxicological concern and, if present in low quantities, they do not adversely affect efficacy. In mass trapping and monitoring, the target insects need to be attracted precisely to a trap, while in mating disruption, it is sufficient to distract males from calling females. Therefore, higher purity may be needed in mass trapping and monitoring than in mating disruption. Because of their chemical similarity, removal of other SCLPs is neither economically feasible nor necessary. By contrast, unrelated contaminants may be of toxicological concern, and they may affect effectiveness. Manufacturers use 'HACCP' (Hazard Analysis and Critical Control Points) procedures to avoid the formation of other contaminants. Information on the manufacturing process is likely to be useful to identify the potential for the formation of substances of toxicological concern.

Currently, Directive 91/414 requires identification of each component present in quantities of 1 g/kg (=0.1%), and validated analytical methods for the detection of these impurities must be provided. It is not differentiated as to whether the components are SCLPs or unrelated contaminants. The REBECA project proposes that information on the manufacturing process shall be used to determine the likely identity of impurities, and that the threshold value of $\geq$1 g/kg is applied only to unrelated contaminants. For impurities which are SCLPs or structurally similar substances, validated analytical methods shall only be required if they are present in quantities $\geq$20 g/kg.

### 15.6.4 Proposal Concerning the Number of Samples to be Analyzed

Directive 91/414 requires analysis of 'representative samples' of the active substance. In practice, analyses from 5 batches are requested in the EU if feasible, while only 3 are requested in the USA, Canada and Switzerland (OECD 2001). Some pheromones are needed only in small quantities and are not produced every year. For these, multiple analyses are nothing other than pseudo-replicated analyses of the same sample. Therefore, the REBECA project proposes that for rarely produced pheromones (e.g. 1 batch/3 years), it should be acceptable to present results from fewer batches (down to 1), together with information on the manufacturing process. In this case, additional analyses must be provided as soon as additional batches have been manufactured, and this approach must be justified by the applicant. Comment: one regulator stated that this is already their current practice.

### 15.6.5 Proposal Concerning Risk Assessment of SCLPs

For SCLPs, the REBECA project proposes facilitations in the requirements for risk assessment in section 3 (human health), section 4 (residues), section 5 (fate and behaviour in soil, water, air) and section 6 (effects on non-target organisms) of the

OECD dossier format. The proposed facilitations depend on the quantities of SCLPs used:

- If SCLPs are used in quantities comparable to natural emission (up to 375 g/ha per year), no data shall be required for OECD sections 3, 4, 5, and 6. Note: Most currently registered uses of SCLPs fall within this category.
- If SCLPs are used in quantities higher than natural emission, or above 375 g/ha per year, data may be required for OECD sections 3, 4, 5, and 6 case-by-case.

These proposals are in line with the US EPA's new proposed rule for biopesticides. The rationale for these proposals can be found in the OECD 12 document (OECD 2001).

### *15.6.6 Proposal Concerning Risk Assessment of Other Semiochemicals*

SCLPs are sexual pheromones of the insect order Lepidoptera (butterflies and moths). They have been intensively studied and are widely used in plant protection. The SCLPs have very similar structure and function, and the currently available knowledge shows that they also have very similar safety profiles.

Other arthropod groups also have pheromones which are structurally very similar within one taxonomic group: the pheromones of beetles (coleoptera) are based on terpenoids, the pheromones of midges (a group of flies; diptera) are based on diacetoxy alkanes and the pheromones of pentatomides (a group of bugs; heteroptera) are based on alkene esters. The REBECA project envisages that many of these pheromones might have similar safety profiles as the SCLPs, and suggests the following approach:

- At the moment, these pheromones are not well known from a regulatory point of view, and have to be evaluated on a case-by-case basis.
- With increasing experience, these pheromones could be regulated in a more generic way, comparably to SCLPs. Such experience could be built up with a literature review carried out by the industry or academia, and/or it could be the result of accumulating experience from registration of individual substances. There would be considerable value in conducting reviews and creating such resources for both regulators and industry to refer to. This would make the use of reasoned cases in lieu of data much easier, particularly when addressing areas of environmental and toxicological risk assessments.
- Although the pheromones of beetles, midges and pentatomides are mentioned together in this paragraph, knowledge on these substances will not progress synchronously, and each group should progress individually from the stage where they are not well known to the stage where they are better known.

As long as information and knowledge of these pheromones remains limited, the following data requirements are proposed: For OECD sections 3 and 4, data requirements shall be determined case-by-case, taking into account natural emissions and the degree of target specifity, as well as the history of exposure to the substance. For OECD sections 5 and 6, no data shall be required, if application rates are comparable to natural emissions. If application rates are higher than natural emissions, data requirements shall be determined case-by-case, taking into account natural emissions.

With increasing experience, however, these pheromones could be regulated in a more generic way. When three to five products have been registered, a working group consisting of the Rapporteur Member States for the semiochemicals in question and the Commission should be established to compare the risk profiles of these substances. If these semiochemicals have been found to have a low risk profile (e.g. similar to SCLPs), no data shall be required for OECD sections 3, 4, 5, and 6. In this case, each new group of structurally similar semiochemicals shall be listed collectively in Annex I (as described above for SCLPs). As with SCLPs, this necessitates that the issue of data protection is resolved. Further, it shall be explored whether it is necessary and possible to make similar relaxations concerning the analysis of impurities as those proposed for SCLPs.

## *15.6.7 Proposals Concerning Efficacy Evaluation*

Semiochemicals acting through mating disruption do not directly reduce the population size of the treated generation, but rather affect the following generations, particularly when used over several seasons. Therefore, a comparison with chemicals for short-term reductions in population size is not appropriate. Furthermore, the use of semiochemicals can involve specialist techniques, which require adapted trial protocols.

The REBECA project proposes that authorities should not be too prescriptive concerning trial protocols for efficacy evaluation (OECD section 7). However, it is important that the rationale for the trial protocol is justified by the applicant (discussed further below). Requirements for efficacy data should be flexible and adapted to the special properties of semiochemicals.

It is recognised that conventional small plots with replicated treatments are not appropriate. Large areas are required for optimal performance, and also to avoid treatment interference. Consequently sites may have to be separated by some distance. It is essential to take into account and explain the pest biology, not only mating but also migration and egg laying behaviour. All these factors will influence the design of trials, including plot size, monitoring of populations and the number and placement of dispensers. The location of the trial sites and conditions in the surrounding area will also affect the design, including the need for any barrier treatments with conventional insecticides. Monitoring population levels during the season can be challenging. Conventional traps often may also use an identical sex pheromone, leading to interference from the mating disruption product with males unable to locate the monitoring traps. Other allelochemicals such as fruit extracts can be used as alternative lures to overcome this. Local and regional monitoring data during the trials season, and historical information on pest attack at particular sites, is very important in providing information on pest pressure and trying to demonstrate reductions in population levels.

Often the crops involved are high value, with only very low levels of damage tolerated. Evaluation and approval of semiochemicals is based largely on demonstrating a reduction in damage, and even small reductions may be economically beneficial and decrease rejection levels of the harvested produce. As there are no direct lethal effects, other assessments can also be very useful in demonstrating the impact on mating behaviour and reducing populations. For example, long term effects accumulating over several seasons could also be taken into account e.g. information on factors such as numbers of overwintering larvae. Assessments of reduced mating capacity using caged or tethered females can also be useful. Even if these effects are frequently observed, they are difficult to quantify. Products with some beneficial effects should be acceptable, provided that the effects can be demonstrated, and the label accurately reflects what benefits the grower may expect, and has clear directions on the optimum use of the product.

PSD (Pesticides Safety Directorate; now Chemicals Regulation Directorate of the HSE [Health and Safety Executive], UK) have produced an efficacy guideline on mating disruption products (Anonymous 2006a). This outlines both the key points for trials design, as well as how to address the 91/414 efficacy data requirements. It identifies those areas which require data and those which can be addressed by reasoned case. The guideline stresses that the quality of information provided on mode of action and pest biology, supported by any studies from preliminary work (e.g laboratory) can reduce the actual number of field trials required. In addition, this can form the basis of reasoned arguments to support other areas of the risk assessment, for example levels compared to natural ones, and the degree of species specificity. Several of the issues raised are also relevant to other types of semiochemicals. This guideline has formed the basis of a new EPPO guideline which was published in December 2008 (Anonymous 2008b). The development of further EPPO guidelines for individual species are also under discussion. It is also recommended to develop EPPO guidelines for evaluation of mass trapping.

In summary, there are practical difficulties in designing efficacy trials for semiochemicals, e.g. the need for large plot sizes and often impracticalities of having replicated treatments. To acknowledge these difficulties, and to avoid unnecessary repetition of efforts, the use of efficacy data from other areas should be actively encouraged (e.g. by mutual recognition). The REBECA project proposes that efficacy data from all areas of the EU or from outside the EU should be acceptable, if they have been generated under comparable conditions. The applicant should justify the comparability of such data with reasoned arguments based on issues such as pest biology, climatic conditions, number of generations and formulations. The arguments must take into account potential differences in agricultural practices, average field size, shape and size of trees/vines and label claims. In the new 'Pesticides Regulation' 1107/2009, the EC is divided into three zones, based on broad similarities of agriculture, plant health and environment. There are provisions for mutual recognition of registrations by other Member States belonging to the same zone. When this chapter was written (June 2008), these provisions were not yet in force, but they were expected to greatly facilitate registrations and reduce the burden of generating data.

One other practical difficulty identified was the need for official recognition of testing organisations conducting field trials, which is a requirement under the Uniform Principles of the 91/414 Directive. In the UK, a new category for biological products has been added as part of the official recognition process called 'biologicals and semiochemicals'. This encompasses micro-organisms, including viruses, as well as pheromones and other semiochemicals. Plant extracts are covered under the existing categories depending on intended use (e.g. agricultural/horticultural). The new category is designed both to identify appropriate contract organisations specialising in these products, as well as directly encouraging those working and researching in this area to apply for recognition themselves. Further information is given in PSD Regulatory Update 11/2006 'New Official Recognition Category – Biologicals and Semiochemicals' (Anonymous 2006c).

Pheromones applied in dispensers need to be applied to large areas, so the requirement for crop destruction would cause very high costs (ca 160,000 Euro per trial), which would effectively preclude doing a trial. The REBECA project proposes that for experiments with semiochemical products which are not directly applied to crops, there should be no requirement for crop destruction. If the products are applied directly to crops, it must be determined case-by-case whether crop destruction is needed.

### *15.6.8 Proposal Concerning Harmonization of Registration for Semiochemicals*

In the semiochemicals working group of the REBECA project, some applicants reported the experience that the registration of pheromones as biocides is far more expensive than the registration as plant protection products. This is due to higher registration fees and to lesser flexibility in the dossier requirements. For example, many SCLPs were submitted in a joint dossier for re-evaluation as plant protection products, while a joint dossier was not accepted in their re-evaluation as biocides. The REBECA project asks that the registration requirements for semiochemicals used particularly as biocides, but also for human and veterinary medicine, are harmonised with the registration requirements for use as plant protection products. In particular, joint dossiers should be permitted for the inclusion on Annex 1 and registration of SCLPs. Also, the registration fees should be lowered proportionately to the volume of work necessary.

**Table 15.2** Evaluation of the REBECA proposals

| REBECA proposal concerning | Implementation | Potential impact: On duration | Potential impact: On costs | Details |
|---|---|---|---|---|
| Collective listing of SCLPs in Annex I | ** | +++ | +++ | *Annex I*: duration reduced to minimal; costs substantially reduced. *National*: reduction depends on local procedures. |
| 'Low risk' status of SCLPs under new legislation | (***) | (++) | + | *Annex I*: no reduction. *National*: maximum 90/60 days (90 days for initial authorization per zone; 60 days for mutual recognition); cost reduction variable. |
| Identification of impurities | ***? | + | +++ | A few 1000 Euro per substance. |
| Number of samples to be analyzed | */**? | +++ | +++ | |
| Risk assessment of SCLPs | ? | +++ | +++ | Duration reduced ca 70%; costs ca 25–50% (several 10,000 Euro per SCLP) |
| Risk assessment of other semiochemicals | ? | +++ | +++ | Similar to SCLPs |
| Flexible trial protocols for efficacy evaluation | *** | + | +++ | |
| Acceptability of minor beneficial effects | ** | ++ | ++ | Variable, depending on country. |
| Acceptability of efficacy data from other areas | *** | +++ | +++ | |
| Harmonization of registration | * | +++ | +++ | Cost reduction ca 100,000 – 150,000 Euro per substance. |

Proposals were evaluated with respect to their implementation and their potential impact. For details, see Section 15.7

*** Implementation of the proposal is easy and/or fast

** Intermediate estimate for implementation

* Implementation of the proposal is difficult and/or slow

+++ The proposal has a large impact (greatly reduces duration of the process or costs for the applicant)

++ The proposal has an intermediate impact

+ The proposal has a small impact (slightly reduces duration of the process / costs for the applicant)

? no estimate was made, or estimate is uncertain

## 15.7 Evaluation of the REBECA Proposals

During the final conference of the REBECA project, the semiochemicals working group has evaluated the proposals described above. On one hand, it was evaluated whether a proposal could be implemented easily (and therefore in a short time-span). This evaluation was mainly based on the expert judgement by the regulators. On the other hand, the potential impact of the proposals was estimated. The impact on the duration of the registration process and on the costs of registration (for the applicant) was assessed separately. This evaluation was mainly based on the expert judgement by the industry. In these assessments, it was only discriminated between small, intermediate and large effects, to reflect the uncertainty associated with these processes. Despite this, the comparison between different proposals gives a clear indication whether or not a proposal is easy to implement, and whether its potential impact is large or small. Where possible, a quantitative or semi-quantitative estimate of the potential impact was made (see last column).

Several proposals were evaluated as relatively easy to implement (score in Table 15.2: ** or ***). Among these, collective listing of SCLPs, identification of impurities, flexible trial protocols and the acceptability of efficacy data from other areas have the greatest potential effect (score in Table 15.2: +++ for duration and/or costs). The other proposals have a lower potential effect (score in Table 15.2: never above ++): low risk status of SCLPs and the acceptability of minor beneficial effects.

Other proposals were evaluated as more difficult to implement (score in Table 15.2: * or ?). All of these were judged as having a great potential effect. These are the proposals concerning: the number of samples to be analysed, risk assessment of SCLPs, risk assessment of other semiochemicals and harmonisation of registration. Harmonisation of registration would reduce costs much more than any other proposal.

**Acknowledgements** We gratefully acknowledge the contribution from many colleagues who provided their expertise within the 'semiochemicals working group' of the REBECA project. The Specific Support to Policy Action REBECA ('Regulation of Biological Control Agents') was funded by the European Commission (SSPE-CT-2005-022709). For details see project website www.rebeca-net.de.

## References

Anonymous (2006a) Data requirements and trial design for mating disruption pheromone products, Pesticide Safety Directorate (UK). Efficacy Guideline 220

Anonymous (2006cs) UK PSD Regulatory Update 11/2006 'New Official Recognition Category – Biologicals and Semiochemicals', http://www.pesticides.gov.uk/biopesticides.asp?id=1849

Anonymous (2008a) Guidance Document on the assessment of new isolates of baculovirus species already included in Annex I of Council Directive 91/414/EEC, European Commission, Health and Consumer Protection Directorate-General. SANCO/0253/2008 rev. 2

Anonymous (2008b) EPPO Guideline 'Mating disruption pheromones' (PP 1/264). Bull OEPP/EPPO Bull 38(3):322–325

Anonymous (2009) Guidance Document on the assessment of new substances falling into the group of straight Chain Lepidopteran Pheromones (SCLPs) included in Annex I of Council

Directive 91/414/EEC. European Commission, Health and Consumers Directorate-General. SANCO/5272/2009 rev. 1

EC (2007) Commission Regulation (EC) No 1095/2007 of 20 September 2007 amending Regulation (EC) No 1490/2002 laying down further detailed rules for the implementation of the third stage of the programme of work referred to in Article 8(2) of Council Directive 91/414/EEC and Regulation (EC) No 2229/2004 laying down further detailed rules for the implementation of the fourth stage of the programme of work referred to in Article 8(2) of Council Directive 91/414/EEC. Off J Eur Commun L 246 (21.9.2007):19–28

EC (2008) Commission Directive 2008/127/EC amending Council Directive 91/414/EEC to include several active substances. Off J Eur Commun L 344 (20.12.2008):89–111

EC (2009) Regulation (EC) No 1107/2009 of the European Parliament and of the Council of 21 October 2009 concerning the placing of plant protection products on the market and repealing Council Directives 79/117/EEC and 91/414/EEC. Off J Eur Union L 309: 1–50

OECD (2001) Guidance for registration requirements for pheromones and other semiochemicals used for arthropod pest control. Paris, Environment Directorate. Organisation for Economic Co-operation and Development

# Chapter 16
# Regulation of Invertebrate Biological Control Agents in Europe: Recommendations for a Harmonised Approach

**Jeffrey S. Bale**

**Abstract** There have been few reported negative effects associated with the import and release of non-native invertebrate biological control agents (IBCAs), yet this practice is subject to stringent regulation in a number of countries including the USA, Canada, Australia and New Zealand. The import and release of IBCAs in Europe is not regulated by an EU Directive, as is the case for microorganisms and semiochemicals. As a consequence, some European countries have regulatory systems, others do not, and among countries with regulation, there is no consistency in the information requirements that biocontrol companies must produce when seeking a licence to release non-native species. Against this background, the REBECA Action: 1. Produced a standardised Application Form for the licenced release of non-native IBCAs in Europe, together with an accompanying Guidance Document. 2. Recommended adoption of a step-wise testing procedure for the environmental risk assessment of insect, mite and nematode agents, and summarised methods to assess establishment, host range and dispersal. 3. Endorsed the reactivation and updating of the EPPO 'Positive List'. 4. Proposed the establishment of an Expert Group to provide advice on the first release in Europe of non-native IBCAs. This chapter reviews the scientific and political dimensions underlying these recommendations and proposals, and provides an update toward the objective of a harmonised regulatory framework for non-native IBCAs in Europe.

## Contents



J.S. Bale (✉)
School of Biosciences, University of Birmingham, Birmingham B15 2TT, UK
e-mail: J.S.Bale@bham.ac.uk

R.-U. Ehlers (ed.), *Regulation of Biological Control Agents*,
DOI 10.1007/978-90-481-3664-3_16, 

## 16.1 Introduction

The regulatory systems for biological control agents that operate in different countries across the world could be seen as a paradox in modern science. As van Lenteren et al. (2006a) have reported, over the last 120 years there have been more than 5000 introductions of around 2000 non-native biocontrol agents for the control of arthropod pests, in 196 countries or islands with very few reports of any negative effects. And yet, 20 countries across the world have regulations on the release of biocontrol agents, and the move toward greater regulation may increase. The backdrop to the 'regulation debate' is complex, involving, in part, a lack of distinction between 'real risks' and 'perceived risks' – although any difference between these cannot be identified without some form of analysis, which therefore requires pre-release screening and post-release surveillance – and in some countries, a legislation-based prohibition or strict control on the import of non-native species.

Regulation in the United States, Canada, Australia and New Zealand is based on national legislation, as reviewed by Hunt et al. (2008 and this volume). The use of invertebrate biocontrol agents (IBCAs) in Europe is not regulated by any EU directive such as EU Council Directive 91/414/EEC that regulates the use of microorganisms, botanical substances and semiochemicals as plant protection products. As a result, there is a 'patchwork of regulation' of IBCAs across Europe, in which some countries have strict controls on the import of non-native species enshrined in national legislation, and other countries, sometimes directly neighbouring countries, have no restrictions on the import and release of so-called 'exotic species'. As insects used in biocontrol are sometimes highly mobile, it is possible, perhaps likely, that an organism will migrate from a country where it has been released without regulation to a different country where its import and release would have been prohibited, a problem exemplified by the predatory ladybird *Harmonia axyridis* (Brown et al. 2008 and Chapter 11 of this volume).

The absence of any EU-wide regulation of non-native IBCAs can be viewed as having both advantages and disadvantages. As an example, the absence of regulation has been cited as one of the main reasons for the success of IBCA-based biocontrol in Europe, and it is the case that there have been relatively few reports of any negative environmental effects arising from such unregulated releases. However, the recent rapid spread through Europe of the predatory ladybird *Harmonia axyridis* (Brown et al. 2008 and Chapter 11 this volume) and concerns about possible local declines in native coccinelid populations has raised awareness among regulators, the biocontrol industry and governmental and NGOs responsible for environmental protection of the need to ensure the safe release of non-native species. Additionally, the fact that countries with regulation have different 'information requirements' within their licence (permit) application forms means that companies have to produce separate dossiers for each country to which an application is made.

Prior to the REBECA project, various organizations (FAO, EPPO, OECD) had produced recommendations and guidelines on the environmental risk assessment (ERA) of non-native biocontrol agents (see Bigler et al. 2005 for details). The content of these documents was recently reviewed by the IOBC-WPRS 'Commission on the harmonization of invertebrate biological control agents' (CHIBCA), which produced an updated review 'Guidelines on Information Requirements for Import and Release of Invertebrate Biological Control Agents in European Countries' (Bigler et al. 2005). Most of the regulators, representatives of industry and scientists who had contributed to the CHIBCA review became participants in the REBECA project, thus providing a continuity of knowledge in this area.

## 16.2 Principles of a Balanced Regulatory System for IBCAs

The primary objective of the REBECA project with regard to IBCAs was to develop a balanced regulatory system, in which the concept of 'balance' reflected the need to minimise the costs on industry involved in the research and commercial production of new agents without compromising environmental safety. Implicit in this objective was the acknowledgement that European biocontrol companies are mainly SMEs with limited R&D budgets. Equally importantly, there was recognition across all stakeholders that although there had been very few reported negative effects of species released to date in Europe (with the exception of *H. axyridis* and a small number of other species), it was unrealistic to propose that the 'release without regulation' approach in countries such as France and Italy should become an EU-wide policy.

These initial conclusions gave rise to a set of guiding principles that became the specific objectives of the REBECA project with regard to IBCAs:

1. The application form (dossier) submitted to 'national competent authorities' should be standardised so that all countries use the same form and seek the same information. Also, the preparation of the dossier, particularly with regard

to the ERA, should be an interactive process between companies and regulators in order that unnecessary and costly work is avoided, and that any studies that are conducted will meet the requirements of the regulatory authority. A 'guidance document' should be produced to accompany the application form. The application form and supporting documentation (guidance on methods for ERA) should be comprehensive, covering inundative as well as classical biocontrol and be equally applicable to all currently used taxonomic groups of IBCAs (insects, mites and entomopathogenic nematodes – EPNs). In the context of this chapter, the term 'inundative biocontrol' is used synonomously with 'augmentative control', acknowledging that inundation is one form of augmentation (Bale et al. 2008; van Lenteren 2008). Whilst inundative control is currently the dominant method in Europe, the use of classical biocontrol may increase, hence the regulatory framework should be 'fit for purpose' for this development.

2. An ERA of novel agents is an essential component of the 'application to release' dossier that is submitted to regulatory authorities; but the ERA should have a hierarchical (step-wise) structure so that 'safe' or 'hazardous' species could be identified quickly, removing the need for unnecessary and expensive tests on organisms that have no prospect of being licensed for release. Also, dossiers could include on an optional basis, and where appropriate, information on the risks and benefits of the proposed release in comparison with alternative controls. Examples of relevant information might include a comparison with chemical control, or situations in which the target is a new exotic pest, and the alternative is a chemical that would undermine existing biocontrol schemes. Such information should be scientifically rigorous and evidence-based. Regulators are responsible for conducting an analysis of the environmental risk assessment (ERA) contained within the application dossier, which would include consideration of information provided on the wider risks and benefits of the release.
3. As some EU countries already had systems of regulation in place, and various, but different, requirements for ERA, it was considered desirable to validate the Application Form and recommendations for ERA with a series of 'case studies', using species with a history of use in Europe and for which extensive scientific information was available.
4. Against the backdrop of EU countries with and without regulation, the apparent absence in some countries of regulatory personnel with training and expertise in biocontrol and ERA, and a regular 'turnover' of people with relevant experience, an important role could be played by an 'EU Expert Group' that could review application dossiers. However, given the absence of an EU directive on IBCAs (and with no desire among stakeholders to introduce one), and thus the 'decision-making' regulatory power lying with individual countries, it was acknowledged that referral of dossiers to such an Expert Group would have to be on a voluntary basis, but that this 'service' would be valuable to countries with limited or no expertise in biocontrol (among the regulatory personnel), or for countries planning to introduce new regulatory systems.

This chapter now summarises the conclusions and recommendations for the implementation for these four components of the proposed regulatory system for Europe arising from the REBECA project.

## 16.3 Standardised Licence (Permit) Application Form and Guidance Document

The format of the application form that would constitute the dossier submitted for the import, shipment, rearing and release of an IBCA in a European country, with particular reference to non-native species is set out in Appendix 1. The guidance document that provides information on how to complete the application form is presented as Appendix 2. The key elements of the application form and accompanying guidance document are that the same form can be used for (i) both native and non-native species, (ii) import for 'research only' as well as commercial release, (iii) all biocontrol situations (inundative, classical, conservation control and weed biocontrol – though the latter was not considered in detail by REBECA), and (iv) all currently used taxonomic groups of agents, including EPNs.

The application form can be customised to comply with specific legislative requirements that may apply in some countries e.g. where it is necessary to seek a licence to release a native species, or if an import is for 'research only' (these requirements do not apply in some countries that have regulation). The documents can also give the contact details of the national regulatory authority, including any website. Whilst the production of a standardised application form reduces the administrative workload for industry, this benefit will only be realized if the form becomes widely accepted and used in the EU (see later in this chapter).

The application form and guidance document are structured in 5 sections (see Appendices 1 and 2). In general, the information required in sections 1, 2, 4 and 5 should be routinely available to any company applying for a licence: contact details and purpose of the release (section 1); taxonomy and origin of species, and nature and formulation of the commercial product (section 2); agreement on general safeguards (section 4); and any supporting documentation (section 5). The part of the form which is potentially problematic for industry is section 3, and particularly section 3B, which constitutes the ERA for the proposed release. The main difficulties likely to be encountered are two-fold: firstly, information on the different aspects of the ERA (overwintering, host range and dispersal) are often unavailable in the published literature, and therefore have to be generated from 'original research; and secondly, even if a company can afford the costs of such experimentation, there has been no consensus among regulators on the methods that should be used to acquire the data, or where waivers (exemptions) might be allowed. There have been a number of recent reviews recommending methods to be used in environmental risk assessment (see Bigler et al. 2006 and van Lenteren et al. (2006a). The REBECA project considered these approaches and produced guidelines on the most appropriate methods to be used for ERA data acquisition as summarized in Appendix 3.

One of main principles of a balanced regulatory system is that it should not place unnecessary costs on industry. This is in turn led to the recognition that the ERA should be 'hierarchical', a concept exemplified by the system proposed by van Lenteren et al. (2006a). This approach (described in Fig. 16.1) provides a good conceptual framework for ERA by including both native and non-native species as well as applications in classical, inundative (augmentative) and conservation biocontrol within the same general process, and thus maps on well to the structure of the standardised application form (Appendix 1). Most importantly, the system is predicated on the basis of a sequential assessment of factors affecting the 'risk' of a candidate biocontrol agent (establishment, host range and dispersal), enabling decisions (and their financial implications for industry) to be made at an early or late stage in the screening process as appropriate.

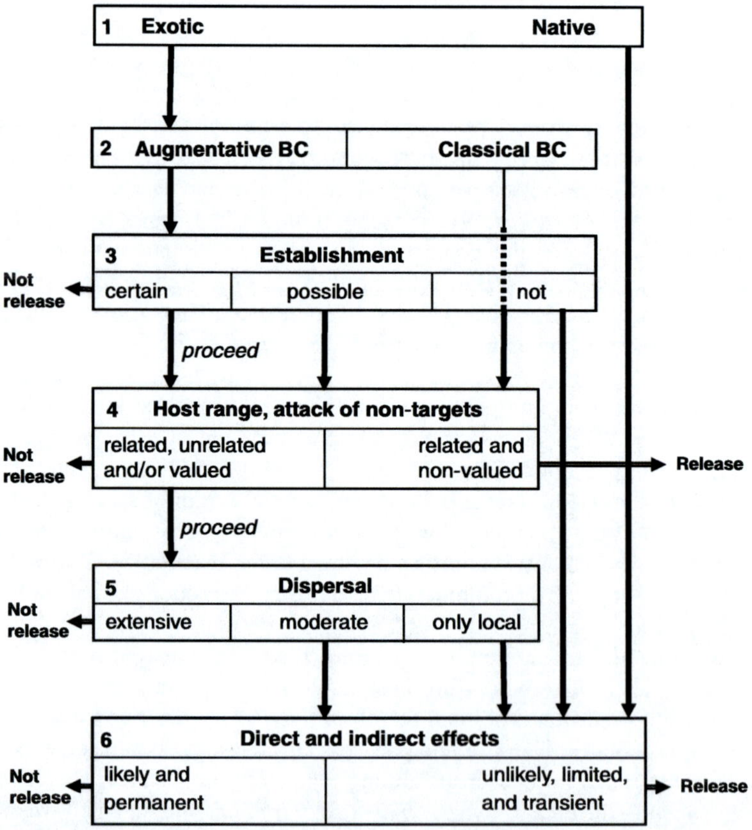


**Fig. 16.1** Flow chart summarising a hierarchical environmental risk assessment scheme for arthropod biocontrol agents (van Lenteren et al., 2006a; see text for additional guidance)

## 16.4 Role of ERA in a Regulatory System

The ERA summarised in Fig. 16.1 should not be seen in isolation from the wider perspective in which it will normally be used – as one part of the application (dossier) submitted for a permit (licence) to release a non-native species and the basis of the risk analysis of information contained within the dossier that will be conducted by the regulator. The ERA is therefore an integral and arguably the most important part of the dossier with the option to include information on the risks and benefits of the proposed release in comparison with alternative controls (see comments above).

If a species is 'Native', it is unlikely that there would be any major negative effects, thus the organism could be 'mass produced' commercially and released in larger numbers than would be normally present (see Fig. 16.1). For an 'Exotic' (non-native or alien) species, the order of testing from the perspective of an ERA would depend on its intended usage. For classical control, 'establishment' is an essential pre-requisite for success; hence, this would be assessed as a necessary requirement for efficacy, and the first ERA consideration would be the likely host range of the candidate agent (note the dotted line by-passing the 'Establishment' test for classical agents). For an exotic species under evaluation as an augmentative agent, if 'establishment' was assessed to be 'certain', the species would not normally be considered any further, unless it was monophagous on the target host or prey.

The information that is evaluated in each stage of the ERA, whilst ideally available from the published literature, may on some occasions require additional experiments to be carried out. Also, the 'order of tests' that make up the ERA is inseparable from the methods by which the tests are conducted (see Appendix 3 for details of recommended methods).

### *16.4.1 Order of Testing in ERA*

In most cases the ERA would follow the order of 'Establishment', 'Host Range' and 'Dispersal' and 'Direct and Indirect Effects' as set out in Fig. 16.1, although there are situations in which some of these assessments could be omitted, by-passed or conducted in a different order (see Sections 16.4.2 and 16.4.3).

### *16.4.2 Flexible Routes Through ERA System*

The order of testing described in Fig. 16.1 should be regarded as the 'default starting point' for an ERA, but there are clearly identifiable situations in which alternative routes through the testing system would be more logical. For example, as many of the non-native agents used in biocontrol originate from tropical, semi-tropical or Mediterranean climates, the climatic conditions experienced outdoors in northern Europe may be an effective natural barrier to permanent establishment. For releases in northern Europe it would therefore be logical to start the ERA with a 'test for establishment', and then consider the need for any host range testing depending on the results obtained. The establishment potential of a range of previously released and candidate non-native agents has been assessed by comparing their laboratory cold hardiness and outdoor winter survival, revealing a strong correlation between these indices (Fig. 16.2), such that in future it may be possible to obtain a rapid guide to likely establishment by laboratory assessment alone (Bale et al. 2009). Conversely, if the release areas are in southern Europe, the climate would most likely support year-round development and reproduction leading to establishment, in which case the ERA should focus on host range tests and data.

The principle of flexibility in the order ERA testing and the granting of waivers (exemptions) for some tests (see next section) highlights a further strong recommendation from the REBECA Action: the preparation of the application dossier and collation of ERA information should be an interactive process between companies and regulators so that unnecessary and costly work is avoided, and so that any studies that are conducted will meet the requirements of the regulatory authority.

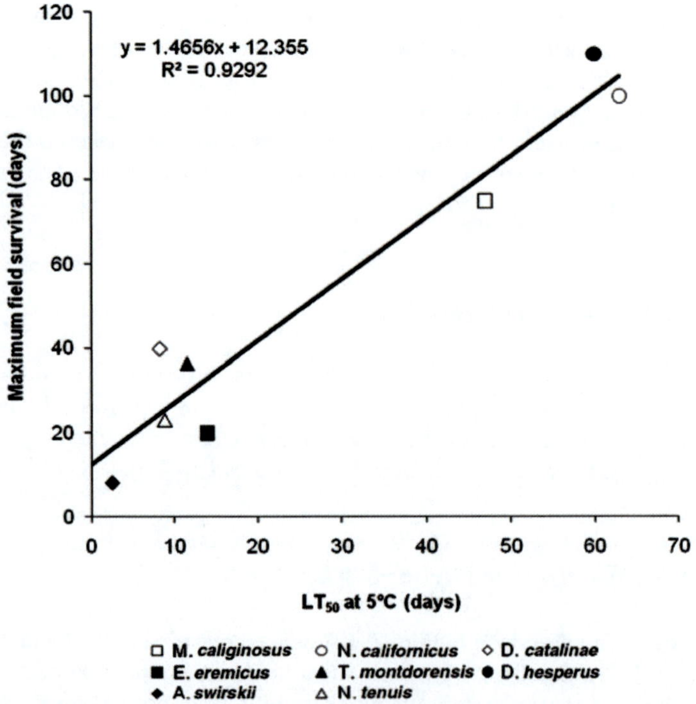


**Fig. 16.2** Relationship between the LTime$_{50}$ at 5°C (days) and maximum field survival in winter of eight biocontrol agents non-native to the UK (*Macrolophus caliginosus, Neoseiulus californicus, Delphastus catalinae, Eretmocerus eremicus, Typhlodromips montdorensis, Dicyphus hesperus, Amblyseius swirskii* and *Nesidiocoris tenuis*; Bale et al., 2009)

In general, the need to assess dispersal will be restricted to a limited number of candidate agents for inundative biocontrol. If it is clear that a species can establish in the release environment, it should be assumed that dispersal will occur – the unknown factors being 'how soon' and 'how far', and these are both difficulty to quantify on a 'pan-European' scale. However, there are circumstances in which dispersal may be limited (flightless species), and such information should be provided in a dossier. If no establishment is predicted, any effects on the wider environment will be transient and generally restricted to the 'summer season'. It is recommended that dispersal should not be assessed in species that are used exclusively in glasshouses where any escapes will involve low numbers of individuals that will have minimal impact on the neighbouring species and ecosystem before they die out. The impact of an 'open field' release where there is no prospect of survival through winter will depend on the numbers released and dispersal distances, and the proximity of the release area to sites of special scientific interest, such as nature reserves. Dispersal data are generally difficult to obtain but a description of methods by which to assess dispersal for inundatively released biological control agents is summarised in Appendix 3 and described in more detail in Bigler et al. (2006) and van Lenteren et al. (2006a). The REBECA Action also recommends that in the longer term, a database of information should be created from the literature and experimental studies to provide 'typical dispersal distances' for different taxonomic groups commonly used in biocontrol. Companies should have the discretion to provide information on atypical species with limited dispersal ability.

Direct and indirect effects are a summary of information gained from the available literature. When such information is not readily available, these effects may be estimated by 'expert knowledge' or generated from the data on establishment, host range and dispersal in the ERA. Examples of direct effects would include effects on non-target species and on other trophic levels (such as intraguild predation and plant feeding damage), hybridization and enrichment and vectoring (van Lenteren et al. 2003; Bigler et al. 2006). Indirect effects are those that occur when there is no direct interaction between the control agent and non-target species, such as competition and competitive displacement (see van Lenteren et al. 2003; Bigler et al. 2006). Indirect effects are difficult to quantify, but are likely to be related to the scale of the direct effects.

### 16.4.3 Waivers (Exemptions) and Taxon-Specific Issues

As part of the interactive discussions between industry and regulators during dossier preparation, there will be opportunities for industry to seek waivers (exemptions) from the need to provide or generate data on every aspect of the ERA. However, it is not possible to list any 'generalised waivers', as each dossier/species has to be considered on a case-specific basis and any 'application for exemption' evaluated in relation to the evidence provided. As an example, the format of the ERA summarised in Fig. 16.1 has been developed mainly from the perspective of insects and mites, but it would clearly be desirable to include entomopathogenic nematodes within the same system, allowing for the development of appropriate methods and modification to the order of testing as appropriate. In general, EPNs have very limited potential to cause non-target effects and, therefore, can be included within the same ERA framework that is applied to insects and mites but with the recommendation that data on establishment, dispersal, host range and indirect and direct effects would not normally be necessary because of the limited potential of EPNs to disperse or persist at the site of application. The remote risk related to the use of *Heterorhabditis indica* can be excluded by a precise identification of its associated symbiotic bacterium (see Appendix 3).

## 16.5 Validation of ERA and Licence (Permit) Application Process

The standardized licence (permit) application form (Appendix 1) is a comprehensive amalgamation of guidelines produced by various international organizations (see Bigler et al. 2005 for details) and incorporates the principles of the systems and guidance that currently applies in European countries that have regulation (e.g. The Netherlands, Switzerland, UK). Whilst wide acceptance of this form as the standard structure for a dossier would undoubtedly benefit industry, the guidelines also emphasise the areas in which provision of information is essential for the analysis of environmental risks, and where such information is not available through published literature, there is an expectation that data will be acquired by experimentation. A number species that had been released in several EU countries, and for which substantial data were available, were 'processed' though the template set out in Appendix 1, with a particular emphasis on the ERA, noting that in some cases important environmental data were acquired some years after the species had been released. This chapter presents brief summaries of the relevant data for some of these species.

The predatory mite *Neoseiulus californicus* is distributed throughout the world in both arid and humid areas of sub-tropical and temperate climates. The mite is not endemic in northern Europe, but has been released for inundative glasshouse biocontrol in many European countries (France, Belgium, The Netherlands, Germany, Italy, UK, Ireland, Denmark, Norway, Finland, Austria and Poland). In 2000 there was a record of outdoor establishment of *N. californicus* in several areas of the UK after it had earlier been released into greenhouses (Jolly 2000); subsequent overwintering field trials showed that non-diapausing adult females could survive outdoors for over 3 months in winter and oviposition occurred during this time (see Fig. 16.2; Hart et al. 2002a). However, the other important discovery with *N. californicus* was the existence of a diapause ability in some strains that seems to have been introduced into commercial production during the 'refreshing' of cultures. Diapausing females are more likely to overwinter than non-diapausing populations, and once this trait had been introduced into the commercial stock, even at a relatively low level, the shorter photoperiod and lower temperatures at the start of a northern European winter would exert a strong selection pressure on glasshouse escapees and, over time, this would result in a dominant diapausing population outdoors (Hart et al. 2002a). The comparatively high level of winter cold tolerance evident in even non-diapausing mites (Fig. 16.2) provides a retrospective ecophysiological explanation for the outdoor establishment of this species. *Neoseiulus californicus* feeds mainly on Tetranychid mites (e.g. *Tetranychus urticae*) and under laboratory conditions can develop and reproduce successfully on several mite species (Castagnoli and Simoni 2003). On other mite taxa, insects and pollen, feeding is common, but development is slower and often no reproduction occurs. This raises the interesting question of whether this 'invasive' species has had any negative effects on the native fauna of the UK but, as yet, this has not been investigated.

A similar analysis of cold hardiness and overwintering dynamics was applied to the predatory mirid *Macrolophus caliginosus* (Hart et al. 2002b) in the knowledge of its polyphagous nature and prey preferences later investigated by Hatherly et al. (2009). Nymphs of *M. caliginosus* were observed to develop slowly through winter (in the UK) with a low level of survival after 4 months when the experiment ended. Thus, *M. caliginosus* is less cold hardy than *N. californicus* (Fig. 16.2) and whilst its establishment potential could be described as 'marginal', it is also evident that in a progressively warmer climate it would be more likely to establish than other recently released species e.g. *Amblyseius swirskii* (see Fig. 16.2). Again, it is clear that if the ERA system set out in Fig. 16.1 had been applied to *M. caliginosus* prior to it commercial release, its possible outdoor establishment would have been detected.

The fragmented nature of regulation of non-native invertebrate biological control agents in Europe provides the explanation as to why the most notorious recent release – the Harlequin ladybird *Harmonia axyridis* – could have occurred (see Chapter 11 for a review). A retrospective ERA was conducted for *H. axyridis* according to the stepwise testing procedure described in Fig. 16.1 (van Lenteren et al. 2008), with the conclusion that it 'fails' every stage of the testing protocol. Thus, the documented establishment in Europe outside the initial release areas is related to its cold hardiness and diapause ability (Berkvens et al. 2010), polyphagous feeding behaviour (Koch et al. 2006) and strong powers of dispersal (Brown et al. 2008).

In general, it appears as though the ERA system summarised in Fig. 16.1, linked to the standardised licence application form (Appendix 1) and guidance on methods (Appendix 3) is robust and 'fit for purpose' and represents a major step forward toward the harmonisation of regulation in Europe – but only if those countries currently without regulation wish to integrate into what would have to be a non-binding pan-European system.

## 16.6 Implementation of a Pan-European Regulatory System

Regulation of IBCAs across Europe is currently fragmented with well organized systems in some EU countries, and no regulation in others. An analysis of the regulatory systems for IBCAs in Australia, New Zealand, Canada and the USA (Hunt et al. 2008 and Chapter 3 this volume) proposed that an Expert Panel should be established in Europe and placed under the jurisdiction of the EU or an 'EU governmental body'. Under this scheme, the Expert Panel would review 'first release applications' received by the EU administrative body and provide a recommendation on release approval. The final decision would then be made by representatives on the EU administrative body. The advantages of this approach are that all countries would be aware of which species had been authorised for release and where and there could be 'mutual recognition' between countries of safe releases. It would also ensure a harmonized process of IBCA regulation across all European member states. However, there are no indications at the present time that the relevant authority that could develop an EU-wide coordination of regulation of IBCAs via a new Directive (e.g. DG Environment) has any plans to introduce such a course of action; the establishment of some form of 'EU Agency' would require legislation and, therefore, may take 5–10 years to achieve. It is interesting to note that the concept of risk assessment of non-native biological control agents is broadly analogous to the 'pest risk analyses' (PRAs) that are reviewed by the Plant Health Panel of the European Food Safety Authority (EFSA), so there is a successful model already in operation that could be adopted for biocontrol. However, with regard to the immediate future, it seems likely that the two main features of the current regulatory environment in Europe will remain unchanged. Firstly, most activity will relate primarily to company dossiers seeking the first release in a European country of a non-native biocontrol agent, or in a second and subsequent country. Secondly, the 'regulatory power' and final decision on releases will be made by individual countries not some centralized body; thus, any coordination of regulation between member states would have to be achieved on a voluntary basis.

Given that the key role in the regulatory system is the scientific analysis of the ERA, a reputable international organisation could provide a viable alternative to a formal EU regulatory agency and this model was explored within the REBECA Action using the European and Mediterranean Plant Protection Organisation (EPPO) Panel on the so-called 'Positive List' of 'safe' biocontrol agents as a model. The EPPO 'List of biological control agents widely used in the EPPO region' ('EPPO Positive List') was first published in 2001 to facilitate decisions by national regulatory authorities on the import and release of IBCAs within EPPO countries, but had not been updated since its second publication in 2002. The concept underlying the Positive List is that because the listing of agents is based on the expert judgment of available information, other EPPO countries can conclude with some confidence that these agents can be introduced and used safely. However, it became apparent that the Positive List was in need of updating and that this would require the reactivation of the relevant EPPO Panel, possibly in collaboration with another body with expertise in biocontrol.

The REBECA Action endorsed the plan for the EPPO Panel to be reactivated in collaboration with the IOBC-WPRS (International Organisation for Biological and Integrated Control of Noxious Animals and Plants – Western Palaearctic Regional Service), and this was achieved in 2007. However, whilst updating the Positive List would have benefits for regulators and biocontrol practitioners, it was recognized that this would not overcome the absence of an EU-wide 'Expert Group' that could provide advice on request (on dossiers and particularly ERAs) to individual member states on the first release in EU countries of non-native species. Therefore, the REBECA Action made a pragmatic recommendation that the EPPO-IOBC Positive List panel should also take on the role of an Expert Group to fulfill this task because (i) it could be developed quickly and without the need for legislation, and (ii) whilst its advice was non-binding, if the aforementioned Dossier Application Form (Appendix 1), Guidance Document (Appendix 2) and 'ERA Methods' (Appendix 3) were published as 'EPPO Technical Reports', they would be distributed to all EU member states within the EPPO region, thus increasing the likelihood of adoption by many countries. In effect, this joint group or 'Panel' would have a dual function: (i) updating regularly the Positive List and (ii) providing 'non-binding' advice on the safety of new (first) releases not only to member states that have no current regulation or expertise in biocontrol, but also to countries with regulation that may wish to seek advice on dossiers from the Expert Group. A schematic description of how such an Expert Group with this dual function could operate is shown in Fig. 16.3, indicating its lines of communication with National Competent Authorities on new applications, and with EPPO with regard to the Positive List. At the present time, the first of these envisaged functions has been achieved: the proposed joint Panel between EPPO and IOBC-WPRS has between set up and updated the Positive List in 2008, 2009 and 2010 and will meet again in 2011 and thereafter at intervals deemed to be appropriate.

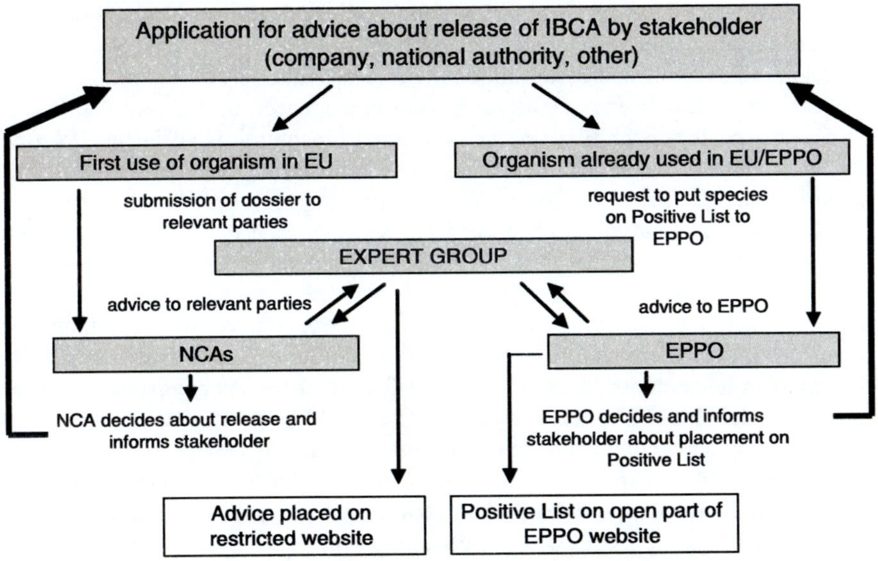


**Fig. 16.3** Schematic description of the role and lines of communication between an Expert Group, international organizations (EPPO) and national competent authorities (based on proposals by the IOBC-WPRS Commission on harmonization of invertebrate biological control agents – CHIBCA)

The issue of how to provide an advisory service on first releases in Europe remains to be resolved. It seems that provision of advice to individual countries would lie outside of the remit of EPPO, and there would be resource implications for any non-governmental organization, such as IOBC, that took on this responsibility. Consultation with EU member states and the biological control industry is now required to assess the likely demand for such a service and how it could be funded. However, it is a helpful step forward that the documents produced by REBECA on applications for the licensed release of non-native species (Appendices 1, 2 and 3) are now being reviewed by EPPO with the aim of including them in the 'EPPO standards', contributing to the harmonization of dossier requirements across Europe and the wider EPPO region and facilitating the exchange of risk assessment information between countries.

## 16.7 Wider Issues Concerning Implementation

An Expert Group coordinated by a reputable international organisation would bring an international dimension, reputation and longer term stability to 'biocontrol expertise' in Europe, acknowledging that final decisions on first releases will remain with individual countries. However, there are other actions that could be taken to achieve a greater consistency in regulatory practice across Europe. For example, countries with regulation and which receive applications for new species on a regular basis should be encouraged to adopt the standard 'Dossier Application form' (Appendix 1) for routine use and disseminate experience to other countries. The UK has already customized this form and uses it for all applications as from 2008; similar versions have been developed in the Netherlands and Switzerland and, subject to review, EPPO will distribute the templates across the EPPO region. The wide applicability of this licence application form is evidenced by the fact that it was used as the basis for the first release of a weed biocontrol agent in the UK (the psyllid *Aphalaria itadori* against the Japanese knotweed *Polygonum cuspidatum*), authorized in 2010.

At the present time the application form and guidance documents are written in English. It is possible that these documents would be used more widely if they were translated into other national languages. A further difficulty is that because of the patchy nature of IBCA regulation in Europe (unlike microbial agents under EU directive 91/414), there was no network and little or no contact between regulators dealing with invertebrate agents in different EU countries. As companies have the most up to date lists of 'national regulators', it would be possible to compile a list of relevant personnel that could be distributed to all member states. Related to this, a common problem experienced by industry is that the 'contact person' responsible for IBCA regulation in individual countries changes at regular intervals and the identity of 'new' people was not communicated to industry or to other countries. This problem would be overcome if each EU country set up a generic website (non-native licensing@), enabling all electronic documentation to reach its intended recipient and office, even if individual personnel had changed.

Finally, a further problem arising from the absence of a communication network between IBCA regulators in different EU countries is the lack of information available to a regulator on organisms for which licence applications had been previously submitted to other member states and the outcome of such submissions. For this reason, the Dossier application form requires a disclosure of this information, enabling a regulator to discuss an application with their counterparts in other countries.

## 16.8 Conclusions

Most herbivorous insects and mites are predated or parasitised by carnivorous species from the same two taxa such that the numbers of plant-feeding species are 'naturally controlled' over long periods of time. Biological control exploits these natural trophic relationships to suppress pest population densities below economically damaging levels. This 'natural' dimension also enables biocontrol to be portrayed as an 'environmentally-friendly' technology with many advantages over the use of pesticides (see Bale et al. 2008 for review). However, some types of biocontrol (classical and augmentative) differ from natural control (and conservation control) in one important respect – they often 'combine' predators with prey and parasitoids with hosts that would not normally encounter each other in their natural distributions. It is around this non-native dimension that the debate over the safety and the risks of biocontrol has become increasingly focused, with two opposing schools of thought. The first view adopts a version of the 'precautionary principle' in which introduced species are deemed to pose a risk to native fauna (and flora in the case of weed biocontrol) which, therefore, must be subject to an environmental risk assessment (ERA) before release. The alternative opinion argues that as a man-managed extension of natural control, there should be a 'qualified presumption of safety' for biological control, unless there is sufficient evidence to prove otherwise. Given that regulatory systems are now well established in North America, Australia and New Zealand (Hunt et al. 2008), it seems that the possibility of 'risks' currently holds sway in countries where biocontrol is widely used in pest management.

The situation in Europe is curious because despite international agreements such as the Convention on Biodiversity, there are still a number of countries (e.g. France and Italy) that have not found it necessary to regulate the import and release of non-native species, whilst others (e.g. Belgium) have recently announced plans to introduce regulation. The situation with France and Belgium is interesting as it was in these two 'non-regulating' countries that *Harmonia axyridis* was first released. The full environmental impact of this predatory ladybird is yet to be known or comprehensively assessed, but it would be hard to disagree that its commercialised release without any risk assessment has damaged the reputation of biocontrol and strengthened the case in favour of more comprehensive regulation in Europe.

Whenever an activity that has little or no regulation finds itself in the 'public spotlight' it has often been found prudent to introduce a 'voluntary code of conduct' as a safeguard against the introduction of a more stringent and bureaucratic system embedded in legislation – witness the problems of Directive 91/414 for microbial biocontrol – and that has been a key aim of the REBECA Action, although the relevant EU authorities confirmed at an early stage that there was no enthusiasm to regulate invertebrate agents. The impact of REBECA must therefore be judged on the basis of the contribution it has made toward its overall objective of 'harmonised regulation' across a continent in which the authority for policy still lies with individual member states.

REBECA benefitted from inheriting from the IOBC 'Commission on the Harmonisation of Invertebrate Biological Control Agents' (CHIBCA) an established network of scientists, regulators and industrialists. REBECA expanded this collaborative group to ensure that the interests of all relevant stakeholders were represented. It was also fortunate that REBECA had available a conceptual framework for environmental risk assessment (Fig. 16.1; van Lenteren 2006a) that laid a foundation around which the practical issues of implementation could be developed. The real merit of this ERA approach is that it encapsulates all biocontrol scenarios within a single system: native and non-native species and classical and augmentative control. Also, there are similarities between classical and weed biocontrol in that establishment is essential for both techniques and the main focus in the ERA is with regard to host range and direct and indirect effects. Whilst endorsing the idea of the 'step-wise' testing of candidate agents, REBECA has also made clear that there should be sensible flexibility in the order of testing and the appropriate use of waivers and exemptions, depending on the intended usage of the organism and relative risks associated with its taxonomy (e.g. entomopathogenic nematodes). A critical component of any risk assessment is the methods by which the data are acquired. Here again, REBECA has taken note of recent developments in this area (Bigler et al. 2006) and produced a succinct statement on recommended methodologies (Appendix 3), recognising that it is not desirable to be over-prescriptive for such a diverse range of potential agents, and also the need to balance the costs of data acquisition born by industry with an assurance of environmental safety.

One of the by-products of the fragmented regulatory system in Europe is that those countries with some form of regulation have developed their own approaches to ERA and associated documentation. As a consequence, companies have to compile separate dossiers for each EU member state that licenses the release of non-native species. In this regard, the production of a standardised 'application form' (Appendix 1) and guidance notes (Appendix 2) is a major step forward. These documents contain all of the relevant information that a 'risk assessing' authority would need to complete an ERA – it is in effect, a 'ready-made' system that could be introduced across Europe with immediate effect, if there was a desire to do so. Furthermore, REBECA has also validated this approach by applying the recommended ERA to a number of well-studied species (to which the assessment of *Harmonia axyridis* by van Lenteren et al. (2008) can be added).

If the intention of environmental risk assessment is to identify both 'safe' and 'risky' species, there is a logic in making known the species that have a high level of safety across designated geographic areas and based on documented evidence. This was the concept of the EPPO 'Positive List' first compiled in 2001, but which had not been revised since 2002. REBECA supported the discussions between EPPO and IOBC to form a joint panel to update the Positive List and this activity has now been placed on a firm footing, with meetings in 2008, 2009 and 2010.

Whilst REBECA has brought a greater clarity, coordination and understanding of the issues surrounding the regulation of invertebrate biocontrol agents in Europe, some of the other problems identified at the outset of the Action have not changed: some major EU countries still do not regulate the import and release of non-native species – there may be another '*Harmonia*-accident' waiting to happen; some countries have limited expertise to implement a regulatory system, even if they wished to do so; and there remains an independence in regulation at the level of individual countries. REBECA recommended the development of a non-bureaucratic, inexpensive and centrally organised system of regulation, with voluntary compliance by EU member states, to deliver the combined benefits of evidence-based advice by experts and the mutual recognition of such advice across different countries. REBECA has established a framework by which that objective can be achieved and it remains an important goal to pursue.

**Acknowledgements** JSB would like to thank the European Commission for funding the REBECA project. The contribution of all those who attended the 'macrobial' workshops and meetings is gratefully acknowledged. Special thanks are given to Anton Loomans, Franz Bigler and Guido Sterk for help with producing Appendices 1 and 2, to Franz Bigler, Joop van Lenteren and Nico van Opstal for comments on parts of this chapter, and to Ralf-Udo Ehlers for his leadership and commitment to the project.

## Appendix 1: Application Form for the Import, Shipment, Rearing and Release of Invertebrate Biological Control Agents in European Countries[1]

### *Using this Form*

This form should be used for the submission of an application to a National Competent Authority (NCA) of the European Union (EU) for a permit to license the import for research, mass-rearing and/or release of an invertebrate natural enemy used for the biological control of invertebrate and plant pests (Invertebrate Biological Control Agent or IBCA) and for other beneficial organisms. Organisms include invertebrates as well as entomopathogenic nematodes,[2] but not micro-organisms. Guidance on the completion of this form is provided in the accompanying Guidance Document[1]. This form is valid for an application relating to a single biological control organism. An organism is characterised as any identifiable and recognisable taxon of the IBCA, either a species, or recognised sub-species, population, strain or biotype.

After the NCA has received your application (administrative forms and documentation/dossier), you will receive an acknowlegdement of receipt within a specified period of time. The application will then be checked for completeness and subjected to a risk assessment in relation to the purpose of your application (e.g. for research under quarantine conditions, or a commercial release). The risk analysis will be conducted by the NCA or – upon its request – by a specified expert or group of experts. The NCA will conduct a risk analysis in the light of the information provided, or any other sources they have available. The NCA may need to contact you to clarify parts of the application or to seek further information. At all times and in all communication, including that with external experts, your application will be regarded as confidential. After the risk assessment has been completed, the NCA will make a decision as to whether to grant a permit within a previously agreed period of time. The licence to permit an import and/or release will be valid for a fixed period of time, assigned by the NCA, after which a renewal may be sought, or a request may be made to place the organism on the EPPO Positive List. In the case of mixed products, an application should be made for each separate component.

[1]Guidance on the completion of this Application form is provided in a separate document – Appendix 2.

[2]See REBECA WP 5 – Recommendations for regulation requirements for entomopathogenic nematodes

## *Information Required to Complete this Form*

This application form and related information requirements for the release of non-indigenous IBCAs contains 5 parts (numbered 1–5) and is structured in a step-wise way: depending on the origin of the organism and the purpose of the application, the sequence of assessments and level of information required is related to the perceived level of risk. An application for any specified organism should include the following information:

Part 1. Application information

A. *Information on the applicant*
B. *Purpose of the application and use*

Part 2. Information for indigenous and non-indigenous IBCAs

A. *Taxonomy and origin*
B. *Product information*

Where an application is made for the import for research and rearing of a non-indigenous species and/or release of a native IBCA, the applicant should proceed to sections 4 and 5 of the form. Where the application is for the release of a non-indigenous IBCA, section 3 of this form must be completed.

Part 3. Information requirements for intentional release of a non-indigenous IBCA with reference to:

A. *Biology and ecology*
B. *Assessment of risks and benefits*
   a. Establishment,
   b. Host specificity
   c. Dispersal
   d. Direct and indirect effects

Part 4. Submission of forms and Signature

A. *Submission details*
B. *Agreement: safeguards and signature*

Part 5. Appendices

## *Sections of the Form to be Completed*

This form can be used for the import and release of all IBCAs. Depending on the purposes of use, either some or all parts of the form must to be completed.

1. Renewal of a previous application Parts 1, 4 and 5
2. First application

- Organism on Positive List Parts 1, 2, 4 and 5
- Import only Parts 1, 2, 4 and 5
- Release of indigenous IBCAs Parts 1, 2, 4 and 5
- Release of non-indigenous IBCAs Parts 1, 2, 3, 4 and 5

*For more information*: Call... or refer to our website...... or consult the Guidance Document

## Part I. Application Information

| A Information on the applicant | | |
|---|---|---|
| *1.1 Who will apply for the permit?*<br>*only a legally authorized person is allowed to apply.<br><br>Include confirmation of the person's authorization and a copy of a valid idenfication card with the application. | **Name of organisation** | |
| | **Name of applicant*** | |
| | **Affiliation of applicant** | |
| | **Address** | |
| | **Post code** | |
| | **City** | |
| | **Phone** | |
| | **Fax** | |
| | **E-mail** | |
| | **Chamber of Commerce #** | |

| | | |
|---|---|---|
| *1.2. Who is the contact person?*<br><br>Contact person, research manager and/or quarantine officer. | Name of contact person | |
| | Affiliation of contact person | |
| | Visiting Address | |
| | Post code | |
| | City | |
| | Phone | |
| | Fax | |
| | E-mail | |

| B Purpose of application and use | | | |
|---|---|---|---|
| *1.3. Information on application* | Application type | Renewal ☐ | First Application ☐ |
| | Renewal (application number and expiry date) | | |
| | Positive List organism | Yes ☐ | No ☐ |
| | Relation with previous/ other applications | | |
| | Application or registration elsewhere in Europe | | |
| | Licence period requested | Mm/dd/year | |
| *1.4. Purpose of use*<br><br>*To include full scale release of a classical biocontrol agent | Import | Research ☐ | (Mass) rearing ☐ |
| | Release | Trials ☐ | Commercial* ☐ |
| | Type of biocontrol programme | | |
| | Area of release | | |

| | | |
|---|---|---|
| ***1.5. Facilities and procedures***<br><br>Describe how the risks, and the extent or probability of escape into the wild will be managed (for import/rearing of non-indigenous organisms only) | Address | |
| | Post Code | |
| | Location | |
| | Facility | |
| | Contingency plan | |
| | Standard Operating Procedures | |
| | Quality control management | |
| | Accreditation | |
| ***1.6. Information on target organism(s)***<br><br>Give a description of the biology and ecology of the target pest(s), including weeds | Target host taxon | |
| | Names of target pests | |
| | Original area of distribution of the pests | |
| | Biology of pests | |
| | Target crops | |

## Part 2. Information for Indigenous and Non-indigenous IBCAs

| A Taxonomy and origin | | |
|---|---|---|
| 2.1. *Identity*<br><br>For what species/organism is the application made? Indicate which species is involved (a single species per application) and full scientific name and taxonomy | Class | |
| | Order | |
| | Family | |
| | Genus | |
| | Species | |
| | Sub-species | |
| | Common names | |
| | Alternative names | |
| | Associated organisms | |
| *ID-Confirmation*<br><br>Indicate means, methods of ID-confirmation and vouchers. | Authority | |
| | Methodology | |
| | Voucher deposits | |

| | | | |
|---|---|---|---|
| *2.2. Characterization of IBCA*<br>Specify life-stages, strains or taxonomic constraints | **Diagnostic descriptions** | | |
| | **Specific characteristics** | | |
| | **Taxonomic characteristics** | | |
| *2.3. Origin and Distribution of IBCA*<br>What is the immediate source of the organism. Include details of the origin and distribution of the IBCA (species or lower taxon) | **Origin** | **Indigenous** ☐ | **Non-indigenous** ☐ |
| | **Field collected** | | |
| | **Laboratory culture** | | |
| | **Producer/Supplier** | | |
| | **Original area and distribution** | | |
| | **Areas introduced before** | | |

| B Product information | | |
|---|---|---|
| *2.4. Product Information* | **Product/Trade name** | |
| | **Producer/Supplier** | |
| | **Method of supply** | |
| | **Life stages** | |
| | **Label information** | |
| | **Storage** | |
| | **Method of use** | |
| *2.5. Product Composition* | **Co-formulants** | |
| | **Contaminants** | |

In the case of a renewal of a previously successful application (Section 1.3), or if the species or population is indigenous to the country or ecoregion, and/or imported for research or rearing only and/or is mentioned on the list of species considered safe for use in the intended area of release, no further information is required and only the submission details in 4A and B and Appendices (Part 5) need to be completed. For other applications, such as the release of a non-indigenous species, the information requirements in Part 3 must be supplied.

# Part 3. Information Requirements for Intentional Release of a Non-indigenous IBCA

| A Biology and ecology | | |
| --- | --- | --- |
| *3.1. Information on Biology and Ecology*<br><br>Give a description of the biology and ecology of the IBCA | Life cycle – generations/year | |
| | Developmental biology | |
| | Mechanisms of survival | |
| | Mechanisms of dispersal | |
| | Climatic conditions | |
| | Habitat range | |
| | Host range | |
| | Natural enemies | |

| B Assessment of risks and benefits | | |
|---|---|---|
| *3.2. Safety and Health Effects*<br>Potential hazards of IBCA, product or any co-formulants, and measures taken to limit operator exposure | Human health | |
| | Animal health | |
| | Measures of prevention | |
| *3.3. Information on Environmental Risk Assessment (ERA)*<br>All fields should normally be completed (but see exemptions listed below), but may be weighted differently in the evaluation of risks | History of previous releases or introductions | |
| | Outcome of previous risk assessments | |
| *3.3.1. Potential for establishment*[a] | Physical constraints | |
| | Resource constraints | |
| | Survival data and methods used | |
| | Evidence of establishment | |
| *3.3.2. Host range assessment*[b] | Wild hosts known | |
| | Organisms tested | |

| | | |
|---|---|---|
| | Procedures used for host range testing | |
| | Target and non-target host plants | |
| *3.3.3 Dispersal*[c] | Ability to disperse | |
| *3.3.4. Direct and/or indirect non-target effects*[d] | Summary of available information and conclusions on risks | |
| *3.4. Efficacy and benefits of the IBCA*<br><br>Assessment of efficacy, economic and environmental benefits | Method(s) to determine efficacy | |
| | Results of efficacy trials | |
| | Economic benefits | |
| | Environmental benefits | |

[a]When outdoor establishment of the IBCA is very unlikely and predicted to die out rapidly (as indicated by the data provided), the subsequent fields need not be completed, and no further risk assessments are necessary;
[b]When outdoor establishment of the IBCA is necessary or likely to occur, host range information is essential for the risk assessment;
[c]Dispersal test results are not required for glasshouse releases, but should be provided when IBCAs are released into open fields or structures that do not prevent escape (e.g. polytunnels) and long term establishment is very unlikely;
[d]A summary of known direct and indirect non-target effects should always be given, irrespective of whether host range and/or dispersal have been assessed.

## Part 4. Submission of Forms and Signature

| A submission details | | |
|---|---|---|
| *4.1. Appendices*<br>Check for completeness of application | **Information requirements** | ☐ |
| | **Literature reference copies** | ☐ |
| | **Identification of applicant** | ☐ |
| | **Chamber of Commerce** | ☐ |
| | **Authorization payment** | ☐ |
| *4.2. Where to submit the application* | **Name organisation** | |
| | **Bureau** | |
| | **Address** | |
| | **Post code** | |
| | **City** | |

B Agreement

## 4.3 General Safeguards

The applicant or authorized user undertaking the release proceeds under the conditions of the authorization for release, taking into account of the following requirements:

- All appropriate safety procedures should be put in place.
- Any relevant information on adverse effects, which might relate to the released IBCA, should be reported immediately to the National Competent Authority (NCA).
- Information on sites and dates of supply or release of the IBCA should be made available to the NCA, if requested.
- Information requirements have been supplied according to the most recent knowledge, and that the conditions made by the NCA will be respected.

| | | |
|---|---|---|
| *4.4. Signature** <br> *completed by a legally authorized person | Date | |
| | Applicant's name | |
| | Signature | |

# Part 5. Appendices

## Appendix 2: Guidelines for the Completion of an Application for the Import, Shipment, Rearing and Release of Invertebrate Biological Control Agents in European Countries[3]

### *Using this Guidance*

The purpose of this document is to provide guidance on how to complete the application form for a permit for the import (including labelling, packaging and storage in transit), mass-rearing and/or release of an Invertebrate Biological Control Agent (IBCA) and other beneficial organisms.[4] The application form and this accompanying guidance document are intended to cover all situations in which a permit (licence) is required: (1) for import and release, (2) for species and strains, (3) for different types of biological control programmes (augmentative, classical biocontrol, weeds) and includes, (4) product and efficacy information. The environmental risk assessment (ERA) and risk/benefit analysis will be based upon the information provided in the application form. It is therefore important that all required parts of the form are completed. It is also recommended that all EU countries should use the same application and guidance documents. The National Competent Authority (NCA) will conduct a risk analysis in the light of the information provided, or any other sources they have available. The dossier to be submitted to the NCA must include information on the organism[5] (IBCA) for import (including shipment), research, rearing and/or release as specified in the following parts of the application form:

Part 1. Information on the applicant (A) and purpose of the application and use (B)
Part 2. Information on the invertebrate biological control agent: identity, specific characteristics, origin and distribution (A), and product information (B)
Part 2. Information relating to intentional release of a non-indigenous IBCA: biology and ecology of the IBCA (A) and an assessment of risks and benefits of the release (B)
Part 4. Information on where to send the application (A) and conditions (B)

---

[3]These guidelines are largely based on Bigler et al., 2005 and redrafted during REBECA workshop discussions in 2005–2007.

[4]IPPC, 2005 – http://www.ippc.int/: Any organism directly or indirectly advantageous to plants or plant products, including biological control agents [ISPM No. 3, 2005, ISPM No. 5, 2007])

[5]Organism = any identifiable taxon of an IBCA; either a species, recognised sub-species, population, strain or biotype. Natural enemy = predator, parasitoid or EPN known to attack and develop on a certain host or prey and intended to be used for the biological control of certain plants, plant pests, stored products; IBCA = product of a certain specified natural enemy; non-indigenous = organism (taxon) orginated and collected ouitside the area of release. For other terminology, the IPPC definitions are used.

Part 5. Appendices

Parts 1–5 of this guidance document are divided into different sections and sub-sections. The title and number of each part, section and sub-section referred to in this document correspond with the same parts, sections and sub-sections of the application form. In the case of renewal of an application, Parts 1, 4 and 5 have to be completed. In the case of a first application, Parts 1, 2, 4 and 5 must be completed by all applicants, including applications for the release of indigenous species, when required by the NCA. For applications to release a non-indigenous species, Part 3 of the application form must also be completed.

*For more information*: Call... Or check our website......

## Information to Be Submitted by the Applicant

## Part 1. Application Information

| A Information on the applicant |
|---|

Provide information (including contact details) on:

1.1. *Who will apply* for the permit[6]; include confirmation of the person's authorization and a copy of a valid identification card with the application.
1.2. *The contact person*, research manager and/or quarantine officer.

| B Purpose of application and use |
|---|

1.3. *Information on the application*:

- Indicate whether this is a first application or a renewal of a previous application. In the case of a renewal, include a dossier reference number and expiry date and highlight any changes introduced since the first application.
- Is the organism on the EPPO 'Positive List of IBCAs'[7]?
- Has an application for this organism been submitted elsewhere in Europe, or has the organism or a product containing the organism been registered elsewhere in Europe? Specify in what country and contact details, when the application was submitted and the outcome.

[6]Only a legally authorized, registered person is allowed to apply.

[7]EPPO (2002). *List of biological control agents widely used in the EPPO region*. EPPO Standard PM6/3(2). *EPPO Bulletin* **32**: 447–461. See full REBECA WP 5 report.

- Is there a relation with other applications currently submitted or previously licensed with other IBCAs or beneficial organism(s) in the same product?
- For what period is the permit requested (within the range allowed by the relevant NCA)?

1.4. *Purpose of Use*:

Indicate the purpose of the application and use of the organism:

- Indicate whether the application is made for (i) import for research and/or (mass) rearing or (ii) direct release.[8] Indicate whether a release is intended in the country of application or not;
- When releases are intended, indicate whether the applications are for trial purposes or for full field releases, in commercial and/or classical programmes;
- Type of biological control programme[9]: classical biological control (CBC), augmentative (inundative) biological control (IBC), weed biocontrol;
- For direct release in field trials or for commercial release, indicate whether permanent establishment is intended (classical release) or not (augmentative release);
- Provide details of area of application (e.g. protected, semi-protected glasshouse, open field, natural environment).

1.5. *Facilities and Procedures*

The research/production facilities and procedures: describe how the risks, and the extent or probability of escape into the wild will be managed (*for import of non-indigenous organisms only*). This can usually be done by means of one or more waivers.

- Address (physical), postal code, location (city);
- For imported material, provide details of labelling, packaging and storage during transit;
- Facility: describe the types of facilities used (greenhouses, laboratories, climate rooms or cabinets);
- Levels of containment: do you have a permit to work with quarantine organisms under the provisions of Directive EC/95/44[10]? If not, justify why the levels of containment proposed for transport, rearing or research are appropriate to avoid escape and spread; where feasible, a contingency plan to prevent undesired environmental effects should be provided.

---

[8]Release: intentional liberation of an IBCA into an ecosystem [see ISPM No. 3, 1996].

[9]Eilenberg J. et al. (2001). Suggestions for unifying the terminology in biological control. *Biocontrol* **46**: 387–400.

[10]Commission Directive 95/44/EC of 26 July 1995 establishing the conditions under which certain harmful organisms, plants, plant products and other objects listed in Annexes I to V to Council Directive 77/93/EEC may be introduced into or moved within the Community or certain protected zones thereof, for trial or scientific purposes and for work on varietal selections: see http://eur-lex.europa.eu/LexUriServ/LexUriServ.do?uri=CELEX:31995L0044:EN:HTML

- Quality control management system: give a description of the measures, methods and intervals to ensure quality and purity of the IBCA (species/strain), and methods for periodic control of purity and identity of mass-rearing, including Standard Operating Procedures for:
  - Life stage and numbers (amount) to be imported;
  - Methods and materials to be used for shipping (e.g. sealed container, host mummies, prey to be included, plant material included, etc.);
  - Procedures to eliminate any contaminants of the imported agent that are of concern;
  - Procedures to dispose of used research materials, including shipping materials;
  - A plan for detecting escape and undesired environmental effects;
  - Any other procedures specific to this importation (i.e. not part of standard procedures).
- Accreditation: is your organization certified and/or accredited for processes and/or activities (ISOs) as developed by the International Organization for Standardization.[11] Relevant standards include ISO 9001 for 'Quality management' (general procedures) and ISO/IEC 17025 for 'General requirements for competence of test and calibration laboratories'. Provide details of the ISO standard(s) and activities for which you have certification and/or accreditation.

1.6. *Information on the target organism(s) and area of application*

- Name(s) of pest(s) to be controlled (order, family, genus, species and author), including weeds;
- Origin of the pest(s)/weeds and the natural occurrence in the area of release;
- Biology of pests: life cycle(s) of pests/weeds released against;
- Crops: damage inflicted on target crops or vegetation; crops or vegetation on which releases will be made.

## Part 2. Information for indigenous and non-indigenous IBCAs

| A Taxonomy and origin |
|---|

2.1. *Identity and ID Confirmation*

For what species/organism is the application made? Indicate which species is involved (a single species per application) and full scientific name and taxonomy. Give an accurate identification of the IBCA or, where necessary, sufficient characterization to allow its unambiguous recognition, such as

[11]For details, see http://www.iso.org/iso/home.htm

- Order, family, genus, species and author, and, where appropriate, sub-species, strain, or biotype; include common names and synonyms;
- Include the name of micro-organisms directly associated with the IBCA, e.g. identity of the symbiotic bacteria in entomopathogenic nematodes.

*ID confirmation*: Indicate means, methods of ID confirmation and vouchers:

- Authority: by which expert or institute has the organism been identified?
- By what method (morphological, molecular): if available, include a letter from a scientific expert, recognized by the NCA, stating the identity of the organism;
- Supply evidence of deposition of voucher specimens, with identity confirmed, in a recognized collection facility (these depositions must be made before the agent is released); include the name and location of institution(s) where voucher specimens are deposited;
- Where cultures are refreshed, confirmation of identity should be sought at regular intervals and additional vouchers should be deposited accordingly;
- Include the accurate identity of the symbiotic bacteria associated with entomopathogenic nematodes used as an IBCA.

### 2.2. *Characterization of IBCA*

Specify life-stages, strains or taxonomic constraints:

- General diagnostic descriptions of all life stages of the IBCA that are relevant for its use in biological control, highlighting details of any taxonomic characteristics and difficulties with the group (e.g. species complexes, cryptic species, poorly studied group);
- Describe specific characteristics of the species/strain(s) (where relevant), such as:
    - cold-hardiness (winter survival, diapausing abilities);
    - known pesticide resistance (if yes: what resistance);
    - information on differences from the parent wild strain.
- Where appropriate, molecular information (e.g. unique micro-satellite markers) used for diagnosis, especially for population identification, species complexes or cryptic species.

### 2.3. *Origin and Distribution*

What is the immeditate source of the organism. Include details of the origin and distribution of the IBCA (species or lower taxon) as follows:

a) Indicate whether indigenous or non-indigenous
b) If field collected, provide information on collection sites and dates, including:

- geographic area (approximate latitude, longitude and altitude of site);
- description of the original habitat(s) and host(s) from which the collection was made.

c) If from laboratory culture or production facility, provide information as indicated in (a) and in addition, the history of the culture stock, including:

- the immediate source of the organism (i.e. where it is produced), giving the name and address of the manufacturer, including the location of the production facility;
- any other source from which the culture has been collected or supplied;
- frequency and origin of additional wild stock used to refresh laboratory cultures.

d) Current distribution, including:

- Known areas of original natural distribution of the IBCA;
- Known areas where the IBCA has been intentionally or accidentally introduced

| B Product information |
|---|

2.4. *Product Information*

For augmentative (inundative) commercial release or classical biocontrol, briefly describe the intended use and potential benefits that may be derived.

- Function of the IBCA (e.g. predator, parasitoid);
- Life stage(s) of the agent(s) to be released (e.g. pupae, adults);

For augmentative (inundative) commercial releases, the following information should be supplied:

- Trade name of the product;
- Method of supply and formulation (e.g. single species, interim prey, mixed species);
- Label and container information;
- Storage conditions (temperature, humidity, expiry date);
- Recommended method of use (e.g. frequency and dosage of release).

2.5. *Product Composition*

Provide evidence that for inundative releases, the product is free from unwanted contaminants i.e. entomopathogens and hyperparasitoids, including:

- Co-formulants: give a description of co-formulants/organic contaminants included with the IBCA (e.g. plant material, live prey or other food materials, carrier material);
- Contaminants: give an assessment of the extent to which these should be of concern; frequency and percentage of hosts used in culture that might be present in the marketed product;
- Any combined or contaminant organism should be separately authorised before import and/or release.

In the case of a renewal of a previously successful application (Section 1.3), or if the species or population is indigenous to the country or ecoregion, and/or imported for research or rearing only, and/or mentioned on the list of species considered safe for use in the intended area of release, no further information is required and only the submission details in Part 4A and B and Appendices (Part 5) need to be completed. For other applications, such as the release of a non-native species, the information requirements in Part 3 must be supplied.

## Part 3. Information requirements for intentional release of a non-indigenous IBCA

| A Biology and ecology |
|---|

3.1. *Information on the biology and ecology (in current area of distribution)*

Information provided below will be the main basis for the environmental risk assessment. Give a description of the biology and ecology of the IBCA, including:

- *Life cycle* and number of generations per year;
- information on *developmental and reproductive biology* (e.g. sexual/asexual reproduction, feeding and parasitisation habits, developmental period, reproductive potential, longevity);
- known mechanisms of *survival* of extreme conditions (e.g. diapause, quiescence, migration);
- known mechanisms of *dispersal* (e.g. flight capability, migratory behaviour);
- describe the *climatic conditions* of areas where the IBCA is known to be native and/or where it has established following intentional or accidental introductions;
- give information on the habitat range, including the *habitat(s)* where the IBCA is known to be native and/or where the IBCA is known to have established following intentional or accidental introductions (e.g. pasture, forest, scrub, etc) and known factors determining habitat selection (e.g. oviposition behaviour);
- Give details of *natural enemies*, including pathogens known to attack the IBCA.

| B Assessment of risks and benefits |
|---|

Information presented in this section forms the basis for the ERA. The ERA should address the whole country within which releases will be made, with reference to regional variation that may affect risk where appropriate. Information required in this section is considered essential to an ERA, and can be acquired from published literature, company reports and/or experimentation. Include details of previous risk assessments for the same species (strain/biotype) with outcomes and other relevant information, including the country of application. The submission of available and/or generated data and subsequent assessment of environmental risks follows a tiered approach: information should be acquired and risks assessed according to the hierarchical system proposed by Van Lenteren et al. (2003, 2006), and further updated in REBECA Work Package 5. When establishment of the IBCA is very unlikely and the organisms released are predicted to die out, the subsequent fields need not be filled in, and no further risk assessments are necessary; when establishment of the IBCA is likely or necessary (e.g. in classical control), host range information is a crucial requirement for risk assessment; dispersal test results are needed when IBCAs are released in open fields and establishment is very unlikely; a summary of known direct and indirect non-target effects should always be given.

### 3.2. *Safety and Health Effects*

Summarize available information on hazards to human, animal and plant health (for example, allergy, skin irritation, disease vectoring etc) by the IBCA, product or any co-formulants and measures taken to limit operator exposure, where necessary.

### 3.3. *Information on Environmental Risk Assessment (ERA)*

All fields should normally be completed (but see exemptions listed below), but may be weighted differently in the evaluation of risks. Summarize the history of previous releases or introductions and the outcome of previous risk assessments, with known consequences, including non-target effects.

#### 3.3.1. *Potential for Establishment*

Indicate any evidence of establishment as a result of previous releases or accidental introductions outside Europe or other IOBC/WPRS countries. Describe conditions (including extremes) affecting the IBCA's survival and reproduction in its current distribution.

Information on *physical constraints*, such as:

- Climatic similarities/differences between area of current distribution and area of intended release (e.g. temperature, altitude, humidity, day length, etc.);
- Probability of temporary survival;
- Ability to survive and reproduce at temperatures and humidities outside the normal range (e.g. cold tolerance, overwintering ability); lower and upper temperature thresholds for development and survival; ability to enter diapause and/or overwinter (include test results);
- Other physiological and behavioural mechanisms for surviving extreme conditions;

- Dispersal potential (where known);

Information on resource constraints, such as:

- Availability and utilization of suitable hosts (target and non-target organisms) for short-term or long-term survival;
- Availability of suitable habitat, vegetation and plant food resources.

Indicate any evidence of establishment as a result of previous releases and/or accidental introductions outside Europe.

When outdoor establishment of the IBCA is very unlikely and the organisms released are predicted to die out rapidly, the subsequent fields need not be completed, and no further risk assessments will be necessary; when outdoor establishment of the IBCA is likely or necessary, host range information must be supplied.

#### 3.3.2. *Host Range Assessment*

When establishment is likely and/or required, provide available information on recorded effects on non-target organisms, including:

- A list of known hosts other than the target pest(s) and potential of the IBCA to utilize non-target host organisms living on wild or cultivated plants;
- A list of non-target organisms that have previously been tested, including unrelated non-target hosts, including pollinators, and threatened and endangered species; indicate hosts that were not accepted in such tests;
- Procedures used to determine host range (e.g. phylogenetic relatedness, experimentation) and methods used for host-range testing (e.g. experimental design, test conditions, rearing methods for non-target species, life-stages tested etc);
- Possible direct effects on plants: describe possible direct effects of the IBCA on the host plant(s) of the target pest and on plant hosts of non-target species.

#### 3.3.3. *Dispersal*

- Indicate potential direct (inundative) effects of mass-releases into open fields to neighbouring non-target hosts and habitats;

Direct effects of dispersal are considered for both indigenous and non-indigenous IBCAs where relevant to the direct environment of release. Dispersal test results are not required for glasshouse releases, but should be provided when IBCAs are released in open fields or structures that do not prevent escape (e.g. polytunnels) and long term establishment is very unlikely.

#### 3.3.4. *Additional Information on Direct and Indirect Non-target Effects*

Describe the history of previous releases or accidental introductions, with known consequences, including non-target effects. Indicate any other possible specific non-target effects, such as:

- Competition with, or displacement of, indigenous natural enemies in the area of intended release;
- Other constraints on the presence of natural enemies, including transfer of pathogens, of the released IBCA;
- Presence of natural enemies, including pathogens, that may affect establishment of the IBCA

A summary of known direct and indirect non-target effects should always be given, irrespective of whether host range and/or dispersal have been assessed. This section should also include conclusions on the risks associated with the intended release.

3.4. *Efficacy and Benefits of the IBCA and Proposed Release*

Provide relevant information on:

- Anticipated contribution to the control of the target pest(s) and weeds;
- Estimated economic benefits (crop specific) of the IBCA;
- Possible environmental benefits, e.g. beneficial effects of release of the IBCA compared with current control methods;
- Method(s) to determine efficacy and, when required by the NCA, results of efficacy trials.

## Part 4. Submission of Forms and Signature

| A Submission details |
|---|

4.1. *Appendices*

Check your application for completeness in the following areas:

- Information requirements (dossier)
- References, other literature and overview of information used in preparation of the dossier: include copies of relevant articles, chapters or reports in an appendix to the application documents;
- Identification of applicant: ID-card or passport;
- Chamber of Commerce copy;
- Authorization for payment of fees;
- Letter from a scientific expert, recognized by the NCA, confirming identity of the organism;
- Evidence of deposition of voucher specimens, with identity confirmed, in a recognized collection facility (these depositions must be made before the agent is released); include the name and location of institution(s) where voucher specimens are deposited

- In case of import for research and/or rearing, include a map of the facilities;
- Any other information that is relevant to the application.

4.2. *Where to Submit the Application*

Address details of the NCA

| B Agreement |
|---|

4.3. *General Safeguards*

The applicant or authorized user undertaking the release proceeds under the conditions of the authorization for release, taking into account the following requirements:

- All appropriate safety procedures should be put in place.
- Any relevant information on adverse effects which might relate to the released IBCA should be reported immediately to the NCA.
- Information on sites and dates of supply or release of the IBCA should be made available to the NCA if requested.
- Information requirements have been supplied according to the most recent knowledge, and that the conditions made by the NCA will be respected.

4.4. *Signature Details*

- Date
- Applicant's name
- Signature

All information and documents submitted for a licence application (dossier) will be regarded as 'commercial in confidence' by the NCA. The Environmental Risk Assessment and decision will be based on data and documents submitted for that specific licence application only.

# Part 5. Appendices

## Appendix 3: Summary of Methods to be Used in the Environmental Risk Assessment of Invertebrate Biological Control Agents

### *Introduction*

This appendix should be read with reference to Fig. 16.1 in Chapter 16 and other information presented in the chapter, particularly the order of testing in an environmental risk assessment, flexible routes through the ERA system, the granting of waivers (exemptions) and taxon-specific issues. In most cases the ERA would follow the order of 'Establishment', 'Host Range' and 'Dispersal', although there are situations in which some of these assessments could be omitted, by-passed or conducted in a different order (for examples, see text in chapter). This system and methods have been devised primarily for arthropod biological control agents, but the principles are applicable to entomopathogenic nematodes, as summarized in the Appendix.

### *Establishment*

1. Long term establishment of a non-native species has two main requirements: (i) ability to survive in the climate in the area/country of release, and (ii) access to a food resource – usually, 'wild prey' (where 'prey' is synonymous with 'host'), which could include established 'exotic' species. It is recommended that both of these requirements are assessed (though not necessarily 'tested'), as this may identify some species that are 'climatically suited' for establishment but unable to establish because of the absence of any acceptable wild prey.
2. In general, an ability to diapause increases the likelihood of winter survival, and in turn, longer term establishment. For this reason, ability to diapause should be investigated as a matter of routine in inundative biocontrol, especially where source populations are collected from different countries or different regions within countries. Information on diapause may be available in the literature or acquired by experimentation. Diapause induction stimuli vary between species but in most cases winter diapause can be induced by a 12:12 LD cycle at 15°C (and often by 12:12 LD at 20°C).

   *Key point*: Ability to diapause should be assessed as a matter of routine prior to other 'tests' for establishment for species intended for inundative release (glasshouse or field), where an inability to diapause would be a desirable feature. Diapause studies are less important for classical control where establishment is normally a requirement for success.

3. Climatic suitability (most often, overwintering ability) can be assessed by the system developed by Bale and co-workers (Hatherly et al., 2005) in which laboratory survival at 5°C is a reliable predictor of duration of field survival in winter, in northern European countries or regions with a winter climate similar to the UK. The system is now based on 8 (mainly predatory) species, which all 'conform' within a strongly correlated relationship. This approach enables species for inundative biocontrol to be categorised as 'safe' (die out within about 4 weeks of release), 'marginal' or 'likely to establish' (can survive for entire winter). Further analyses should be conducted to identify as far as possible 'survival time limits' for each category, and the extent of the ecoregion to which the data could be applied. Also, as most of the species examined so far are predators, further studies are required to assess the wider applicability of this system to parasitoids. It should also be noted that whereas as some predatory species have a cold hardy stage of the life cycle that can survive throughout winter, there are likely to be some parasitoid species that are not so long-lived, but are sufficiently cold hardy to complete one or more generations in winter and thus have the same or greater establishment potential. As those species which 'die out quickly' are usually unable to survive below their developmental threshold (often in the range of 8–10°C for species of tropical origin), the developmental threshold might be an additional predictor of establishment. This could be investigated as the data are usually available in the literature, but some caution is required as reported thresholds may vary depending on regional variations in different source populations, stage specific differences, and differences related to different prey-host plant combinations used in experiments and in commercial production.

   *Key point*: When experimental data on establishment are required, it is proposed that a laboratory assessment of survival at 5°C is appropriate to predict field survival, particularly for weakly cold tolerant species. Companies could usefully indicate to which countries or ecoregions such data would apply.

4. The species that pose interpretational difficulties are those in the 'marginal' zone that can survive for 1–2 months but not entire winters, and cannot reproduce in winter. For such species, it would be relatively easy to assess their acute lethal temperature and compare this with regularly occurring minimum temperatures in areas of intended release. However, as the effect of cold stress is determined by both the temperature and the duration of exposure, the reliability of this 'quick test' for 'marginal' species requires further evaluation.

   Any 'climate survival' test should include different life cycle stages (unless there is a known overwintering stage), with and without an acclimation treatment, and where appropriate, with access to a food (host or prey) resource.

   *Key point*: For species that are predicted (or shown) to die out after brief periods of winter low temperatures, no further risk assessments are necessary, other than a consideration of direct and indirect effects, as for a native species.

5. The second requirement for establishment is availability of one or more species of wild prey (which, depending on the climate, may be target or non-target species). The ability of the candidate agent should be assessed on one, or a small

number, of commonly available wild prey that are phylogenetically related to the target species. With the benefit of experience it may be possible to produce a 'recommended list', but as an example, if the glasshouse target was a species of whitefly, then the cabbage whitefly *Aleyrodes proletella* would be an appropriate wild prey. For most non-native biocontrol agents there are likely to be suitable wild prey, but if a candidate species did not feed on one or more close relatives of the target, this might be an indication of host specificity, and would therefore be valuable information in the overall risk assessment. In this part of the establishment assay, the response of the control agent should be recorded in terms of attack (attempt to feed or oviposit), death of the prey, and ability of the agent to develop on the wild prey and produce reproductively viable adults.

## *Host Range*

6. The second aspect of risk assessment of inundatively released agents is host range, but this would be the first area of investigation for a classical control agent (see Introduction to this appendix). There have been a number of studies and recommendations on host range testing. It is recommended that the testing scheme for arthropod biocontrol agents proposed by van Lenteren (2006b) should be adopted and non-target species for host specificity testing selected as described by Kuhlmann et al. (2006). Testing schemes for weed biocontrol have been reviewed by Sheppard et al. (2005), and the selection of non-target species follows recommendations made by Wapshere (1974).

   *Key point*: Species selected as 'test' prey and hosts are used to obtain an indication of the likely host range, not a precise list of non-target species that are accepted or rejected. For this reason, the selected list should be representative of different taxonomic groups rather than a particular country. Ideally, the same list, or at least a similar one, should applicable across Europe.

7. Host range testing can be an expensive exercise, beyond the financial limitations of even the largest companies. For this reason, it is proposed that host range testing should be conducted in two stages. It is unwise to be prescriptive about the exact number of species to be used in each stage, but typically this could be 3 species in stage 1 and a further 6 species in stage 2 of an arthropod biocontrol programme. This would allow companies to decide at stage 1 whether to continue with further host range testing.

   *Key point*: The identity and number of species to be included in host range tests should ideally be discussed with experts and agreed with the regulator prior to any experimentation.

   Stage 1 assessment should include a phylogenetically close relative of the target prey or host (such as the species used in the establishment assay above), a second close relative, and a third species that is taxonomically distinct but commonly available outdoors, including during winter when appropriate to the seasonal biology of the agent. Data recorded should be attack, death of prey or host and development to adult as with the establishment assay.

8. Where the Stage 1 test indicates some level of specificity (e.g. only the phylogenetically related species are accepted as prey or hosts), testing should proceed to stage 2. For arthropod biocontrol, the system proposed by Kuhlmann et al. (2006) is recommended in which non-target species are selected from three categories: 1. Phylogenetically related; 2. Occurs in the same ecological niche; 3. Unrelated 'safeguard' species.
9. A number of studies have compared the physiological ('apparent') host range of some parasitoids, and the 'ecological' host range that is observed in nature. Invariably, laboratory assessments in which hosts are offered to natural enemies in 'no choice' tests overestimates the natural host range. The stepwise procedure proposed by Van Lenteren et al. (2006b) is recommended as the method that should be used for arthropod biocontrol to make an estimation of the range of non-target species attacked under field conditions.

   If the host/prey is accepted in the first two steps (conducted in small arenas), the step 3 test should be carried out in contained environments such as large cages, in which prey or hosts feed on growing plants and the agent is able to move freely around the cage. It is recommended that three treatments are compared with appropriate replication: 1. Target species alone (control); 2. Non-target alone; 3. Target and non-target together.

   *Key point*: If acceptance of non-target hosts is observed in no-choice tests, a further test needs to include direct comparison of the acceptance and development on non-target species when the target species is simultaneously available.

## Dispersal

10. In general, the need to assess dispersal will be restricted to a limited number of candidate agents for inundative biocontrol. If it is clear that a species can establish in the release environment, it should be assumed that dispersal will occur – the unknown factors being 'how soon' and 'how far', and these are both difficulty to quantify on a 'pan-European' scale. However, there are circumstances in which dispersal may be limited (flightless species), and such information should be provided in a dossier.

    *Key point*: Dispersal should not be tested in species that can establish in the release environment.
11. If no establishment is predicted, any effects on the wider environment will be transient and generally restricted to the 'summer season'. It is recommended that dispersal should not be assessed in species that are used exclusively in glasshouses where any escapes will involve low numbers of individuals that will have minimal impact on the neighbouring species and ecosystem before they die out.

*Key point*: Dispersal should be tested only when agents are released into open fields or structures that do not restrict escape.

12. The impact of an 'open field' release where there is no prospect of survival through winter will depend on the numbers released and dispersal distances, and the proximity of the release area to sites of special scientific interest, such as nature reserves. Dispersal data are generally difficult to obtain but a description of methods by which to assess dispersal for inundatively released biological control agents is provided by Mills et al. (2006). It is also recommended that a database of information should be created from the literature and experimental studies to provide 'typical dispersal distances' for different taxonomic groups commonly used in biocontrol. Companies should have the discretion to provide information on atypical species with limited dispersal ability.

## *Direct and Indirect Effects*

13. Direct and indirect effects are a summary of information gained from the available literature. When such information is not readily available, these effects may be estimated by 'expert knowledge' or generated from the data on establishment, host range and dispersal in the ERA. Examples of direct effects would include effects on non-target species and on other trophic levels (such as intraguild predation and plant feeding damage), hybridization and enrichment and vectoring (van Lenteren et al. 2003; Bigler et al. 2006). Indirect effects are those that occur when there is no direct interaction between the control agent and non-target species, such as competition and competitive displacement (see van Lenteren et al. 2003; Bigler et al. 2006). Indirect effects are difficult to quantify, but are likely to be related to the scale of the direct effects.
14. In situations where winter survival of the candidate agent for inundative biocontrol has been demonstrated in the establishment experiments (or seems likely to occur) and where the species is known or shown to be polyphagous, a company may decide that further investment in host range or other forms of testing would not be cost effective, as the dossier may not lead to a successful licence application. In such situations, a company could prepare a dossier describing a 'worst case scenario' that might arise from a release and provide relevant information for a 'risk-benefit' analysis compared with other available methods of control. In effect, although the biological control agent may pose some risk, this may be less than for other control options. A problem with this approach is that there may be difficulties in obtaining reliable comparative data for the alternative method(s) of control. However, there are examples of previously released species that have survived in the northern European climate, and are known to be polyphagous, but as yet, have not had any detectable impact on native species or ecosystems. A risk assessment for such species evaluated under current regulatory guidelines would almost certainly lead to a 'licence rejection' when

considered in isolation, but the species might be the best option in comparative terms.

*Key point*: For polyphagous agents with establishment potential, companies should have the option to submit a dossier containing information on the risks and benefits of the proposed release compared with other possible controls. This information would be evaluated by the regulator as part of the ERA.

15. Direct and indirect effects of classical biological agents should be addressed in pre-release studies, because establishment of such species is essentially irreversible. Additionally, negative direct effects of classical biocontrol agents on non-target prey or hosts have become a major issue in this method of control.

## *Nematodes*

16. The proposed ERA should include entomopathogenic nematodes (EPNs), allowing for the development of appropriate methods and modification to the order of testing as appropriate. On the basis of available information EPNs (i) have very limited potential to cause non-target effects, and (ii) should be included within the same ERA framework that is applied to insects and mites, but with the recommendation that data on establishment, dispersal, host range and indirect and direct effects would not normally be necessary because of the limited potential of EPNs to disperse or persist at the site of application. The remote risk related to the use of *Heterorhabditis indica* can be excluded by a precise identification of its associated symbiotic bacterium.

## *Related Issues*:

### Efficacy Trials

17. It is likely that companies will want to carry out efficacy trials and ERA experiments simultaneously to minimize the time between product development and commercial release. Some of the efficacy work needs to be conducted under commercial or semi-commercial conditions (to determine effectiveness of agent on different crops, release rates etc), but this would pose some risk in species with the potential to establish. It was recommended that 'establishment potential' should be assessed before any commercial scale efficacy trials. In situations where there is no prospect of establishment in the local environment, companies should be able to conduct efficacy trials under outdoor or open field conditions. Where some establishment is possible or likely, the location and biosecurity of efficacy trials should be discussed with the regulator. As a general principle, companies should conduct such trials in a contained facility (large cage, glasshouse), taking all reasonable effort to prevent escape, in sites that are geographically isolated from areas of 'scientific sensitivity', and with

regular monitoring in the immediate vicinity of the trial to detect any occurrence of the agent outside of the enclosed environment. When such escapes are observed, the trial should be terminated immediately and all plants and invertebrate material destroyed. Similarly, at the end of the trial, all plants and pests/control agents should be destroyed. These conditions should be applied to all researchers involved in biocontrol research, including universities and research institutes.

*Key point*: Establishment potential of inundative biocontrol agents should be assessed prior to commercial scale efficacy trials. For species with no ability to establish in the climatic area of the trial, experiments can be conducted under 'open field' conditions if appropriate. When establishment is possible or likely, an appropriate level of biosecurity should be adopted in discussion with the regulatory authority.

## References

Bale JS, Allen CM, Hughes GE (2009) Thermal ecology of invertebrate biological control agents: establishment and activity. In: Third international symposium on biological control of arthropods (ISBCA) Christ Church, 8–13 February 2009 pp 56–65

Bale JS, van Lenteren JC, Bigler F (2008) Biological control. In 'Sustainable Agriculture', Phil Trans R Soc 363(special issue):761–776

Berkvens N, Bale JS, Berkvens D et al (2010) Cold tolerance of the harlequin ladybird *Harmonia axyridis* in Europe. J Insect Physiol 56:438–444

Bigler F, Babendreier D, Kuhlmann U (2006) Environmental impact of invertebrates for biological control of arthropods. Methods and risk assessment. CABI Publishing, Wallingford, UK

Bigler F, Bale JS, Cock MJW et al (2005) Guidelines for information requirements for import and release of invertebrate biological control agents in European countries. Biol Control News Info 26:115–123

Brown PM, Adriaens T, Bathon H et al (2008) *Harmonia axyridis* in Europe: spread and distribution of a non-native coccinellid. BioControl 53:5–21

Castagnoli M, Simoni S (2003) *Neoseiulus californicus* (McGregor) (Acari Phytoseiidae): survey of biological and behavioural traits of a versatile predator. Redia 86:153–164

Hart AJ, Bale JS, Tullet AG et al (2002a) Effects of temperature on the establishment potential of the predatory mite *Amblyseius californicus* McGregor (Acari: Phytoseiidae) in the UK. J Insect Physiol 48:593–599

Hart AJ, Tullett AG, Bale JS et al (2002b) Effects of temperature on the establishment potential in the UK of the non-native glasshouse biocontrol agent *Macrolophus caliginosus*. Phys Entomol 27:112–123

Hatherly IS, Hart AJ, Tullett AGT et al (2005) Use of thermal data as a screen for the establishment potential of non-native biocontrol agents in the UK. BioControl 50:687–698

Hatherly IS, Pedersen BP, Bale JS (2009) Effect of host plant, prey species and intergenerational changes on the prey preferences of the predatory mirid *Macrolophus caliginosus*. BioControl 54:35–45

Hunt EJ, Kuhlmann U, Sheppard A et al (2008) Review of invertebrate biological control regulation in Australia, New Zealand, Canada and the USA: recommendations for a harmonized European system. J Appl Entomol 132:89–123

Jolly RL (2000) The predatory mite *Neoseiulus californicus*: its potential as a biological control agent for the fruit tree spider, *Panonychus ulmi*. BCPC Conf Brighton Pest Dis 1:487–490

Koch RL, Venette RC, Hutchinson W (2006) Invasions by *Harmonia axyridis* (Pallas) (Coleoptera: Coccinellidae) in the western hemisphere: implications for South America. Neotrop Entomol 35:421–434

Kuhlmann U, Schaffner U, Mason PG (2006) Selection of non-target species for host specificity testing. In: Bigler F, Babendreier D, Kuhlmann U (eds) Environmental impact of invertebrates for biological control of arthropods. Methods and risk assessment. CABI Publishing, Wallingford, pp 15–37

Mills NJ, Babendreier D, Loomans AJM (2006) Methods for monitoring the dispersal of natural enemies from point source releases associated with augmentative biological control. In: Bigler F, Babendreier D, Kuhlmann U (eds) Environmental impact of invertebrates for biological control of arthropods. Methods and risk assessment. CABI Publishing, Wallingford, pp 114–121

Sheppard AW, van Klinken RD, Heard TA (2005) Scientific advances in the analysis of direct risks of weed biological control agents to nontarget plants. Biol Control 35:215–226

van Lenteren JC (2008) IOBC internet book of biological control. http://www.unipa.it/iobc/view.php?pg=services

van Lenteren JC, Babendreier D, Bigler F et al (2003) Environmental risk assessment of exotic natural enemies used in inundative biological control. BioControl 48:3–38

van Lenteren JC, Bale JS, Bigler F et al (2006a) Assessing risks of releasing exotic natural enemies of arthropod pests. Ann Rev Entomol 51:609–634

van Lenteren JC, Cock MJW, Hoffmeister TS et al (2006b) Host specificty in arthropod biological control, methods for testing and interpretation of the data. In: Bigler F, Babendreier D, Kuhlmann U (eds) Environmental impact of invertebrates for biological control of arthropods. Methods and risk assessment. CABI Publishing, Wallingford, pp 38–63

Van Lenteren JC, Loomans AJM, Babendreier D et al (2008) *Harmonia axyridis*: an environmental risk assessment for Northwest Europe. BioControl 53:37–54

Wapshere AJ (1974) A strategy for evaluating the safety of organisms for biological weed control. Ann Appl Biol 77:201–211

# Chapter 17
# Proposals on How to Accelerate Registration of Biological Control Agents

**Anita Fjelsted and Ralf-Udo Ehlers**

**Abstract** The chapter presents a compendium of proposals on how the current regulation system for biological control agents can be improved. The thematic areas are fees, communication, generic approaches, timelines, centralised regulation, legislative framework and efficacy evaluation. The aim of the REBECA project was to propose improvements in the registration process of biological control agents, plant extracts and semiochemicals (here collectively called "BCAs"). Previous chapters of this book contain proposals for specific groups, e.g. baculoviruses, botanicals and semiochemicals. This chapter presents a number of proposals that relate to more general aspects of the regulation and registration of BCAs. The proposals have been discussed and/or elaborated in the REBECA project. The recommendations and discussions are based on a questionnaire, which was discussed during REBECA workshops and circulated for comments to all stakeholders to highlight advantages and disadvantages of each proposal. Special emphasis is placed on two aspects that have emerged in all discussions:

- Implementation. Whether or not a proposal can be implemented in the short term. Short-term implementation is likely for proposals which are not controversial and require no changes of legislation.
- Potential impact. Here, we have tried to take into account expected impact on both the speed of the evaluation process and the impact on the costs of registration (for the applicant).

## Contents



A. Fjelsted (✉)
Ministry of the Environment, Danish Environmental Protection Agency – Pesticides and Genetechnology, 1401 Copenhagen, Denmark
e-mail: anfje@mst.dk

R.-U. Ehlers (ed.), *Regulation of Biological Control Agents*,
DOI 10.1007/978-90-481-3664-3_17, © Springer Science+Business Media B.V. 2011

This chapter provides an inventory of possible measures to change the regulation and registration process of BCAs that can stimulate further discussions. In order to allow a thorough decision-making, this inventory includes measures that we consider advantageous, together with measures that we consider less suitable, disadvantageous or unlikely to be implemented. The position of the REBECA project is explained in the "REBECA conclusions" at the end of each sub-chapter. It cannot be assumed that all project partners, or that all experts, who participated in the REBECA workshops, fully agreed with all conclusions.

## 17.1 Introduction

Improvement of the regulation process can either be achieved by modification of the specific regulation requirements for the different groups of BCAs or by changing the administrative organisation of the regulation process. In this chapter we will present recommendations and discuss several proposals on how this administrative process can be improved in order to accelerate registration of BCAs.

These recommendations are based on information provided by the participants of several REBECA workshops. In addition a questionnaire on "main obstacles and proposals" was sent to all stakeholders and their responses were utilized to produce a draft version of the proposals, which was then circulated between regulators, industry and scientists for comments. The comments were used to further improve and specify the proposals and recommendations. Thus, this chapter reflects much of the outcome of the REBECA project.

## 17.2 Fees and Financial Support

Microbials, botanicals and semiochemicals are still mostly products for niche markets. In addition, most of these products have very low risk profiles and are therefore particularly in line with relevant EU policies on reduction of pesticide use. This justifies indirect subsidies in the form of reduced registration fees or other means of support/subsidies, as is already the case, e.g., in Canada, the USA and many EU member states. Registration fees can make up a significant proportion of the total costs for product development. Fee structure varies greatly among EU member states. It is not within the authority of the EU Commission to decide on the size of the fees required by the regulating authorities in the member states. In some member states, there are no fees for products for minor use, or for products containing new active substances, and some have reduced fees for BCAs. In Denmark no fees are requested for national authorisations of plant protection products. Instead there is a tax system, with 3% tax on microbials, 25% on chemical herbicides and fungicides and 35% on chemical insecticides. The tax revenue is partly used to finance the risk assessment and the authorisation carried out by the Danish regulatory authorities.

A survey on fees charged by member states (MS) is presented in Table 17.1, which contains information about the fees that were requested by rapporteur member states (RMS) for the evaluation of new active micro-organisms and Table 17.2, presenting the amount charged by RMS for the micro-organisms on the 4th list (existing active substances, which had been in the market in MS 2 years after the date of notification of Directive 91/414/EEC and which had to undergo a re-registration to be listed in Annex 1). It is evident that fees are not harmonised. Regulators acknowledge that this may be a problem for notifiers. However, they also believe that harmonisation is not possible. It is up to each MS to decide on the amount of the fee. Several regulators mentioned that fees charged for new active substances only covered a small part of the actual expenses for the evaluation

**Table 17.1** Fees requested for the evaluation of the new active microorganisms (the information was collected from member states in 2007)

| Type of micro-organism | Species | Rapporteur member state | Requested fee (Euro) |
|---|---|---|---|
| *Fungi* | *Paecilomyces fumoseroseus (1)* | Belgium | 10,000 |
| | *Coniothyrium minitans* | Germany | 0[a] |
| | *Gliocladium catenulatum* | Finland | 840[b] |
| | *Ampelomyces quisqualis* | France | ? |
| | *Paecilomyces lilacinus* | Belgium | 10,000[c] |
| | *Pseudozyma flocculosa* | The Netherlands | 5,000 |
| | *Paecilomyces fumoseroseus (2)* | Belgium | 10,000 |
| *Bacteria* | *Pseudomonas chlororaphis* | Sweden | 0[d] |
| | *Bacillus subtilis* | Germany | 0[a] |
| *Virus* | *Spodoptera exigua NPV* | The Netherlands | 5,000 |
| | *Zucchini Yellow Mosaic Virus* | UK | 42,000 |
| | *Adoxophyes orana GV* | Germany | 0[a] |

[a]Not specific to micro-organisms; all new active substances were exempted from fees for the RMS work. Such applications were only accepted in conjunction with a national product application
[b]In 1998 RMS's national application fee was 5000 FIM (corresponds to 840 euros). RMS's national legislation has been updated since then
[c]The importance of the fees is not related "mathematically" to the work that has to be done. The importance of the fee for micro-organisms and substances of the 4th list is also linked to the fact that the market for these products is small. It is also an incentive for organic farming
[d]By the time Sweden got the application, 1997, there were no fees established for inclusion in Annex I of Directive 91/414/EEC

**Table 17.2** Fees requested for the 4th list microorganisms (2007)

| | |
|---|---|
| Denmark | 110,000 € |
| Germany | 86,000–143,400 € |
| Netherlands | Cost-recovery basis |
| Estonia | 11,610 |
| Sweden | Maximum 215,000 € |

process, whereas most regulators expect the fees, which had been requested for the evaluation of the 4th list micro-organisms, to cover all expenses related to the registration effort by the RMS. However, in most MS these fees are still much lower than the fees requested for the evaluation of chemical active substances. For example, in Denmark the fees for the evaluation of existing chemical active substances are twice as high as for the 4th list microbials (220,000 € versus 110,000 €). In the UK the fees for microbials were 22,500 £ for national authorisation plus 7,500 £ for the rapporteur service for Annex I inclusion. The fee for chemicals was 110,000 £.

In the US, EPA (Environmental Protection Agency) charges fees for microbials of up to 20,000 € and in Canada no fees are requested for microbials.

## *Proposal 1: Reduction of fees*

Description: National registration fees as well as fees for Annex I inclusion to be lowered substantially for microbials and semiochemicals.

Advantages: Lower fees will make it easier for companies to register new BCAs. This would demonstrate the intention of governments to encourage submission of files for authorisation of BCAs.

Disadvantages/Problems: Today, many evaluating agencies are under financial pressure. Reduction of fees is only possible if governments bear the expenses. By contrast, following this policy would cause a disparate treatment of BCAs compared to synthetic compounds. Some chemicals might have similarly low risk profiles and it would be difficult to justify why they are not given the same favourable treatment.

Implementation: Likelihood and degree of implementation varies greatly between member states.

Potential impact: This proposal will lower the costs of product development. However, its potential impact is limited, because registration fees make up only a minor part of the total development costs.

## *Proposal 2: Support of SMEs for Registration*

Description: SMEs (Small and Medium-sized Enterprises) applying for registration of new microbials, botanicals or semiochemicals should be financially supported by specific programmes (e.g. financial support for specific requested studies) and should be given detailed guidance on the regulatory process by the regulatory authority (including help for the compilation and presentation of data in the required applications). Funding could come from various sources, such as rural development actions, IPM and organic action plans, promotion of SMEs or from taxes on pesticides. In The Netherlands, the project GENOEG has used such an approach with success. In the UK, the Biopesticide scheme provides guidance to applicants.

Advantages: This proposal is more flexible than proposal 1. Such programmes could provide support for production of studies or dossiers, or for covering registration fees. Support is only given to SMEs and it might be adjusted to the degree of exigency by the growers.

Disadvantages/Problems: Such programmes need financial sourcing. A clear definition on environmental benefits obtained from such product is essential to decide on subsidies. This concept might not be in line with a free-market economy.

Implementation: Likelihood and degree of implementation varies greatly between member states.

Potential impact: This proposal will lower the costs of product development and speed up the registration process. Its potential impact is higher than in proposal 1, because it is not limited to registration fees.

**REBECA conclusions concerning fees/financial support**

REBECA recommended lower fees and support programmes for SMEs to bring BCAs to the market

## 17.3 Improve Communication Between Regulators and Applicants

In order to accelerate the registration and evaluation process and to reduce data requirements, communication between regulators and applicants should be modified and intensified. Arranging pre-submission meetings is a first pre-requisite to ease communication. Applicants and evaluators gain a better understanding of the matter and of the procedures relevant for the evaluation of risk data, and clarify which data are likely to be required for the evaluation and risk assessment. Many countries have established pre-submission meetings as a routine. This is, for instance, the case in the UK and the experiences are positive throughout. Applicants avoid producing unnecessary data, and regulators save time because dossiers better address those points that the regulators consider important. In addition, further improvement of the communication is also needed later during the evaluation process, which can be organised, e.g., during expert meetings.

Regulators report that some companies have not been interested in adopting the offer for a pre-submission meeting. Some regulators report that meetings to check completeness also provide immediate feedback for applicants and improve quality of future dossiers.

### *Proposal 3: Pre-submission Meetings*

Description: Pre-submission meetings shall be established as a routine in all EU Member States.

Advantages: Today, expertise and experience in SMEs on regulation is limited. Applicants would get a better understanding on how to prepare requests for data waivers, how to address data requirements more effectively and how to avoid producing unnecessary data. If personal contacts have already been established, communication between applicants and regulators would be easier. SMEs will have an easier approach to regulators with questions while preparing the data package. Better understanding of the regulation process, data requirements etc. by industry will likely result in submission of higher quality dossiers. Regulators gain better understanding of company's products and target markets. All these measures will speed up the evaluation/authorisation process.

Disadvantages/Problems: Realisation of this proposal will impose a greater work load on regulators and some authorities may lack resources for such meetings. Much effort can be avoided if the applicant commissions a qualified consultant to support the application.

Implementation: This proposal can be easily implemented.

Potential impact: Realisation of this proposal would save costs for the applicant for producing studies which will not be required. It improves dossier quality and thereby speeds up the evaluation process.

## *Proposal 4: Pre-submission Information Package*

Description: A pre-submission information package can provide additional guidance. This proposal can be considered as a further elaboration of proposal 3.

Applicants are encouraged to contact the RMS at an early stage of the product development and before preparing a dossier. Each MS can appoint a contact person (a "BCA champion") to provide guidance. The main objectives of pre-submission meetings are to determine the appropriate test substances, study protocols and data that are required for the dossier of a particular active substance and plant protection product, and determine the information required to support a justification for non-submission of data (waiver). Before consulting the RMS, the applicant should familiarize himself with the data requirements.

When an applicant has asked the RMS for a pre-submission consultation meeting, the RMS will ask the applicant to send a pre-submission information package. A pre-submission meeting will take place no later than 90 days after the submission of the information package. The information package should contain the following:

- A cover letter requesting a pre-submission meeting (for which a template could be made available).
- A proposed agenda of the issues to be discussed (a template should be made available).
- Completeness check tables (document O) containing details of (a) the information included in the pre-submission information package, (b) the studies that have already been carried out (if any), and (c) the justification for non-submission of data.
- Proposed use pattern (Table of Good Agricultural Practise), proposed label, international regulatory status.
- Characterization of the active substance (for microbials also information on mode of action).
- Short summaries of available information regarding manufacturing processes, product specifications, safety to the environment and human health.
- Scientific justifications for non-submission of data (waivers).
- Proposed study protocols (if available).

After the pre-submission meeting, the completeness check table will be updated by the RMS. A copy of the completeness check table must be enclosed in the dossier.

The applicant will be reminded that, depending on the outcome of the risk assessment, additional data/information may be required. Since most studies are unlikely to be available at the pre-submission meeting, the regulators will at this stage not be able to guarantee that no further data will be necessary.

Advantages: This procedure gives guidance to the applicant, which facilitates the preparation of the dossier, avoids the applicant carrying out unnecessary studies, and improves the quality of the dossier.
Disadvantages/Problems: see proposal 3
Implementation: see proposal 3
Potential impact: see proposal 3

### *Proposal 5: Contact with Applicant During Evaluation*

Description: When appropriate, the applicants could be given the opportunity of attending part time at the EU evaluation/expert meetings during discussions of their specific product. It must be clear that they are only invited for clarification of questions, and not for introducing new data or for lobbying. The applicants should not attend throughout the discussions, so that regulators may have additional discussions in the absence of the applicants.
Advantages: Minor issues/mistakes can be resolved much faster. The applicant will have a better understanding of the procedure and the comments/reasoning made by other MS.
Disadvantages/Problems: In the presence of applicants, it is not possible to refer to previous discussions on other compounds due to confidentiality. Some regulators feel that the presence of applicants would compromise the independence of the expert meetings. Some regulators also fear unwanted pressure/lobbying from applicants, or criticisms of the RMS and MS experts. Finally, this system would cause additional costs for the applicant, particularly if there are many evaluation/expert meetings.
Implementation: This proposal can be implemented quite easily.
Potential impact: This proposal will help to clarify certain questions and misunderstandings rapidly, which speeds up the registration process.

**REBECA conclusions concerning communication between regulators and applicants**

REBECA supported proposals 3–5.

## 17.4 Improve Communication Among Regulators of BCAs

Only few dossiers on BCAs have been reviewed until today. The major expertise in current regulation and risk assessment panels is on synthetic plant protection products but not on BCAs. Acceleration of the regulation process would be possible, should the existing expertise in Europe or elsewhere be better exploited and made use of for the regulation process of BCAs. Networking among regulatory authorities can help to disseminate experience and information on risks and safety of BCAs. The EU expert/evaluation meetings should be attended by experts, who can provide specific knowledge on the BCA under review.

## *Proposal 6: Networking and Involvement of Experts*

Description: Further and more regular EU expert and evaluation meetings should be arranged and further outside resources allocated to such meetings.

Advantages: This will speed up procedures. Regular meetings also contribute to building a network among member states, and improve expertise and harmonisation between member states.

Disadvantages/Problems: Depending on the issue, it might be difficult and in particular expensive to organise such specialised meetings.

Implementation: This proposal can be implemented immediately if the Commission and/or the member states are willing to cover the additional costs.

Potential impact: This proposal will help to speed up the registration process.

## *Proposal 7: Appointment of Lead Rapporteurs*

Description: The member states that had been appointed as lead Rapporteurs for BCAs in the 4th list review process should, after the finalisation of the review of these substances, have a function as lead Rapporteurs for new, but similar, BCAs and thereby contribute to the harmonisation and consistency in the evaluation process. The aim should be to facilitate communication and close cooperation among regulators as well as between regulators, experts, EFSA and the Commission.

Advantages: This approach would increase the communication, harmonisation and consistency between member states, and would facilitate and speed up procedures. It is a simple way to make use of the experiences gained in the 4th stage.

Disadvantages/Problems: This approach places an additional burden on a few member states. Some stakeholders expressed fears that the involvement of "lead rapporteurs" could make the process more complicated than necessary (although it is meant to achieve the opposite). The possibility of free choice for the applicant to choose a rapporteur authority and competition between regulation authorities would be eliminated.

Implementation: This proposal can be implemented easily, if the RMSs concerned are willing to take over this task.

Potential impact: The guidance and harmonisation resulting from this proposal will in the long term reduce the costs for dossier preparation, and speed up the registration process.

## *Proposal 8: Establishment of Expert Groups to Support Risk Assessment*

Description: EU expert groups are established for each of the following types of active substances: microbials, botanicals and semiochemicals. For each expert group, one member state can be appointed as chair. The chair facilitates a high level of information exchange and is responsible for the coordination of two annual meetings. The groups comprise a representative from the Commission, an EFSA expert, national regulatory authorities and national experts with experience in evaluating the particular type of active substances (ideally 10–15 experts in total). The expert meetings should be hosted by the Commission, EFSA (European Food Safety Authority) or by a MS. Travel expenses should be covered by the Commission. The minutes of the meetings should be made available to all MS (reported at meetings of the WG legislation).

The purpose of the expert groups is to give guidance to the RMSs, other MS, SANCO (European Directorate General for Health and Consumer Affairs) and EFSA and the applicants. The groups should discuss both risk assessment and risk management issues. Discussions in these groups will facilitate the peer-review process. The group can also discuss issues raised during pre-submission meetings with applicants. RMSs may ask the expert groups for an opinion on specific issues. The groups can also develop draft guidance documents, which are subsequently discussed and finally agreed upon by all MS. To reduce travel expenses, the expert groups should try to organise their meetings jointly with other meetings or use modern communication media (video conferences and e-mail).

Advantages: Increases the communication, harmonisation and consistency in the risk assessment and risk management throughout the EU. Better guidance to the applicant facilitates preparation of the dossier, and increases its quality. Better guidance to the RMS facilitates preparation of the Draft Assessment Report (DAR) and the subsequent peer-review process.

Disadvantages/Problems: Lack of time among regulators/experts to attend these meetings. With regular meetings, there may not be enough issues to discuss, or the timing of the meetings may not fit into the schedule of the evaluation process. High cost to cover expenses for these meetings.

Implementation: This proposal can be implemented relatively quickly. The workload can be adjusted to the needs, which facilitates its initiation and its acceptability.

Potential impact: The guidance and harmonisation resulting from this proposal will in the long term reduce costs for dossier preparation and speed up the registration process.

**REBECA conclusions concerning communication between regulators of BCAs**

REBECA supported all initiatives that improve communication between regulators of BCAs. Of the three proposals, proposal 8 will probably have the highest impact and was particularly supported by REBECA.

## 17.5 QPS Approach in Risk Assessment

The "precautionary principle" is a fundamental element of Directive 91/414 (see also Chapter 6). Its assumption is that all potential risks have to be excluded, before a substance can be included into Annex I of the directive. A practical consequence in the registration of microorganisms is that most data are required at the strain level and not at the species level. In areas other than plant protection, other strategies of risk management are discussed. For microorganisms entering the food chain, EFSA considers the QPS concept (Qualified Presumption of Safety). QPS is based on scientific evidence and experience. The development of a QPS concept was initiated in 2003 by a working group consisting of members of several former (EC) scientific committees. The work was continued within an EFSA working group. The aim was to develop a scheme that would harmonise the risk assessment of microorganisms throughout the various EFSA panels and a scheme developed as a tool for setting priorities within the risk assessment of micro-organisms used in food/feed. Wherever possible, a more generic approach is taken instead of a full case-by-case assessment. It allows the generic listing of microorganisms, provided that certain criteria are met, e.g. absence of acquired antibiotic resistance factors. QPS is similar in concept and purpose to the GRAS (Generally Recognised As Safe) concept used in the USA, but is not identical to GRAS.

QPS based its safety assessment of defined taxonomic groups on four pillars: establishing identity, body of knowledge, possible pathogenicity and end use. If the taxonomic group did not raise safety concerns or, if safety concerns existed but could be defined and excluded, the group could be granted QPS status. Thereafter, any strain of the micro-organisms given QPS status would be freed from further safety assessments other than satisfying any qualifications specified. The final opinion of the Scientific Committee was adopted on 19 November 2007. Table 17.3 contains groups of micro-organisms included in the concept. The committee explained that filamentous fungi could not be recommended for the QPS

**Table 17.3** List of taxonomic units proposed for QPS status

| Genus/species | | | Qualification |
|---|---|---|---|
| Gram-positive non-sporulating bacteria | | | |
| ***Bifidobacterium*** | | | |
| *B. adolescentis* | *B. bifidum* | *B. longum* | |
| *B. animalis* | *B. breve* | | |
| ***Corynebacterium*** *glutamicum* | | | QPS status applies only when species is used for production |
| ***Lactobacillus*** | | | |
| *L. acidophilus* | *L. farciminis* | *L. paracasei* | |
| *L. amylolyticus* | *L. fermentum* | *L. paraplantarum* | |
| *L. amylovorus* | *L. gallinarum* | *L. pentosus* | |
| *L. alimentarius* | *L. gasseri* | *L. plantarum* | |
| *L. aviaries* | *L. helveticus* | *L. pontis* | |
| *L. brevis* | *L. hilgardii* | *L. reuteri* | |
| *L. buchneri* | *L. johnsonii* | *L. rhamnosus* | |
| *L. casei* | *L. kefiranofaciens* | *L. sakei* | |
| *L. crispatus* | *L. kefiri* | *L. salivarius* | |
| *L. curvatus* | *L. mucosae* | *L. sanfranciscensis* | |
| *L. delbrueckii* | *L. panis* | *L. zeae* | |
| ***Lactococcus*** *lactis* | | | |
| ***Leuconostoc*** | | | |
| *L. citreum* | *L. lactis* | *L. mesenteroides* | |
| ***Pediococcus*** | | | |
| *P. acidilactici* | *P. dextrinicus* | *P. pentosaceus* | |
| ***Propionibacterium*** *freudenreichii* | | | |
| ***Streptococcus*** *thermophilus* | | | |
| Gram-positive spore-forming *Bacillus* spp. | | | |
| *B. amyloliquefaciens* | *B. fusiformis* | *B. mojavensis* | Absence of emetic food poisoning toxins with surfactant and enterotoxic activity[a] |
| *B. atrophaeus* | *B. lentus* | *B. pumilus* | |
| *B. clausii* | *B. licheniformis* | *B. subtilis* | |
| *B. coagulans* | *B. megaterium* | *B. vallismortis* | |
| ***Geobacillus*** *stearothermophillus* | | | |
| Yeasts | | | |
| ***Debaryomyces*** *hansenii* | | | |
| ***Hanseniaspora*** *uvarum* | | | |
| ***Kluyveromyces*** | | | |
| *K. lactis* | *K. marxianus* | | |

**Table 17.3** (continued)

| Genus/species | | | Qualification |
|---|---|---|---|
| ***Pichia*** | | | |
| *P. angusta* | *P. anomala* | | |
| ***Saccharomyces*** | | | |
| *S. bayanus* | *S. cerevisiae* | *S. pastorianus (synonym of S. carlsbergensis)* | *S. cerevisiae* subtype *S. boulardii* is contraindicated for patients of fragile health or a central venous catheter in place. A specific protocol concerning the use of probiotics should be formulated |
| ***Schizosaccharomyces** pombe* | | | |
| ***Xanthophyllomyces** dendrorhous* | | | |

[a]When strains of these QPS units are to be used as seed coating agents, testing for toxic activity is not necessary provided that the risk of transfer to the edible part of the crop at harvest is very low

status. Further more, all strains belonging to the *Bacillus cereus sensu lato* group (e.g. *Bacillus thuringiensis*) should not be given a QPS status either, since it is known that the vast majority of strains within this group are toxin producers and thus can not meet the required qualifications. The Scientific Committee wrote as follows: "The Scientific Committee is of the opinion that the use of strains from the *B. cereus* group should be avoided whenever there is a possibility of human exposure whether intended or incidental. The *B. cereus* group is therefore excluded from consideration for QPS status. There is an artificial distinction held between *B. cereus* and *B. thuringiensis* (used for plant protection) which has little scientific basis. The plasmid encoding the insecticidal enterotoxin, which provides the phenotypic distinction for *B. thuringiensis,* is readily lost, particularly when grown at 37°C, leaving an organism indistinguishable from *B. cereus*. Consequently it is likely that *B. thuringiensis* has been the causative organism of some instances of food poisoning but identified as *B. cereus* because clinical investigations would have failed to recognise the distinguishing features characteristic of *B. thuringiensis*. However, the Scientific Committee recognises that *B. thuringiensis* has value to the industry as a means of biological pest control and that its widespread use for this purpose may not lead to significant human exposure." (EFSA 2007).

Bacteria directly consumed by humans only qualify for QPS status if they are free of acquired resistance to antibiotics of importance in clinical and veterinary medicine. Furthermore, all bacteria capable of toxin production should be demonstrated to be free of any toxigenic potential. It is important to stress that QPS does not carry any legal status.

Since neither *B. thuringiensis* nor any of the filamentous fungi are included on the list of species proposed for QPS status, the QPS, in its present form, does not offer a *generic approach* to the safety assessment of most micro-organisms used as biological control agents. Nevertheless, the EFSA Scientific Committee considers that it may be possible to devise robust use qualifications, which would allow a QPS approach for further groups of micro-organisms relevant for biological control in the future. The system is developed in order to provide a generic assessment system for use within EFSA that can be applied to all requests for the safety assessment of micro-organisms deliberately introduced into the food chain or used as producer strains for food/feed additives. This implies that when industry applies for Annex I inclusion of micro-organisms belonging to microbial taxonomic units, which are now included in the list of organisms for which a QPS status is proposed (e.g. *Bacillus subtilis* and *B. pumilus*) with the intention to market these in plant protection products, the industry can in their dossier argue that the species are given QPS status, and that the risk for consumer health (due to exposure from residues on crops) is likely to be low when these strains are applied as plant protection products. This information can be used to ask for a waiver for residue data for micro-organisms given QPS status. The applicability of the QPS approach for broad use of micro-organisms as plant protection products needs to be discussed further, possibly also in the context of definition of low-risk BCAs (see Section 17.7.).

Experience gained during the EU evaluation of the microorganisms in the 4th stage of re-evaluation may be taken as a basis to determine in which cases a generic approach is justified.

## *Proposal 9: Generic Approach to Risk Assessment*

Description: Establish risk management strategies taking a generic approach wherever possible, and restricting case-by-case evaluations to those cases where this is necessary and justified. I.e., evaluate microorganisms at species level whenever possible and evaluate other substances as groups as well (e.g. certain botanicals and semiochemicals). However, this approach can only be followed if there is enough experience/scientific evidence about a certain group.

Advantages: Saves costs for producing studies, and speeds up the registration process. Proposal provides incentives to better exploit progress and innovation in biocontrol. Currently, registered strains are not replaced by more suitable strains as registration costs would incur again.

Disadvantages/Problems: Data protection must be ensured for those applicants who have provided data on which regulatory experience is based. This approach cannot be followed if there is not enough experience/scientific evidence or only unsubstantiated claims about a certain group of BCAs.

Implementation: Such an approach is now being taken for baculoviruses and maybe for another couple of groups of substances. However, in order to expand this proposal to further groups, this proposal will require considerable discussion.

Potential impact: This proposal will greatly reduce the costs for dossier preparation, and speed up the registration process.

**REBECA conclusions concerning precautionary principle vs. QPS approach.**

REBECA considered a more generic approach as promising. Experience from the 4th stage registration should be used to determine groups which are amenable to such an approach (e.g. baculoviruses, straight chained lepidopteran pheromones (SCLPs) and certain microbial species).

## 17.6 Define Low Risk Substances for Fast Track Authorisation

One possibility to accelerate authorisation of BCAs would be to differentiate between low risk and high risk active substances. A fast track regulatory system would be possible for low risk substances. It is considered that a great number of BCAs could be placed into the low-risk category. For such a classification it is necessary to establish a definition or criteria of low risk substances. In Directive 91/414/EEC no differentiation was made between low risk and higher risk active substances. However, Regulation (EC) 1107/2009 contains separate paragraphs relating to "low-risk substances", "basic substances" and "substances of concern". Article 22 of the regulation extends the period of approval from the normal 10–15 years for low risk active substances (European Commission 2009).

Of low risk could be substances that are of a type considered inherently unlikely to cause an adverse effect on humans, animals or the environment. The active substances cannot be regarded as low risk if they are classified as carcinogenic, mutagenic, toxic to reproduction, very toxic or toxic, sensitising or explosive. In addition, substances that have the following characteristics cannot be regarded as low risk either: persistent (half life of less than 60 days), endocrine disrupter or bioaccumulative and non readily-degradable (Annex II of Regulation 1107/2009).

Timelines for the authorisation of plant protection products based on low risk substances are shorter. The member state shall within 90 days decide whether to approve an application for authorisation of a low-risk plant protection product. This period should only be 60 days if an authorisation has already been granted for the same low-risk plant protection product by another Member State located in the same zone. However, if the Member State needs additional information, it shall set a time limit not exceeding 6 months for the applicant to supply the information.

Article 23 of the regulation provides criteria for basic substances and extends the period of their approval to an unlimited time. The basic substances will have to be included in a separate list. Basic substances are substances that are placed on the market for purposes other than plant protection, e.g. for food, fertilizers etc. These will not be regulated at national level.

The criteria for low risk substances seem to be made with chemical active substances in mind. First of all, there is a general risk of micro-organisms being sensitisers. This may disqualify them as low risk substances. However, so far no proper guidelines are available that can be used to carry out studies in order to investigate the sensitising properties of micro-organisms. In the data requirements for micro-organisms (Annex IIB to Directive 91/414/EEC), which are listed in Dir. 2001/36/EC, it is mentioned that it is not necessary to present data on sensitisation, due to this lack of guidelines, unless the micro-organism is considered to be sensitising. Secondly, the three terms: persistence, bioaccumulative and non readily-degradable, and endocrine disrupters are all terms originating from the classification of chemical active substances. These criteria do not take into account that, e.g., micro-organisms are naturally occurring substances.

In the Biocide Directive 98/8/EC the active substances regarded as being of low risk are included into a specific list: 1A. The criteria for substances to be included into this list are quite similar to the criteria that are now suggested to be included in the new regulation on plant protection products. However, the biocide criteria do not include: toxic, very toxic, explosive and endocrine disrupters.

A questionnaire was sent to all REBECA participants, in which they were asked to list active substances (BCAs consisting of micro-organisms, botanicals, semiochemicals or macrobials), which they would regard as being of low risk and to give their reasoning for such proposals for low risk substances. Furthermore, the participants were asked to give a definition and/or criteria for low risk substances. 46 persons replied to the questionnaire (9 regulators, 12 persons representing industry, 22 from the scientific community and 3 from consultancies).

The participants representing the industry and the scientific community all gave lists of active substances which they regarded as of low risk. In particular, the participants gave a long list of macrobials. However, the semiochemicals, in particular the SCLP were also mentioned by representatives from both the industry and regulatory authorities as a category of low risk. It was mentioned by several participants that if SCLP were applied in concentrations similar to the background concentration occurring in areas with high densities of the pest insect, they should definitely be regarded as of low risk. Baculoviruses was another group of active substances that was mentioned by many participants as being of low risk. A number of botanicals were listed as well. These were products which are also used for human consumption.

Arguments for listing these as low risk were:

- Long history of safe use
- Micro-organisms frequently causing natural epizootics in presence of the host pest

- Micro-organisms that are ubiquitous in soils around the world
- Micro-organisms that do not grow at 37°C
- Narrow host range/very specific
- Low persistence
- Substances used for human consumption (e.g. rapeseed oil, garlic oil, olive oil)
- Substances used as household products (e.g. for cleaning)

The general opinion expressed by regulators and the European Commission (DG SANCO) was, that it would not be possible to establish a list of substances of low risk prior to a risk assessment, i.e. a list of substances that would not need a risk assessment. However, all regulators seemed to agree, that there was a need for a definition/criteria of low risk substances for the new EU regulation of pesticides, but such criteria will be applied only after the risk assessment has been carried out and will determine which substances will get an Annex I inclusion of a longer period (15 years) and an easier/faster process for national registration.

REBECA participants discussed the possibility of the national authorities giving priority to low risk products during the evaluation and authorisation process. These issues had been discussed in Sweden, The Netherlands, Belgium, Denmark and in the UK. The purpose in all three countries was to increase the number of such products at their market, e.g., by reducing the fee requested for low risk products and in The Netherlands, Belgium, Denmark and the UK also to provide further guidance to applicants in order to speed up the preparation of dossiers and the subsequent evaluation of those dossiers. However, none of the regulatory authorities in these countries found the term "low risk" very helpful, simply due to the difficulties in defining such a category. In the UK the Pesticide Safety Directorate (PSD) has not used the term *low risk products* in their BioPesticide Scheme but instead the term *alternative products* (however, also without a specific definition). For this product group they have lowered the fees and are arranging pre-submission meetings; in addition, they have increased the web-site information of the regulatory process, established a specific contact point in PSD for these product types (a champion) and the applicants can be guided throughout the process of putting together an application.

A somewhat similar project has been taking place in The Netherlands, where the project was called GENOEG. It was also aiming at getting further low risk products on the market. In The Netherlands they have used the term *natural pesticides* rather than *low risk products*. Via this project the applicants can get up to 100,000 € co-finance for registration fees and extra studies needed for the risk assessment, and the regulatory authority here also help applicants put together good dossiers and invite applicants for pre-submission meetings. A similar project is under way in Denmark.

In the USA, there is a list of substances that can be used as pesticides without any registration, however, they still need a residue limit, or exemption, for food or feed uses. These substances are called *Minimal Risk Pesticides*, as described in the

US Code of Federal Regulation, 40CFR 152.25(f). The list contains many essential oils.[1] All inerts must be on EPA's 4A inert list, all ingredients must be identified on the label, and the label may not contain false or misleading claims. This regulation was developed by an EPA workgroup in 1994 and revised in accordance with public comments for a final Federal Register publication in 1996. The EPA has experienced a problem since it has been difficult identifying exactly which chemical substances are included under the names listed. Currently, CAS numbers (Chemical Abstracts Service) are used to describe the substances on the EPA inert substance classification lists.

The BCAs that could possibly be listed as low risk products are summarized for baculoviruses in Chapter 12, for bacterial and fungal products in Chapter 13, for botanicals in Chapter 14 and for semiochemicals in Chapter 15. REBECA recommendation were also based on: (i) a case by case evaluation of microbial biocontrol agents, assessed by international experts, recognised by the REBECA consortium; (ii) the safety data fact sheet published by the US Environment Protection Agency (EPS); and (iii) publications of the European Council regulations, reporting the opinion of the safe use of micro-organisms listed in Annex I. Criteria, which could be used to define low-risk micro-organisms for a comparative analysis, are described by Laengle and Strasser (2008).

## *Proposal 10: Define Low Risk BCAs/Substances for Fast Track Authorisation*

Description: Establish lists of low risk BCAs after a risk assessment. This list can contain species/genera of micro-organisms or groups of substances, which have been generally accepted as safe. These BCAs would receive an authorisation as defined in the Regulation (EC) 1107/2009.

Advantages: Saves costs for producing studies, and speeds up the registration process. The proposal provides an incentive to exploit progress and innovation in biocontrol more effectively, as mentioned under proposal 8. Those BCAs, which target smaller markets, could be listed to avoid registration for SMEs. Macrobials could be included in such a list. Currently regulation of macrobials is in the hands of national authorities only and prevents European-wide approaches to regulation of macrobials (see Chapter 16). The

[1]Currently, the list includes the following substances: castor oil, cedar oil, cinnamon and cinnamon oil, citric acid, citronella and citronella oil, cloves and clove oil, corn gluten meal, corn oil, cottonseed oil, dried blood, eugenol, garlic and garlic oil, geraniol, gernanium oil, lauryl sulfate, lemongrass oil, linseed oil, malic acid, mint and mint oil, peppermint and peppermint oil, 2-phenethyl propionate (2-phenylethyl propionate), potassium sorbate, putrescent whole egg solids, rosemary and rosemary oil, sesame (includes ground sesame, plant) and sesame oil, sodium chloride (common salt), sodium lauryl sulfate, soybean oil, thyme and thyme oil, white pepper and zinc metal strips.

new Regulation prioritises the use of less hazardous plant protection products. The definition of low risk BCAs can provide one basis for the decision on priority within a comparative assessment.

Disadvantages/Problems: As a risk assessment is necessary prior to listing, protection of data must be ensured should the application for listing come from a private enterprise.

Implementation: Such an approach is now being taken for baculovirus species. It could easily be implemented for other groups, provided that comprehensive data are available for the risk assessments. Expert panels could perform a risk assessment and provide recommendations for a listing in order to avoid problems with data protection.

Potential impact: This proposal will reduce efforts for dossier preparation and significantly speed up the registration process.

**REBECA conclusions concerning fast track authorisation for low risk BCAs**

REBECA supported the idea of an introduction of fast track authorisation for low risk BCAs

## 17.7 Guidance Documents Based on Experience from the 4th Stage Evaluation

A large number of microbials, semiochemicals and botanicals have been reviewed in the so called 4th stage evaluation of BCAs, a re-evaluation obligatory for products that had been on the market in MS for 2 years after the date of notification of Directive 91/414/EEC. The 4th list consisted of substances that were regarded as less problematic and included the BCAs. This process is well ahead by now, but for the microbials it is unfortunately not finalised by EFSA yet. Further EU expert meetings regarding the microbials will take place. However, by now the EU regulators have obtained more experience in assessing BCAs. In a number of reports/draft guidance documents on "lessons learned from the 4th stage" regulators could summarise their experiences with these substances. Of course, data protection has to be respected. For microorganisms the production of such documents has already been discussed.

### *Proposal 11: Introduce Experience to Decide on Data Requirements*

Description: The "lessons learned documents" should be used by applicants and regulators in general, and in particular during future pre-submission meetings to determine data requirements/waivers for new substances in analogy to substances evaluated during the 4th stage. In the pre-submission meeting, it must be clarified in which way the applicant has to address the data requirements.

Advantages: This approach will result in better/more focused dossiers. It improves consistency across member states, reduces data requirements, lowers the costs for applicants and it results in a faster procedure.

Disadvantages/Problems: Data protection may be a problem when writing the lessons learned guidance documents. The lessons learned documents would have an uncertain legal status.

Implementation: As soon as such documents have been prepared this proposal can be implemented easily by those member states that are open to this approach.

Potential impact: This proposal may reduce the costs for dossier preparation and speed up the registration process.

### *Proposal 12: Use Experience to Further Develop Generic Safety Profiles*

Description: The "lessons learned documents" could be used to justify a generic approach, and as a basis for determining generic safety profiles.

Advantages: see proposal 7

Disadvantages/Problems: see proposal 7

Implementation: see proposal 7

Potential impact: see proposal 7

**REBECA conclusions concerning guidance documents based on experience from the 4th stage approach**

The 4th stage of re-evaluation has imposed enormous work and costs both on applicants and on evaluators. The preparation of "lessons learned guidance documents" is a way to utilise the experience gained through these efforts. The REBECA project recommended the preparation of such documents. However, data protection must be respected. The "lessons learned guidance documents" will be particularly useful to determine data requirements/waivers (see proposal 11) and they may also be useful for developing a more generic approach (see proposal 12).

## 17.8 Timelines

Most applicants of BCAs are SMEs and only have resources to apply for national provisional authorisation of their products in very few (1–2) member states during the process of Annex I inclusion of their BCA. However, due to the large investments in the preparation of the dossier etc., it is crucial for the industry to reach the market as soon as possible, either by provisional authorisations, or by obtaining authorisations immediately after the Annex I inclusion. It is thus important for the applicants that the Annex I inclusion is obtained as quickly as possible.

In the past, Annex I inclusion of microorganisms has taken several years. This is a hurdle for the industry compared with the authorisation system in the USA, where the products often reach a large part of the US market within 1–2 years after the application is submitted, because authorisation by the US EPA is quite fast and most states do not require any further evaluation and authorisation. In the USA and Canada, strict timelines are in place for the registration of BCAs, and the industry reports good experience with this. With the implementation of Regulation (EC) 1107/2009, which will replace Dir 91/414, fixed and short timelines will be introduced.

## *Proposal 13: Strict Timelines*

Description: Strict and short timelines for the EU risk assessment as well as for national registrations should be the rule in the EU regulation. The timelines should be as short as is practicable to enable the appropriate risk assessments to be checked, and to ensure they have been supported by robust information.

Advantages: Gives the applicant the opportunity for more adequate planning, since strict timelines would provide better predictability on the length of the evaluation/registration process.

Disadvantages/Problems: There are already strict deadlines, but member states have difficulties in keeping to them due to lack of personnel. What sanctions should be applied if the deadlines are not met? If additional information or clarifications are needed, these may have to be provided within a very short time, otherwise the application must be rejected. In such a case, a "clock-stopping" mechanism may be more useful than strict and short timelines. However, strict and short timelines could also be combined with the possibility of obtaining Annex I inclusion with the requirement of submitting, e.g., one or two confirmatory studies within a short timeframe.

Implementation: With the implementation of EC 1107/2009 part of this proposal is already implemented. Lack of resources within member states and EFSA may have to be taken into account.

Potential impact: This proposal will help to ensure short duration of the registration process and, in particular, increase security about its duration.

**REBECA conclusions concerning timelines**

REBECA considered strict and short timelines to be useful.

## 17.9 Centralised Registration Authority

Microbials, semiochemicals and botanicals, until today, are only a small fraction of all plant protection products registered in the EU. Consequently, many regulators have relatively little experience with the evaluation of these substances. Taking into account the diversity of BCAs and their multifunctional and versatile attributes, environmental and human health risk profiles of BCAs may be equally diverse and consequently require specialised expertise for evaluation. If there was a centralised and specialised authority for the evaluation of such substances, it could build up more expertise, it would speed up the evaluation process, reduce the costs and potentially improve the quality and consistency of risk assessment and DARs. Within the US Environmental Protection Agency (EPA) a separate unit specializes on the evaluation of BCAs and industry reports good experience with this.

### *Proposal 14: European Agency for Registration of BCAs*

Description: Establishment of a new and centralized authority (similar to US EPA) for the evaluation of the BCAs: microbials, botanicals and semiochemicals (and potentially also macrobials).

Advantages: Many applicants consider a centralized authority to be the most effective means to speed up procedures and secure consistency in risk assessments. Applicants have experienced severe problems with regulators new in the field of BCA risk assessment. Within a centralized agency expertise would be handed over to newcomers.

Disadvantages/Problems: Considering the subsidiary principle of the political organisation of the EU (delegate only those tasks to Brussels, which cannot be better organised by MS) it will be difficult to give up national evaluation of BCAs and hand it over to a centralized organization. As a result of this proposal, member states will gain less experience with these specific groups. This lack of experience will cause problems when MS have to decide on national authorisation of the BCA-product. It is not evident that such a centralised authority would require fewer studies/accept further waivers than in the present system. The resources needed to create and run such a new authority will be relatively high compared to the small number of active substances to be dealt with. What will this authority do, if only few applications are submitted, or if applications are submitted at irregular intervals?

Implementation: This proposal conflicts with the sovereignty of member states and seems to meet strong opposition from this side. Within the new regulation, plant protection products are still authorised at MS level, but simultaneous application to several member states ("zones") is introduced and mutual recognition is facilitated.

Potential impact: The impact of this proposal is difficult to estimate. The industry claims that a centralised registration authority would greatly reduce their costs for dossier preparation, and might speed up the Annex I listing process as has been experienced in the USA. On the other hand, this might be offset by a lack of expertise in the member states, which could affect national registration of products.

**REBECA conclusions concerning a centralised registration authority**

The opinion of the REBECA project was that the evaluation of BCAs should be harmonised as much as possible, and the evaluators should have as much expertise as possible. This may be achieved through further communication among regulators (in particular by implementing proposal 8 (Establishment of expert groups). REBECA did not support the establishment of a centralised authority, mainly because this was regarded as politically unachievable.

## 17.10 Optimal Legislative Framework

Some stakeholders believe that even though it has been attempted, the Directive 91/414 (Regulation 1107/2009) is still not adequately adapted to the special properties of microbial biocontrol agents and semiochemicals, which have completely different modes of action than the conventional pesticides, as well as completely different protocols for production, methods for characterization and environmental and human health risk profiles. Proper evaluation of microbials and semiochemicals requires a different approach with different data. However, microbials and semiochemicals also differ greatly from each other. For these reasons, separate legislation for these two groups of substances seems relevant. Chemical as well as microbial biocides are regulated according to Directive 98/8/EC. The data requirements and format for submission of biocide dossiers are different from those for plant protection products.

### *Proposal 15: New Directive/Regulation for Registration of BCAs*

Description: Microbial biocontrol agents and semiochemicals to be taken out of EU Regulation 1107/2009 and regulated by a separate directive (or regulation) for each group.

Advantages: More "tailored" regulation/requirements.

Disadvantages/Problems: It will be a long and time consuming process to prepare and agree on a separate regulation/directive. Separate regulations may then fall under the responsibility of different (national) authorities, which have otherwise no experience with regulating plant protection products. They may also fall under different General Directorates of the Commission.

Implementation: This proposal will require scientific, political and administrative efforts and long discussions before it can be implemented.

Potential impact: This proposal will have great influence on the registration process. It may change procedures and data requirements considerably and speed up the registration process. However, there are also great uncertainties involved in this proposal, particularly if a new set of agencies become involved.

## *Proposal 16: Specify Requirements for BCAs Within a New Annex*

Description: Keep microbials and semiochemicals in EU Regulation 1107/2009, but specify separate data requirements for semiochemicals in a new Annex, and revise the Annex with the data requirements for microbials.

Advantages: Some stakeholders stated that it would be an advantage to keep the regulation of all kinds of plant protection products within one EU legislation.

Disadvantages/Problems: The preparation of separate data requirements as an Annex to a regulation is a long, time-consuming process (although certainly less than proposal 15). If the data requirements are too prescriptive, there may be a loss of flexibility, which the current system allows.

Implementation: This proposal will require long discussions before it can be implemented (although probably less than proposal 15).

Potential impact: This proposal may change data requirements considerably. It may increase predictability of the process.

## *Proposal 17: Produce BCA-Adapted Guidance Documents*

Description: Keep microbials and semiochemicals in Regulation 1107/2009, but specify separate data requirements for semiochemicals with new guidance documents.

Advantages: Some stakeholders stated that it would be an advantage to keep the regulation of all kinds of plant protection products within one EU legislation. The approach with guidance documents is faster and more flexible than changing legislations.

Disadvantages/Problems: Guidance documents are not legally binding.

Implementation: This proposal can be implemented quite easily and changes are possible thereafter.

Potential impact: This proposal may change data requirements considerably, increases predictability of the process and facilitates preparation of the dossier.

## *Proposal 18: Harmonisation of Data Requirements with Biocides*

Description: Harmonisation in the regulation of plant protection products and biocides based on BCAs as well as increased communication among national authorities of the two product types. Comment: Harmonisation has already started, also facilitated by the OECD Pesticide Steering Committee (Organisation for Economic Co-operation and Development). Some stakeholders stated that there is now a high level of harmonisation of data requirements and uniform principles for microbial biocides and pesticides. Other stakeholders stated that there is an urgent need to harmonise the data requirements and the structure of dossiers. It is clear that it is more difficult to obtain harmonisation in situations where products have different fields of use.

Advantages: Saves costs in the preparation of dossiers.

Disadvantages/Problems: It may be a time consuming process to harmonise between different DGs (General Directorate) within the Commission. However, a review of the data requirements for the microbial biocides is currently taking place with regulators involved in both biocides and plant protection products and this process is thus expected to increase harmonisation.

Implementation: Given the involvement of two DGs within the Commission, this proposal would probably require long discussions before it could be implemented.

Potential impact: This proposal is expected to have a great influence on the registration process. If the data requirements for biocides are adapted to those for plant protection products, this will result in considerable improvements. Harmonisation in dossier format will facilitate dossier preparation for those active substances which are used both as plant protection products and as biocides.

**REBECA conclusions concerning the optimal legislative framework**

REBECA considered separate data requirements for BCAs to be adequate. This goal can be achieved fastest with guidance documents (proposal 17). REBECA also recommended harmonising dossier format and data requirements for BCAs, which are used both as plant protection products and as biocides (proposal 18).

## 17.11 Efficacy Evaluation

Compared with conventional chemical substances, some BCAs may have a lower efficacy. There is some uncertainty as to what levels of efficacy are required for BCAs as their effects are often not only biocidal. Effects are also achieved by promoting plant health and growth. Additional indirect effects are achieved by preservation of the natural antagonistic potential. BCAs often have a different mode of action compared with conventional chemical substances, which often makes it necessary to adapt trial protocols. This is particularly the case for semiochemicals, where it is often impossible to use replicated trial designs.

It should be noted that within Dir. 91/414/EEC efficacy has only been an issue at member state level, not in the EU review system for Annex I listing. However, with the new Regulation 1107/2009 the efficacy data will have to be included in the EU dossier.

## *Proposal 19: Modification for Requirements on Efficacy Testing*

Description: Authorities should accept modified trial protocols, provided that the applicant can justify the modification. Rationale: Efficacy testing of BCAs may involve specialised techniques, which require modification of trial protocols (e.g. plot size, replicates, parameters for assessment). BCAs may be more variable in their performance than conventional chemical pesticides, but provided a demonstrable and consistent benefit is achieved, approval should still be acceptable. The product label should accurately reflect the levels of performance that may be expected, as well as provide guidance on how to achieve these. United Kingdom PSD efficacy draft guideline 220 on mating disruption products provides an example of such a flexible approach.

Advantages: Efficacy testing and evaluation can be tailored to the specific properties of each BCA.

Disadvantages/Problems: No serious disadvantages obvious.

Implementation: This proposal can be implemented easily.

Potential impact: This proposal will simplify the preparation of efficacy data.

## *Proposal 20: Acceptance of lower level efficacy*

Description: Even products with only minor beneficial effects should be acceptable, provided the effect is shown to be reproducible and the label accurately reflects the likely benefits.

Advantages: This proposal facilitates the registration of products with minor beneficial effects. Such products may be particularly useful for organic farming and as components of IPM (Integrated Pest Management) programmes or resistance management strategies. It would facilitate the registration of BCAs for use on a wider range of crops (e.g. minor uses), on which they have only a partial efficacy.

Disadvantages/Problems: No major disadvantages or problems are obvious.

Implementation: For most member states, it is not a question whether this proposal can be implemented, but to what extent it should be implemented.

Potential impact: This proposal would facilitate the registration of BCAs considerably, and result in a wider use of BCAs.

## *Proposal 21: Waiver for Efficacy Data*

Description: The applicant can choose the option that the authorities do not evaluate efficacy during registration. In this case, there must be a disclaimer on the product label saying that efficacy has not been evaluated. It must be ensured that regulators are not liable for failures in efficacy of the products. The registration procedure in the USA by EPA does not require data on efficacy as they are not considered to be part of the risk assessment.

Advantages: This proposal makes registration faster, cheaper and more flexible, and delegates the responsibility for selecting efficient products to the market.

Disadvantages/Problems: End users who do not carefully read the label might have wrong expectations concerning efficacy. This option might also be used for marketing products/uses that do not provide control effects and thus economic benefit, which is not in public interest. The BCA industry fears that this might compromise the reputation of BCAs.

Implementation: It is not clear under which circumstances this proposal is consistent with the requirements of Dir. 91/414 (Art. 4 requires that substances must be "sufficiently effective"). It is also not clear whether the BCA industry would support this proposal, because of fears that their reputation might be adversely affected.

Potential impact: This proposal would considerably reduce the applicants efforts needed for product registration. It is likely to increase the use of BCAs considerably, including possibly some products/uses which have little or no effect.

## *Proposal 22: Time-limited Authorisation Until Efficacy Data are Provided*

Description: The same proposal as 21, but limited to a period of 5 years. After 5 years, efficacy data would need to be submitted, otherwise the authorisation of the plant protection product would be revoked. This option would therefore only be available for a transitory phase. This might be useful under the following circumstances:

- To register a BCA against an emerging pest or for minor uses;
- In cases where there are some, but not sufficient, data supporting the efficacy claims, e.g. trials carried out under unfavourable field conditions, or by institutions which are not officially recognized;
- To facilitate product development and market entry of SMEs with very limited research funds.

Advantages: This proposal facilitates market entry of all BCAs, but limits the possibilities for misuse and market entry of ineffective BCAs.

Disadvantages/Problems: During the period of 5 years, products with low or no efficacy might be on the market, and might compromise the reputation of BCAs to some extent.

Implementation: See proposal 21.

Potential impact: This proposal would facilitate market entry of BCAs.

**REBECA conclusions concerning efficacy evaluation**

REBECA supported proposals 19 and 20. REBECA regarded proposal 22 as an interesting potential approach, but did not support proposal 21.

## 17.12 Perspectives

The aim of the REBECA project was to propose procedural improvements which result in easier market access for BCAs, while ensuring their safety. The present proposals focus mainly on the EU review, and to a lesser extent also on the national registration processes. Table 17.4 sheds a brief light on the priorities with which these proposals should be followed up. Some proposals were rated as being easy (***) or relatively easy (**) to implement. It can be expected that a number of these proposals will be implemented within the next few years. Several of these proposals can be implemented at member state level, and therefore require consensus only within one member state. Some of these have already been implemented in certain member states or are likely to be implemented soon.

REBECA proposals that are relatively easy to implement at member state level:

1. Pre-submission meetings
2. Pre-submission information package
3. Use of "lessons learned" guidance documents from the 4th stage in pre-submission meetings
4. Flexible efficacy trial protocols
5. Acceptance of products with minor beneficial effects

REBECA proposal that are more difficult to implement at member state level:

1. Reduced registration fees
2. Financial support and guidance to applicants
3. Strict and short timelines for national authorisation

Other proposals require action at EU level. These require willingness of the member states and the Commission (DG SANCO) to consider specific needs for registration of biological control agents.

**Table 17.4** List of proposals that can help to accelerate the registration process, estimated time span and difficulties for implementation, potential impact on the duration of the registration process and on the costs for the applicants

| Chapter/proposal | Implementation | Impact on duration | Impact on costs |
|---|---|---|---|
| *17.3. Fees and financial support* | | | |
| P1 Lower fees | ** | – | ** |
| P2 Financial support and guidance | ** | – | ** |
| *17.4. Improve communication between regulators and applicants* | | | |
| P3 Pre-submission meetings | *** | *** | *** |
| P4 Pre-submission information package | ** | *** | *** |
| P5 Contact with applicant during evaluation | ** | ** | * |
| *17.5. Improve communication among regulators* | | | |
| P6 Networking and involvement of experts | * | ** | *** |
| P7 Appointment of lead rapporteur | * | *** | ** |
| P8 Establishment of expert groups to support risk assessment | ** | *** | ** |
| *17.6. QPS approach in risk assessment* | | | |
| P9 Generic approach to risk assessment | * | *** | *** |
| *17.7. Define low risk substances for fast track authorisation* | | | |
| P10 Define low risk BCAs for fast track authorisation | ** | *** | *** |
| *17.8. Guidance documents based on experience from the 4th stage evaluation* | | | |
| P11 Use experience to decide on data requirements | ** | ** | ** |
| P12 Use experience to further develop generic safety profiles | * | ** | ** |
| *17.9. Timelines* | | | |
| P13 Strict timelines | * | ** | ** |
| *17.10. Centralised registration authority* | | | |
| P14 European agency for registration of BCAs | * | */*** | */*** |

**Table 17.4** (continued)

| Chapter/proposal | Implementation | Impact on duration | Impact on costs |
|---|---|---|---|
| *17.11. Optimal legislative framework* | | | |
| P15 New directive/regulation for registration of BCAs | * | */*** | */*** |
| P16 Specify requirements for BCAs within a new Annex | ** | ** | ** |
| P17 Produce BCA-adapted guidance documents | *** | ** | ** |
| P18 Harmonisation of data requirements also with biocides | * | */*** | */*** |
| *17.12. Efficacy evaluation* | | | |
| P19 Modification on requirements on efficacy testing | ** | ** | * |
| P20 Acceptance of lower level efficacy | ** | ** | * |
| P21 Waiver for efficacy data | * | *** | *** |
| P22 Time-limited authorisation until efficacy data are provided | * | *** | *** |

***: implementation easy and/or fast or large impact; **: implementation possible and will take time or with impact on duration and costs; *: proposals which require a change in EU legislation or which conflict with the sovereignty of MSs and are thus difficult to implement or with small impact on duration and costs; -: without impact

REBECA proposals that are relatively easy to implement at EU level:

1. Applicants to attend evaluation/expert meetings
2. Establishment of expert groups of BCAs (a microbial expert group already exists) as well as further and more regular expert-/evaluation meetings on BCAs
3. Strict and short timelines for EU risk assessment
4. Definition of low risk BCAs

REBECA proposals that are more difficult to implement at EU level:

1. New specific data requirements for microbials and semiochemicals
2. Generic approach in risk assessment
3. No efficacy evaluation prior to a 5 year registration period

REBECA proposals that are very difficult to implement at EU level:

1. Centralised registration authority
2. No efficacy evaluation prior to registration

Biological control of pests and diseases can provide a viable alternative to synthetic plant protection compounds specifically under the perspective of increasing problems with resistance development to synthetic compounds, residues of synthetic plant protection products in agriculture produce and political efforts to reduce pesticide use. It is generally accepted that one of the major hurdles hampering market access of products based on BCAs is the currently existing registration process. It is now in the hands of governments, science and the private sector to continue the discussions which have been started by the REBECA policy support action. Member state authorities, SANCO, EFSA and the OECD are possible platforms to continue the process. As OECD member countries have already adapted common rules and many non-OECD countries copy what is recommended by the OECD, a reform of the data requirements and registration process for BCAs can only be a global approach, which makes the implementation of innovation an even more time-consuming and difficult process. However, the potential of biological control agents is not well exploited world-wide. If authorisation is not improved within the next decades, biological control will probably continue to exist only in niche markets and its impact on plant protection will be of minor importance.

In autumn 2009 the new EU directive on sustainable use of pesticides (Dir. 2009/128) was in place. The directive requires that member states will ensure that all professional use of pesticides will be in accordance with IPM principles before January 2014 and that before growers decide on using chemical plant protection products, they should first investigate the possibilities of using alternative methods and products – thus using BCAs.

## References

EFSA (2007) Opinion of the Scientific Committee on a request from EFSA on the introduction of a Qualified Presumption of Safety (QPS) approach for assessment of selected microorganisms referred to EFSA.

European Commission (2009) Regulation (EC) No 1107/2009 of the European Parliament and of the Council of 21 October 2009 concerning the placing of plant protection products on the market and repealing Council Directives 79/117/EEC and 91/414/EEC. Off J Eur Union L 309/1

Laengle L, Strasser H (2010) Developing a risk indicator to comparatively assess environmental risks posed by microbial and conventional pest control agents. Biocontrol Sci Technol 20: 659–681

# Index





R.-U. Ehlers (ed.), *Regulation of Biological Control Agents*,
DOI 10.1007/978-90-481-3664-3, © Springer Science+Business Media B.V. 2011

## B



## C